FREE ENERGY PIONEER:
John Worrell Keely

Theo Paijmans

Adventures Unlimited Press

Other Books in the Lost Science Series:

FREE ENERGY PIONEER:
John Worrell Keely

Adventures Unlimited Press

Free Energy Pioneer: John Worrell Keely

Copyright 2004 by Theo Paijmans

ISBN: 1-931882-33-9

Adventures Unlimited Press
One Adventures Place
Kempton, Illinois 60946 USA

www.adventuresunlimitedpress.com

Other Books in the Lost Science Series:

The Free Energy Device Handbook
The Anti-Gravity Handbook
Anti-Gravity & the World Grid
Anti-Gravity & the Unified Field
The Time-Travel Handbook
The Fantastic Inventions of Nikola Tesla
The Tesla Papers
The Energy Grid
The Bridge to Infinity
The Harmonic Conquest of Space
Tapping the Zero-Point Energy
Quest for Zero-Point Energy

To my parents, my brother Robin and his Charity. To my little niece Anouk, my beloved Anouk Helder and her parents.

My grateful thanks go to the following persons for their help: to George Andrews; to Dale Pond of SVPvril.com; to Jerry Decker of Vanguard Sciences and KeelyNet; to Mark Chorvinsky of *Strange Magazine;* to Gerry Vassilatos; to Paul Theroux of the Borderlands Research Association; to P.G. Navarro; to Peter Bahn; to Jan Alderich of Project 1947; to Barry Greenwood; to Jerome Clark; to Adolf Schneider; to senior curator John Alviti of the Franklin Institute; to head librarian Jeannette Rowden of A.R.E.; to museum specialist William Worthington of the Smithsonian Institute; to Roberto M. Rodriguez and Lori Hood of the American Precision Museum; to librarian Peter Drummey of the Massachusetts Historical Society; to librarian Ineke Vrolijk of the Theosophical Society Arnhem; to the staffs of the Bibliotheca Philosophica Hermetica in Amsterdam and the Koninklijke Bibliotheek in Den Hague; to Glen Houghton of Weiser Books antiquarian dept.; to Maarten Beks; to my friends Mark-Paul Vos, Mariëlle, and Carina van der Snee; and of course to Ron Bonds of IllumiNet Press, who gave me every opportunity to make this book the best I possibly could.

Needless to say, any conclusions and opinions in this book, except where otherwise stated, or any errors are entirely my own.

Contents

Introduction

How would you like to be able to install a spherical tank in your back room and make five million dollars?

That's the legend that has long surrounded a gentleman in Philadelphia named John Keely. Many learned scholars (and a number of total crackpots) have claimed for a century that Mr. Keely really used a hidden spherical tank of compressed air to run the phenomenal machine he invented and spent a large part of his life building and rebuilding.

Actually, the famous Keely machine was a very complicated device...so complicated, in fact, that maybe he did not understand it himself. Leading scientists, engineers and journalists visited his workshop to see demonstrations, eagerly expecting to expose him. But after most demonstrations they enthusiastically whipped out their checkbooks, all wanting a piece of the machine Keely had convinced them would revolutionize the world.

This book, the result of years of research by European author Theo Paijmans, tells the whole story for the first time. It is an amazing, heavily documented contribution to the mysterious land of basement inventors, and others who have spent their lives trying to develop free energy.

Keely found the secret, and it lies somewhere in this book. Mr. Paijmans has sifted through everything ever published about John Keely, and assembled a collage of stunning detail. He proves once and for all that the tales of compressed air are nothing but hot air. But the carefully drawn descriptions of the actual machines also seem to eliminate the recurrent theories about atomic energy. Water appears to have played a key part, as did sound waves and harmonics. Keely could produce thousands of pounds to a square inch and produce a power that could bend bars of steel.

As time passed, Keely's machines grew smaller. At first they weighed tons, later they were the size of dinner plates. Unfortunately, whenever the inventor was asked to describe the principles behind it all, he resorted to his own non-technical neologisms, only adding to the confusion. Mr. Paijmans has laboriously turned such passages into basic English, and perhaps some astute reader can even build one of these machines.

Another part of the Keely legend is exploded in this monumental work. He was not a charlatan and swindler. John Keely lived modestly, giving large sums of money to metal-workers, foundries and manufacturers who constructed parts of his machines (eventually he built over 120 of them). If anything, he was frequently conned out of money by numerous companies and individuals, according to Paijmans' research.

The author leaves no cult unturned. John Keely was a contemporary of many of the great occultists, like Rudolph Steiner, Madam Blavatsky, even Jules Verne. Some were ardent supporters of Keely's work, and their ghostly presence's are woven throughout this book.

Mr. Paijmans has carved his own permanent niche among them by producing an epic book that will be studied for generations to come.

John A. Keel
New York, 1997

Foreword

Almost a decade before Charles Jeantaud invented the electric automobile, an inventor in America named John Worrell Keely claimed to have made an even more important discovery. Not a combination of two known devices, the carriage and the electromotor, nor the development of a crude fossil fuel burner, but the discovery of an entirely new form of energy.

When the first electric taxi appeared on the streets of London, Keely was still experimenting and working to construct the ideal machine that would operate on his force, and he was still fathoming the incredible secrets of the energy he discovered. In the process, many machines and wonderful devices were built. Many people attended his demonstrations. Keely had visions and he had dreams. Cars, ships, aeroplanes, even spaceships would someday be propelled with his force. Man could travel to the stars and would have an unlimited, everlasting and clean source of energy at his disposal. A golden age dawned for all!

However, in 1901 when Edison stated that he would build a super battery, Keely, who at one time invited him to inspect his inventions, had already been dead for three years. Today, Edison is well remembered, as are his inventions. Keely fared less well, as did his discovery. What was the fate of Keely's inventions and his many devices? What was the essence of that mysterious energy that he claimed to have discovered that could have changed the course of history? Why was the technology of the enigmatic inventor so quickly forgotten?

With these questions and many others in mind, I embarked on my search. I began my attempt to reconstruct the life of John Worrell Keely, one of the most puzzling and mysterious inventors of the 19th century.

While studying Keely's life and inventions it was necessary to research other little-known areas, sifting through archives, collections and specialized libraries on two continents and in five countries. During my research I encountered strange tales and histories, and the obstacles I encountered were many.

All of Keely's writings, except for a few fragments here and there, have disappeared. What remains is contemporary newspaper articles, testimonies and memoirs of people who knew him personally, or who were lucky enough to meet him at one time or another. And as in any collection of sources, there is a certain percentage of impurity, contradiction, misunderstanding, and embellishments on tall tales or outright yarns.

It appeared that people did not always grasp what Keely was exploring, what his aim was, or what the nature of his enigmatic source of energy was. Although the accounts do paint something of his life, his explorations and the controversy he unwillingly evoked, these form only the visible part - his public life, the surface.

But the Keely history has a much deeper layer. To my surprise I uncovered the outlines of a strange tradition where the early free-energy inventors and the esoteric underground meet, a tradition involved in the creation and construction of what I have termed 'occult technology.' Lured into the center of Keely's labyrinthine history, I found myself delving into the heart of the 19th and 20th century occult undercurrents, and was amazed to find Keely's heritage there, and in some quarters still very much alive.

There was an obscure current that I have termed 'the early 19th century anti-gravity underground.' There were ambiguous hints in certain 19th century futuristic novels. There were other free-energy inventors. There were connections and influences crossing lands, continents, philosophies and times...all of which will be offered to the reader in this book.

The information that I was able to find that has survived into this century, is in this book. The Keely history is also a Keely mystery; his personal papers do not seem to have survived. What happened to these is a tantalizing question that haunts us even today.

Much of the information in these pages is published here for the first time in more than a century, or for the first time in connection to each other. In many of the notes you may find the source for these materials and you may see how a particular line of thought in the Keely history, erroneous or not, developed from a specific source.

At each turn of my research, care was taken to use the most thorough studies in the fields, and as many primary sources as possible. When I found an anecdote in several clippings, I preferred the articles published in the Philadelphia newspapers, and not in a newspaper miles away from the event or incident, although I compared them for discrepancies. When I found similarities or dissimilarities in content, I described so in the notes.

Those who wish to follow the strange trail where dreams and reality meet, may be especially interested in the last chapter which —I must warn beforehand —is quite speculative in nature.

One of the pleasures of writing this book was in the people that I came to know. They are mentioned in my acknowledgments. These times witness an upsurge of interest in Keely's life and inventions. I hope that this book will offer enough materials to all those persons to satisfy their interest. Please feel free to contact me for the useful exchange of materials, if you have additional information that may shed light on the subject, or to discuss the materials in this book.

Theo Paijmans
Postal Box 213
6800 AE Arnhem
Holland
th.paijmans@wxs.nl

PART I

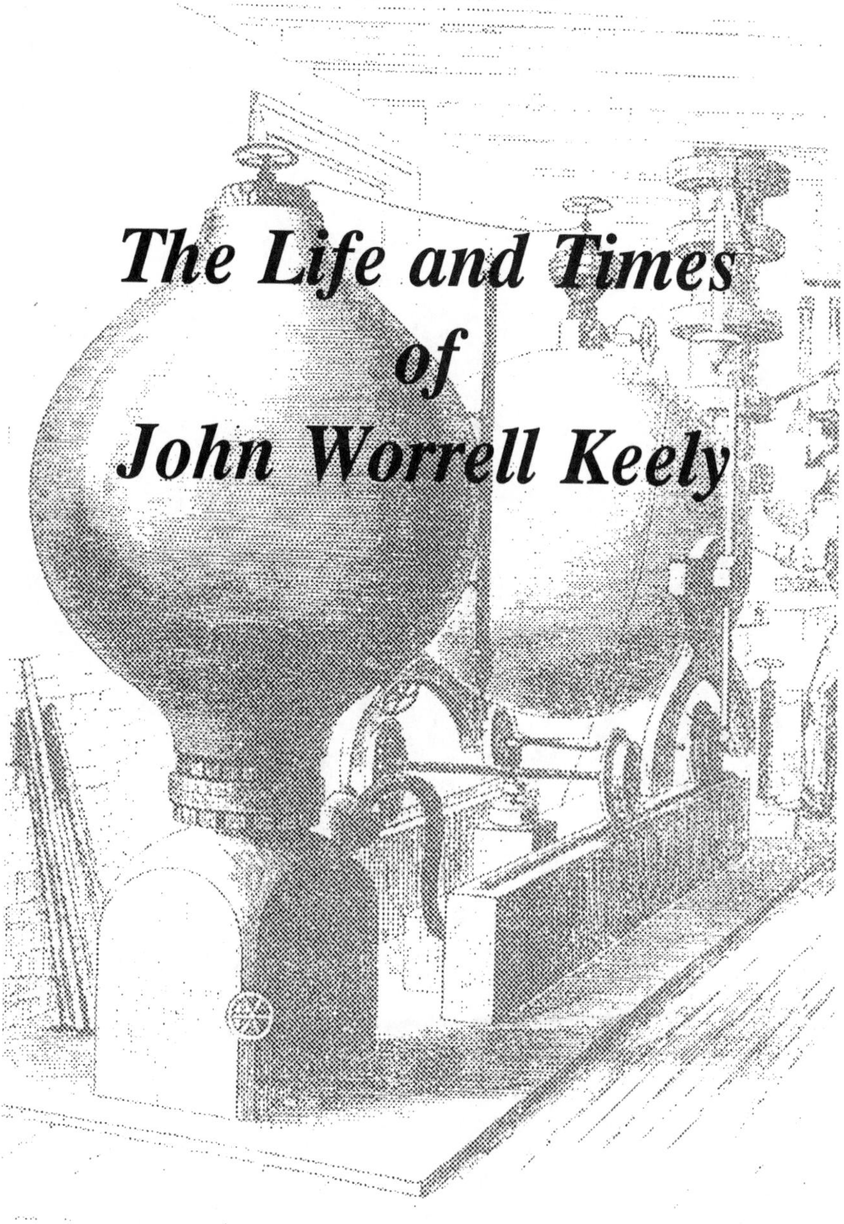

The Life and Times
of
John Worrell Keely

1

Discoverer of the Ether
The Early Life of John Keely

"If all, or even one-half, of what is reported of the Keely Motor is true, the world is on the eve of the most tremendous revolution it has had since it began to revolve at all...."

The Evening Bulletin, July 8, 1875

"Let a note be struck on an instrument, and the faintest sound produces an eternal echo. A disturbance is created on the invisible waves of the shoreless ocean of space, and the vibration is never wholly lost. Its energy being once carried from the world of matter into the immaterial world will live forever."

H.P. Blavatsky, *Isis Unveiled*, 1877

One of the most enigmatic persons in the history of free-energy research is inventor John Ernest Worrell Keely. He stood at the very base of the history of free-energy research, and he possibly made the most important discovery of all time. Not only are the principles of the new form of energy that he claimed to have discovered still not completely understood, but almost all of his incredible machines and inventions are now lost.

John Worrell Keely was born September 3, 1837, either in Philadelphia in a two-story frame house that stood on the corner of Jacoby and Cherry Streets, or in the old town of Chester, Pennsylvania.[1] Sometimes his birthyear is given as 1827.[2] Keely was twice married, the second time in 1887, but left no children.[3] His first wife and his only child died many years before his death.[4] Of his second wife, only her name, Anna Keely, is known. Keely died in the same city on November 18, 1898, and he was buried in the West Laurel Hill Cemetery, 1305 Arch Street, Philadelphia. He has plot number 313 in the River Section. His grave is unmarked. A hard maple tree stands over the plot, which is covered with ivy vines.

The parents of Keely's mother were of English and Swedish descent and his father's parents were of German and French descent.[5] His father was an iron worker[6] and it is alleged that his paternal grandfather was the German composer, Ernst, who led the Baden-Baden orchestra in his day.[7]

Keely had one brother, a J.A. Keely.[8] Since he was orphaned in early childhood, they were probably separated early. Keely lost his parents in his infancy, his mother never recovering from his birth, and his father dying before he was three years old. After his father died, he came to live with his grandmother and an aunt, but his aunt died before he was sixteen, and his grandmother died a year later. Thus, Keely was thrown early upon his own resources. His educational opportunities were limited to the city schools of Philadelphia, which he left at the age of 12.[9]

It is also claimed that he went to live with his grandfather, the German composer, who was unable to do much for him. After he left school, he was apprenticed to the carpenter's trade.[10] Keely, then 12 years old, allegedly worked at this trade until 1872.[11]

Somewhere in these early years, Keely made his astonishing discovery and decided to follow that road. Contemporary sources have their own, sometimes contradictory, versions of how this happened. The nucleus of these tales is that Keely had his moment of illumination while observing peculiar effects of sound on certain objects.

Keely himself claimed that in his childhood he started his research, leading him to the discovery that would haunt him for the rest of his life: "Before I had reached my tenth year, researching in the realm of acoustic physics had a perfect fascination for me; my whole organism seemed attuned as if it were a harp of a thousand strings; set for the reception of all the conditions associated with sound force as a controlling medium, positive and negative; and with an intensity of enjoyment not to be described."[12]

But while he was drawn to this line of research before he was 10 years of age, an incident several years later set him on the course that he decided to follow: "...the first manifestation of which to me when I was twelve years old drew my attention to the channels in which I have since worked..."[13]

Keely never told what that manifestation, incident or discovery exactly was, although elsewhere a poetic picture is presented of a young Keely who "held the sea-shells to his ear as he walked the shore and noted that no two gave forth the same tone...," with which his work of discovery commenced.[14]

Since he never explained the nature of this incident, it was simply written by another that, "...since he was ten years of age [Keely] has been interested in the study of tones and resonances; of those rapid and incessant vibrations which underlie all we see in the world around us, and to which all the energies of the acting universe are primarily due. It is this study which he still continues, and the power which he has developed is claimed to come from a control of these vibrations...."[15]

And elsewhere it is confirmed that Keely not only was a "poor lad reared in obscurity and privation, in early childhood drawn to these unique researches...," but that "from his earliest recollection...was drawn to the study of sound as relating to force, and commenced his first systematic investigation when hardly 10 years of age, making his first encouraging discovery at thirteen. As a child he noticed how powerfully windows were often agitated by the heavy tones of an organ, and this led him to place various objects about the room, suspending glass dishes, etc., and then watching for any effect that might be produced by the various chords he was able to secure by the combination of different tones. He soon found that certain chords invariably resulted in the forcible agitation of objects at a distance."[16]

This was further explained by stating that, "The discovery of the fact that objects composed of a material such as glass could be made to vibrate at a distance only in response to one particular chord to which their mass seemed to respond led to the discovery on which his work is based—the finding of the so-called 'chord of mass' of any material body, and the application of this discovery to the production of vibrations at will. The utilization of this chord produces disintegration, and this disintegration in turn is, of course, capable of being converted into motion."[17] Although the consensus was that Keely's experiments in vibration had their origin in his knowledge of music, and were commenced in his childhood,[18] sometimes a different story arose: "While he was working as a carpenter the vibrations of windows and glass dishes in response to the soundings of the various musical chords first set his mind upon the subject of vibrations and the curious sympathy between distant waves vibrating in harmony. He became interested in speculations concerning physical forces and originated many theories."[19] It is even alleged that he simply became interested in music, and claimed that the tuning fork had suggested to him the idea of a new motive of power.[20]

Apart from working as a carpenter's apprentice, Keely also held other jobs to sustain his living in his pre-motor-fame days. A Philadelphia newspaper, while admitting that of Keely's early life, "little is publicly known," wrote: "He was born in this dear town of ours, and when he was 10 years of age, was thrust out in the world to battle alone. He had two hobbies, music and mechanics. He was a cabinet-maker and musician by turns. The shop in which he worked at the former, some say, was on Market Street, while others claim it was on Jayne Street...."[21]

During these early years Keely also worked from time to time as a physician, a pharmacist, "and in other occupations...,"[22] as an upholster or a cabinet-maker.[23] Keely "worked at many trades to which he had not been apprenticed. He was a plumber, plasterer, carpenter, mason and many other things combined...,"[24] and elsewhere it was added: "He is said to have been employed in various business establishments in this city, where it was noticed that he had an inventive genius and gave much more attention to mechanical problems than to his employment...."[25]

When Keely was about 20 years of age, and it is asserted that he showed in his youth "great interest in physics and chemistry and a wonderful desire for adventure," he went to work in a drugstore in Philadelphia, where he remained for a few years. He then left this job to become a locomotive engineer on the Pennsylvania Railroad, where he "passed several years in an engineer's cab." In 1850, he returned to be a druggist's clerk, but "When the Indian troubles in the West excited the country he joined the army..."[26] There he led a life of adventure for a time, until he was wounded during an uprising of the Native Americans and was sent to a hospital. After his recovery, he returned to Philadelphia and played in an orchestra.[27] Possibly this was around 1856, when Keely was "a varnisher by trade, and in addition a flutist and writer of music..."[28]

That Keely at one time was a flutist, but not the leader in an orchestra as is so often alleged, stems from the recollection of a certain W.D. who wrote about the time "long before the Keely motor was known."[29] The author claimed to have spent "many nights in a musical way" with Keely: "He appeared to me to fall into deep thought during the intervals of our performances, and it was no uncommon thing to have to rouse him from his abstraction. At times he would turn to me, who sat next him, and with great animation explain some improvement, or satisfactory result about an experiment regarding machinery. As I knew nothing of the subject, I could only politely congratulate him."

If the memory of the writer was correct, this again is a hint that Keely, long before his invention of the device that brought him fame, was deeply involved in similar pursuits. The writer gives other interesting details of his early life: "I remember that our party performed at Atlantic City on one occasion, and the next day, seated in the shade with a friend, I saw Keely in a black frock coat buttoned across his breast and a high silk hat, walking along in the sun in an absorbed manner. The heat was terrible—I think about 98 degrees—and when I made a remark about Keely to my companion, he said: 'Oh, John don't know anything about the heat: his brain is busy with some jimcrank about an engine.' His instrument was the flute, upon which he had a remarkable degree of execution; in fact, he was constantly introducing turns and trills, to the annoyance of our professor. On one occasion we were about to play the 'Immortal Waltzes,' and our leader took occasion to speak of the beautiful simplicity of the first movement; but Keely was deep in thought, and when he played his part introduced the most florid turns, at which the professor rapped fiercely and shouted: 'For heaven's sake, John Keely, stop your infernal frills!' This reproof was taken calmly, and the offense was not repeated. A most pleasant and agreeable man he was socially, although subject, even in those days, to fits of deep and intense thought."[30]

There is also a rumor that in his early life Keely was a sleight-of-hand performer and a circus performer.[31] That he worked in a circus was already doubted in his time: "...Keely in earlier life was a cannon-ball tosser and went around the country with a circus throwing cannon-balls in the air, and catching them in his hands.... That story has since been denied."[32] But although denied, doubts remained: "I think there must be truth in it, for away back in the early seventies...it was my great privilege...to have an introduction. ... I was told by the friend who did me this favor, and who had a close intimacy with Keely as a neighborhood-born companion for years, that for a certain time he was connected with a travelling circus...."[33]

This doubt has remained ever since, and current conventional historical sources consider Keely's circus career as more or less apocryphal.[34] The general confusion at that time, the lack of biographical data and the various anecdotes, similar or different, substantiated or unsubstantiated, would also influence later writers about him.[35]

To reconstruct Keely's early life, we must therefore include a biographical sketch of him, that—although colorful—is probably loosely based on such sources as referred to above, and remains unsubstantiated as well: "He worked as a carpenter, played violin in a small orchestra, showed amazing dexterity with card tricks and other paraphernalia of the magician's trade. John longed for adventure and joined a group of trappers who spent three years in the Rockies. Badly wounded by an Indian arrow, John came home to the relative safety of Philadelphia." It was there that he befriended mechanics and professors and "within a few years had acquired smatterings of the fundamentals of both science and machinery."[36]

On only very rare moments, Keely himself confided details of his early life to a reporter. While he denies ever having been a carpenter, we also perceive details of what must have been a most unusual career at times: "'Are you a spiritualist, Mr. Keely?' I asked him. 'That is one of many lies propagated about me,' he answered. 'It has been said I started life as a carpenter (though that is not a slander), but I didn't: I never was a carpenter. Instead of being a Spiritualist myself, I once exposed Spiritualistic mediums in St. Paul, Minnesota, in 1857, 1859 and 1861, and I was nearly run out of town for doing so. Everything their mediums did in the dark, I did in the light, and that naturally enraged them. I do not believe in Spiritualism or in anything of the kind. I am, I hope, a Christian, and a regular member of the Methodist Church....'"[37]

While it is asserted that Keely made his discovery in his childhood, long before he announced it to the public, it is also claimed that this was accompanied by the construction of a number of curious devices or prototypes of engines. A theosophist wrote that Keely "...has worked since he was a boy, at times, upon various inventions before his discovery of ether...."[38]

There is a description of what may be one of Keely's very first devices, but unfortunately no date or time period is given: "His earliest mechanism for noting the uniform force of sound vibrations was a steel bar set full of pins of various lengths, while his first 'resonator,' or 'intensifier,' consisted of a shingle screwed to two hollow wooden tubes. Keely's first rudimentary engine was 'a simple ring of steel with 300 pins set into it, and this first wheel ran in an open box, into and through which an observer was free to look while the wheel was in motion...'"[39]

In 1856, Keely "had running and was experimenting with a toy engine, the boiler of which was fired by a 'burning fluid lamp' in his home on Fifth Street, five or six doors above Queen Street."[40] In 1863, Keely became employed as a furniture varnisher at $10 per week in the furniture shop of Bennet C. Wilson.

Wilson financed Keely's experiments and provided him with a furnished shop on Market Street in which to work. There Keely experimented "year after year" with an engine which he called a "reacting vibratory motor." This led to the construction of a device invented in 1869 that can be considered the forerunner of his Globe Motor.[41]

It is alleged that Keely's motor began to attract attention as early as 1865,[42] but around 1871 he started to try to put his discovery into practical, commercial use, for it is claimed that he announced his invention to the public in that year. Keely, who at the time was "a varnisher of furniture on Market Street, above Seventh," put an advertisement in a Philadelphia evening paper "relating to a new motor, or motive power, which he alleged he had invented or discovered and was prepared to exhibit." This new motor was named the "Globe Motor," presumably because it consisted of a hollow sphere, which revolved at great speed and, as Keely declared, automatically. "It was seen in motion at different times by different persons, but always while in the presence and under the control of Keely. ...The motor attracted the attention of two gentlemen of capital in our city, who consented to advance money with the view of developing this invention and others which Mr. Keely claimed to have made...."[43]

The recollection of another man who was also drawn to Keely's Globe Motor and who retold the incident 30 years later,[44] seems to confirm other allegations that Keely had a globular motor in operation in 1871. In fact, if the 30-year period would be correct, it would mean that Keely had already constructed his device prior to 1869.

The man who came forward after all these years was one Joseph Repetti, decribed as an "expert Vineland machinist and inventor" and "a Mason of excellent standing in the local lodge and a highly respected citizen, whose veracity and capabilities are well known in Vineland." Repetti stated that he knew Keely's much-talked-about secret, and that he was "the only person to whom Keely ever showed his complete motor." Of course this was nonsense; Keely built several devices, of which the Globe Motor was only one, but again it reflects something of the confusion that has always surrounded historical data concerning Keely, even in his day.

Repetti, who lived in Philadelphia, remembered how he met Keely around 1869. According to him, "It had become noised about that a Mr. Keely, who then occupied rooms on Market Street, had accomplished perpetual motion. I became very much interested and, securing permission from my employer, decided to take a half-day off and visit the much-talked-of Keely at his place of work and have a talk. Accordingly I made a call. Mr. Keely was, to all appearances, carrying on a second-hand furniture business. I said: 'I am interested in the report that you have solved the problem of perpetual motion. I am a machinist employed in the city, at the shops of Vere, Camp & Leopold. I have also been working for some time on a machine for the same purpose and am interested in the subject. This is why I have called.' Mr. Keely replied: 'The machine to which you refer is not in running order, but if you will wait for a few minutes I will put it together and show it to you.' In a few minutes, Mr. Keely, who had retired to his workshop, called me to that apartment showing me a peculiarly constructed machine in the form of a globe, of about eight inches in diameter. The ball had a vertical rotary motion. I inquired: 'How long would the machine continue to revolve?' 'Until worn out,' replied Mr. Keely. I asked: 'Does the machine develop any power?' Mr. Keely, placing his finger on the revolving globe, stopped it, saying: 'This machine is too small to develop much power. I shall make a larger machine that will have power to run other machines.' Keely then stated that he had not begun the actual construction yet, but that he had decided the size of the machine, and that the ball would be 'between five and six feet in diameter.' Keely also stated that he didn't know when he would start the construction; that would depend on the help 'from the outside' that he could get. When Repetti asked how much help it would take, Keely's answer was, 'Something like one thousand dollars.'"

At first, Repetti saw something in Keely's device, for he offered to build certain parts of his machine at the workshop of his employers, where he had access to the "proper tools." Keely's reply was enigmatic: "I cannot answer just now, but if you will call in a few days I will let you know what I will do."

Repetti returned and Keely stated that he would go along with his offer, but only if Repetti would pay for the making of the machine, for which he was to receive "a half interest for doing so." Repetti agreed on the condition that he could see all the parts of the Globe Motor "in order to know if I can make them." Keely did not object to this, but told Repetti: "I cannot show them to you now. You will have to call some other time," which Repetti did, and within a few days. The machine had been taken apart and the parts were spread out on the table in the workshop. Keely said: "You can examine the parts and determine if you can make them." Repetti did this and said that he saw no difficulties. But then Keely said that he had a secret part that he could not show without Repetti's promise to oblige himself by oath not to tell anyone. Repetti agreed and Keely "went into another room and brought out the part and showed it to me, after which I said: 'I am very sorry, Mr. Keely, that I have put you to so much trouble, but I think that I will not go into this business any farther....'"

What Repetti saw that changed his mind, he kept secret for 30 years offering no further explanation. Repetti's tale might be truthful in some respects: There is evidence that Keely indeed was living at Market Street around that time; he lived there in 1866, as he admitted.[45]

In 1871, Keely pursued his investigations in the effort to work out his discovery, using water and air in connection with sound vibrations.[46] In 1872, he allegedly made his discovery of an energy that he called "the force" by accident, while experimenting on vibrations. He then "imprisoned the ether" the same year and "commenced his experiments with ether in the winter of 1872-73...."[47] At first he had no idea what he had found, as he readily admitted; it would take him 12 years before he realized that he had "imprisoned the ether."[48]

Between 1871 and 1875, he also constructed six different devices. At that time, 34 documents were in existence relating to the transfer of interests in inventions which were called the independent flywheel, the hydro-pneumatic-pulsating-vacuo engine, the Globe Motor, the dissipating engine, the multiplicator or generator and the automatic water lift. The first assignment was dated July 11, 1871, the last, February 15, 1875.[49]

Keely's Globe Motor, although exhibited in operation around 1871, was never patented; what was found during an investigation in 1875 was "on record in Liber L, 18, page 370, of transfer of patents, U.S. Patent Office, an assignment of this so-called Globe Motor by Mr. Collier to the Keely Motor Company, this assignment bearing date February 15, 1875, and being recorded May 8, 1875." The patent office at that time also had an abstract of "all assignments, agreements, licenses, powers of attorney and other instruments in writing on record in the patent office in the name of John W. Keely since January 1, 1871."

However, there does exist a patent by Keely, granted August 15, 1871, for his flywheel, an arrangement of gearing for causing a wheel to revolve at a greater speed than the shaft to which it is hung.

In 1872, Keely constructed a new motor at his place in Market Street.[50] This second motor was variously called the "hydraulic motor," the "hydro-vacuo engine," the "hydro-pneumatic vacuum engine" and the "hydro-pneumatic-pulsating vacuum engine."[51] A New Yorker who claimed to have known him during that time later wrote that, "I was with him when the idea first entered his head that he could combine steam and water to run an engine. At that time he made a crude machine, which he actually ran for some time; and this was the model of the Pneumatic-Pulsating-Vacuo-Engine. ...In those days I have known him to sell and pawn everything of value in his house to obtain means to continue his investigation with the money thus acquired."[52]

This model was subsequently located at 1010 Ogden Street in Philadelphia, where Keely was then living. The model was described as an engine placed in a bathtub and run by a stream of water that passed through a goose quill. This device "...soon grew into the machine which he called a 'generator,' and which the world named the Keely motor, and in which power was produced from the vibratory qualities of water and air."[53] Elsewhere it is claimed that the generator not so much evolved out of his hydro-pneumatic-pulsating-vacuo engine, but that after its construction, Keely "took a new departure," which culminated in the so-called Keely motor, or, as it has been termed, "a dissipating engine and multiplicator and generator."[54]

While working with his generator one day in 1873, Keely "suddenly felt a cold vapor blow in his face. He tried to wipe away the moisture, but was surprised to find there was none upon his countenance...The curious phenomenon of a vapor that was absolutely dry caused him to take up a new line of experiment."[55]

This mysterious vapor was described as "a heretofore unknown gaseous or vaporic substance,"[56] and it was the power on which the generator—also termed the dissipator or the Keely motor—worked. Keely, being "a poor man, but, having a wonderful degree of natural mechanical skill...devoted all his time for the past fourteen years to experiments with water with a view of procuring a motive power from it. He was engaged upon an idea of his own regarding the force of columns of water when he accidentally discovered the vapor which he has harnessed. He studied the subject, ascertained how it was generated, learned its power, and thenceforth applied himself solely to the perfection of this idea, working night and day for a number of years, until his efforts were crowned with success."[57] Since the above quote is taken from an article written in 1875, this would imply that Keely was involved in this line of research since 1861.

That it is successively claimed that he discovered the force in 1872 and 1873 is explained thus: Keely, while "experimenting with a hydraulic engine...according to his own statement...accidentally made his discovery of the tremendous and mysterious energy which he afterwards pronounced to be etheric force. Over a year passed in various experiments...before he was able to repeat its production at will."[58]

Elsewhere it is confirmed that in 1873 Keely became known as the discoverer of a new power, "which he had not then been able to utilize, to operate machinery, but which could be supplied in limitless quantities at practically no cost. ...He said himself that he made the discovery in 1872, but then had no idea of its origin or laws. He gave no indication of its character, but kept the secret within his own breast until such time as patents could be secured...."[59]

Keely did attempt to secure a patent on his device, which was filed on November 14, 1872, titled: "Specification describing a new and useful Hydro Vacuo Engine, invented by John W. Keely of the City and County of Philadelphia and state of Pennsylvania." The purpose of the machine was described as follows: "The end and design of the invention is an engine wherein the actuating power is produced by a vacuum in connection with water pressure."[60]

Yet nothing came of his application. When information was gathered around 1875, a search of the records of the patent office brought to light an abandoned application for a patent for the hydro-vacuo engine, "the said application having been filed November 14, 1872...At the request of the applicant's attorney, a model was dispensed with by the authorities in the first instance; but on November 26, 1872, a working model was demanded before the examination could be completed. Whenever an application for a patent is of doubtful practicability, or based on what is believed to be a fallacy, it is the practice of the patent office authorities to demand a working model, and to refuse to examine the case until the demand is complied with. Nothing was done in this case of Keely's until March 20, 1874, when he appointed Mr. J. Snowden Bell, now the mechanical associate of Mr. Collier, to prosecute the application; but as two years elapsed without any action, the application was thereby under the law abandoned."[61]

There are rumors of subsequent patents and the assignment of rights. Wilson for instance claimed that Keely had assigned to him "one full half ownership" of the principles or machine that he built in 1869.[62] The other rumors stem in all probability from the 1871 patent, which was granted directly to the assignees, two unidentified persons who at first advanced money for the further development of the Globe Motor and the hydro-pneumatic-vacuo engine, or his patent application in 1872.[63]

A mysterious patent that Keely supposedly requested on November 26, 1873, and of which it was remarked that the accompanying drawings "are now lost," never existed.[64] In contemporary sources there is no further reference to other patents. It is stated that Keely did obtain a patent in 1874 on the hydro-pneumatic-pulsating-vacuo engine, the original design of the Keely motor.[65]

Around 1874 Keely moved to a building at 1420 N. Twentieth Street above Ridge Avenue in Philadelphia, where he established his workshop. There he would construct, over a period of a quarter of a century, various other motors and devices. The workshop was a modest building that was formerly used as a stable. He used the first floor as a general storeroom, and there he conducted his "rough experiments." The second story consisted of three apartments, the first being the office of the inventor, the second the workshop for his globe engine, with an adjoining room where he "religiously guards his latest creation."[66]

Several stories exist of the next important phase that was to be his break-through, and that would lead to the establishment of a company to fund his researches.

One variant has it that Keely's new invention quickly drew the attention of a group of New York bankers and businessmen, and they asked Charles B. Collier, a well-known patent attorney in Philadelphia, to investigate Keely's invention. Prior to this request, Collier alleged that he had never met Keely, although this is not certain, since he would contradict himself a number of times on this point over the years.[67]

Another variant tells that around 1874 two persons, one in New York and one in New Jersey, held contracts with Keely, whereby they were entitled to certain rights in his inventions to be patented thereafter. By mutual consent they agreed to merge their respective rights into a corporation that would be known as the Keely Motor Company. Collier was asked to be their counsel. "At that time," Collier wrote, "I knew but little of Keely's invention. I had seen in his workshop—a room say ten feet square—a 'receiver' charged with a vapor or gas having an elastic energy of 8,000 lbs. to the square inch. I interrogated Mr. Keely critically as to how he had produced this substance. Pointing to an inoffensive looking machine, which stood in close proximity to the receiver, he said to me that he introduced a certain quantity of air into that machine under no greater pressure than was the capacity of his lungs, a certain quantity of water under no greater pressure than was the ordinary hydrant pressure at his residence, and then, by a simple manipulation of the machine, unaided by any chemical substances, heat, electricity etc., he converted a small portion of the introduced water and air into the cold vapor then contained in his receiver."[68]

Collier enquired into Keely's character, consulting Rutherford, chief engineer of the U.S. Navy, and Boekel. Collier was favorably impressed, signed an agreement with Keely and went to New York. There he met with some of the most influential citizens, among whom was Charles H. Haswell, who also visited

Keely's workshop and had seen and reported on the receiver charged with the enormous vaporic pressure.[69]

The meeting resulted in $10,000 of the stock being subscribed. Collier made Keely's written declaration a part of his contract with them. Collier returned to Philadelphia with $3,000 and gave the money to Keely, who then paid the constructors of his machine $2,850. But the term of the agreement was that Keely was obliged to explain the principle of his invention, so Collier went to Keely's workshop with his engineering assistant, J. Snowdon Bell. But even with the sectional drawings of the machine—made by Bell—in front of them, Collier and Bell could not understand why the result would follow from its operation. Collier therefore requested Keely to put the machine together and give a demonstration.[70] Collier and 10 others witnessed the experimental test run of the Keely motor that fateful night of November 10, 1874, in his workshop at N. Twentieth Street.[71]

The demonstration took place in a gas-lit room, with Collier holding a lighted candle.[72] Keely proceeded to make an expulsion which meant that he would develop a force or pressure from the multiplicator. The force was sufficient to exert a pressure of 1,430 lbs. per square inch.[73] He did this by blowing from his lungs for about 30 seconds into the nozzle upon the multiplicator. He then shut the cock and turned on the water from the hydrant. The water that was poured into the multiplicator had a pressure of twenty-six pounds and a quarter to the square inch. The operation was completed in about two minutes after the attachment of the hydrant was made by simultaneously opening two cocks upon tubes connected with the first and second drums, when the lever and the force register were raised.[74]

The multiplicator—or the generator—was not an engine or a motive power motor in the strictest sense; it served as an apparatus for containing the high pressure vaporic substance. The tube for the discharge of the vaporic substance was "about the size of an ordinary knitting needle," with about one-tenth of an inch bore. The first engine that was connected to the generator was a reaction wheel with two arms, each 2.5 inches long. There is no data on the size of the openings through which the vaporic substance escaped in contrary directions in the air, thus causing the arms to revolve.[75]

The reaction wheel was screwed to the reservoir and was put into rotation at high velocity by the manipulation of two cocks. After two minutes, the reaction wheel was removed and connections were made to a small beam engine, which was rotated at 400 revolutions. After three minutes, the reaction wheel was once again rotated. After a minute, the wheel was stopped, and the gaseous fluid was allowed to escape against a candle flame and blow it out. Then the small beam engine was run again for a few turns. Two minutes later the reaction wheel was run again, then the experiments were concluded. The entire experiment lasted 17 minutes. Clocker, who constructed both the multiplicator and the engine, described the engine as having cylinders of three-inch bore and three-inch stroke, with a flywheel of 200 pounds weight, which revolved at 300 revolutions per minute.

It was also stated that the development of the force caused no noise and that no chemical compounds were used or could have been used without detection. Neither heat, galvanism nor electricty was used. The water was "as pure after it left the multiplicator as it was when introduced to the same." The gas had neither smell nor taste and would not ignite. When the multiplicator was dismantled, it showed no traces of chemicals or explosive substances.[76]

Collier then wrote a report for the New York group, which included John J. Cisco, a wealthy banker and former United States Sub-Treasurer in New York; Charles G. Franklyn of the Cunard Line of steamships; Charles H. Haswell, the author of Haswell's tables and a leading authority among mechanicians—who already had visited Keely's workshop; Henry C. Sargeant, president of the Ingersoll Rock Drill Company; W.D. Hatch and Enos T. Throop of the Hatch Lithographic Company; John S. Smith of Baker, Smith & Co., the large manufacturers of steam heating apparatus; and William B. Meeker, a banker.[77]

Collier met with this powerful group of wealthy bankers, merchants and businessmen in the Fifth Avenue Hotel in New York. With Sidney Dillon in the chair, they listened patiently to Collier's report, which had also been made in writing. Collier told his audience of Keely's claims to transform soundwaves in energies exceeding those of a hurricane. This new physical force, when properly applied, would generate immense powers. Keely also claimed that he would be able in a short time to take a train of cars from the Pennsylvania Railroad depot at 32nd and Market Streets to Jersey City in 60 minutes, with a power that could be stored in a teacup or in the hollow of his hand. With as much ease, he would also send a Cunarder across the Atlantic in 48 hours. With a bucket or a barrel of his "etheric force" he might move a continent! All the glories of the age of steam... "would be obsolete long before the close of the century."[78]

Collier convinced the New York businessmen, bankers and orthodox scientists, for the day after his lecture he was to receive a check for $10,000 for the purchase of stock of the Keely Motor Company and to pay the debts that he had incurred in the construction of his machine. The New York group was also given an option for $40,000 more of the stock. The option was kept open until Collier's report was confirmed. In the meantime Collier negotiated a sale of the rights for six New England states for an option of $50,000, the purchasers to invest half a million dollars to introduce the invention, and then to pay another half a million dollars. For the benefit of both groups of investors, a public exhibition was given in Philadelphia that was attended by 300 persons, including engineers and scientists from all parts of the country. The day after this exhibition, the investors handed over their checks for $40,000 and $50,000. Collier had now raised capital of $100,000.[79]

Yet another variant of this important phase has it that there were two meetings of the wealthy group, and that Keely contacted Collier. When Keely succeeded in again producing this dry vapor, he called Collier's attention to it. Keely had built a machine which he called a multiplicator, the forerunner of his liberator and disintegrator. After getting an expert draftsman to make drawings of all its parts, Collier called a meeting of capitalists at the Gilsey House in New York in the fall of 1874 and laid the matter before them, telling them what he had personally witnessed. The result was that $50,000 was "almost immediately" subscribed, and Boston parties also took $50,000 worth of stock. Sidney Dillon presided over the second meeting of the group.[80]

Keely's demonstration on November 10 sparked the controversy regarding his claims that would mark his career for the rest of his life. From that date on Keely was to be a public character, supported in his livelihood by those who believed in the reality of his claims and the grandiose future of his inventions based on his discovery.

Before the demonstration in 1874, the general public at large had heard nothing of Keely's generator, although, as we have seen, it is suggested by Repetti and others that prior to this time, the Philadelphians did hear rumors of Keely and his Globe Motor as far back as 1865.

This 1874 demonstration probably accounts for the fact that most contemporary and later sources date his discovery and subsequent invention of the Keely Motor around 1874, and not three years earlier, when Keely apparently constructed his Globe Motor, obtained a patent on his flywheel and applied for a second patent in 1872. Why he never resolved this issue is one of the countless minor mysteries that riddle his life and career.

In 1874,[81] Collier founded the Keely Motor Company together with Keely, his workman Beckel who "had worked for Keely a number of years," and Sergeant.[82] Sergeant had been skeptical at first, but after a visit in July that year to Keely's workshop at the request of unidentified parties, "some of whom were pecuniarily interested in the discovery," he changed his mind. The interested parties desired a careful investigation to be made, and asked him to go to Keely's workshop as an expert. Sergeant found Keely in his workshop, being "very jealous of his secret. He would not for a long time admit me to see what he was doing that I might investigate the matter. It was only after a strong pressure had been brought to bear on him that he consented to do as much as let me see him work, and when this was at length accomplished by the intervention of his friends who were interested in the invention...so great was his anxiety to keep the secret that the very stockholders were kept out of the room while we were together, and their counsil coming to the closed doors knocked, and was answered by Keely, who would not admit him." Keely not only demonstrated the enormous pressure that he could obtain with his device, but also let Sergeant handle the device himself: "At length he allowed me to work his machinery myself, and I found I could do so as he did; and the machinery for producing this vapor is so simple that a child eight years old could work it. Deception was impossible under such circumstances."[83] Sergeant's reference to "stockholders" is an indication that already-interested persons had begun to invest money in Keely's invention before the foundation of the Keely Motor Company.

When the Keely Motor Company was established, it began selling stock to trusting shareholders throughout the United States. Not only businessmen, speculators and rich investors bought stock; there were thousands of stockholders in Philadelphia and other cities, who were clerks, shopgirls, widows and orphans, "all looking for the day when the increased value of their stock would make them independent."[84]

The headquarters of the Keely Motor Company were at Collier's office. The purpose of the company was to fund the research, construction and subsequent manufacture and marketing of the Keely Motor. But its organizers received stock without paying for it, and about three-fourths of the whole amount was thus given away by Keely. He kept about one-seventh, and "was cheated out of a good portion of that before he had gone far."[85]

When the by-laws of the Keely Motor Company were published a year later, the title page proudly announced that it had amassed a capital of $1,000,000 in 20,000 shares, each with a value of $50.[86]

The company would ultimately have a capital of $5,000,000.[87] This money went to Keely's salary, which was $200 a month, the construction of his workshop and the building of other demonstration devices. Unscrupulous stock speculators were to cause Keely great difficulties however, and Keely saw very little of the proceeds. Nevertheless, he was now a public figure, and articles about him and his mysterious invention, or one should say discovery, began to appear at regular intervals in various newspapers and magazines. The Keely history was launched.

2

Where the Molecules Dance
The First Decade

"A molecule of steel, a molecule of gas, a molecule of brain matter are all of the one primeval substance—the Ether."

Clara Bloomfield-Moore
Keely and his Discoveries, Aerial Navigation, 1893

"What ether is, no one knows."

Thomas H. Burgoyne, *The Light of Egypt,* 1900

With the funds obtained through the sale of stock, Keely began the construction of a much larger multiplicator with which to decompose or disintegrate water, and with the vapor obtained he promised to run an engine. The multiplicator became a huge structure, weighting 80,000 lbs. and costing $60,000. Although the device was crude and unsatisfactory, the machine served another purpose; demonstrations were given to "satisfy popular clamor and boom the stock of the company." Its shares rose considerably and were bought and sold in the stock markets throughout the United States. Keely was by now one of the most talked-of men in the country, and thousands of people visited his workshop at 1422 N. Twentieth Street, admirers as well as scoffers and sensation seekers.[1]

But with fame came scorn; with belief, disbelief and rebuttal. Keely would meet his first major setbacks in 1875. *Scientific American* published a scathing article in which it was alleged that he and Collier were nothing more than frauds, "whose chief purpose appears to be the wriggling of money out of silly people." Keely replied in writing that he had by, "the introduction of atmospheric air into my machine, a limited quantity of natural water direct from the hydrant at no greater than the hydrant pressure, and the machine itself, which is simply a mechanical structure" produced, "by a simple manipulation of the machine, a vaporic substance, at one expulsion of a volume of ten gallons, having an elastic energy of 10,000 pounds to the square inch."[2]

Collier wrote a long letter in return, including several statements by witnesses of Keely's demonstrations. Amongst these was the statement from G. F. Glocker, an employee in charge of the tool room for 26 years at the Port Richmond Iron Works in Philadelphia. Glocker had constructed the multiplicator for Keely "which he operated on the 10th of November, 1874." Furthermore, Glocker stated that "in said multiplicator, there are no secret chambers or recesses in which chemicals or compressed air could be contained, and no spaces not fully accessible to a stream of water passed through the apparatus; further, that, in said apparatus, there are no pistons or moving parts other than valves. I have also constructed for Mr. Keely a vertical direct-acting double cylinder engine, having cylinders of 3 inches bore and 3 inches stroke, and a flywheel 24 inches in diameter and 4 inches face, weighing 2,000 pounds, which engine I have seen rotated at a speed of not less than 300 revolutions per minute with vapor generated in said multiplicator."[3]

All this was published in the *Scientific American;* nevertheless, the compressed air story would follow Keely doggedly in his shadow for the rest of his life and beyond. In the meantime, articles in the press rapidly grew in number and news of his invention was being published by most of the leading newspapers in the country.

One newspaper printed a long article mentioning some of Keely's plans for the commercial use of his inventions: "Mr. Keely says that the first public exhibition will be upon the Pennsylvania Railroad, when he proposes to take a train from this city to New York and return. He will have the 'generator' stationed at West Philadelphia, fill the 'receiver' which accompanies the engine and take vapor enough to draw twenty cars to New York and back. The passage of the train will be silent. There will be no cinders, no escaping steam, or dropping of coals to set fire to bridges. The engine will be smaller than those now in use, but will be of greater horsepower. He says that the generator can either be carried on the train or left at a depot, according to the wishes of the engineer. It is small and compact and takes up very little room. For street cars, as a motive power, this invention, it is claimed, will undoubtedly become popular."[4]

Another newspaper wrote that Keely's invention, "if applied to navigation, the propelling power, it is said, would not cost more than $5 to run a steamer from Savannah to New York; and, as the necessary machinery will not take up one-fourth part of the weight and room of the boiler, engine, and fuel of equal power, another advantage would be additional carrying capacity and space for freight."[5]

But Keely's discovery would have many more applications; "Guns are to be fired by the same power that drives the ship that carries them; explosions are to be rendered comparatively harmless; engines of 5,000 horsepower are to be constructed so as to occupy no more space than an ordinary steam-engine, and all the marvels which are accomplished by steam are to be performed with infinitely greater ease by the cold vapor evolved from water and air." Notwithstanding all these grandiose schemes, some of the stockholders were becoming impatient, and, "not being men of science, are unable to see why we cannot patent what we have, and then patent the improvements that we are able to make."[6]

More philosophically, and probably with its tongue firmly lodged in its cheek, a Philadelphia newspaper had room enough to wonder and to point to the inherent dangers of such a discovery: "Such being the force of the new agent, the United States government ought to buy and keep a monopoly of it. ...With such a possession the country could defy the assembled powers of the earth. ...There is one thing that may save us—the one thing that seems to have thus far delayed the practical use of the new power; it is so subtle that no vessel of any kind of material will long contain it. The vapor from a pint of water penetrated through the pores of an iron receiver three and a half inches thick, forming a damp circle of three feet radius on the floor around."[7]

In the winter of 1875-1876, Keely constructed two metallic spheres, one was about thirty inches in diameter that hung like "an ordinary terrestrial globe." This device, Keely claimed, would revolve with a force equal to two horsepower and would continue to run when started "as long as the Centennial Exhibition would be open, and until the device would be worn out by friction." The device produced power through what was described as "a peculiar-shaped hole in a sphere of iron."

A very sketchy report of the workings of the device has survived. Keely started explaining its working by writing some figures in chalk on a blackboard: "Keely pretended to explain this phenomenon by a string of unintelligible jargon, but the point of all this was that he said the thing ran in consequence of its internal mechanical arrangement—or, in other words, that by combining pieces of metal in a certain way power was generated without any other expense than that required to construct the apparatus. Naturally he refused to show the interior construction which did the miracle, but if his statements were true, it existed inside that globe, and could be produced indefinitely with the result of producing an indefinite amount of horsepower without current expense."

The stock of the Keely Motor Company rose considerably, about 600 percent, and stockholders expressed their wishes that Keely would patent the device. However, he refused to do so. Disappointed, some stockholders sold their stock and left, and the stock gradually declined.[8]

In 1877, a committee of the Franklin Institute witnessed some of Keely's experiments[9] at his invitation, but were not permitted to see anything but the results. Many people, scientific and otherwise, would visit Keely's workshop by invitation, but there was no agreement regarding the character of the experiments. Machinists, physicists and engineers generally were disinclined to believe that any new force had been demonstrated by the experiments, but the vast majority of those who were permitted to see the experiments were convinced that his invention was "destined to revolutionize the world."[10]

That year, Keely also completed an improved machine and was "about to submit it to a thorough scientific test." This new machine was described as being "placed on an iron bed-plate, two cylinders, or upright tubes, on each end, 8 feet high, the right-hand one 12 inches in diameter, that on the left 9 inches in diameter, each three inches bore on the inside. Wrought-iron rings are shrunk on these tubes to strengthen them. At different distances, graduated to a gauge to show pressure on the top and bottom, are compressing valves. In the center between these cylinders, at the top, is a sphere of fourteen gallons capacity, with three tubes connecting it with a middle one, smaller, which rests on a chamber, tea-cup shaped, and in turn rests on an octagonal box, which is fixed to the bed-plate, known as the expulsion-box of the machine. On each side of the central portion is a shell fourteen inches in diameter, seven inches inside, resting on a base through which is a three-quarter inch tube. One tube connects the shell with the expulsion-box, and all are bolted strongly together. On a line with the side shells rests what is called the expulsion-lever, and, singular as it may sound to those who have never seen the operation, this lever opens communication with all parts of the machine, producing vibrating action and the vapor, that impalpable thing about which there has been so much speculation. Connected by small copper tubes, not over three-fourths of an inch in diameter, are two very large spheres of steel—one, termed 'the register of force,' with a capacity of 12,5 gallons and 9 inches thick: the other, 60 gallons and 6 inches thick—the two becoming reservoirs of the power."[11]

The operation was apparently simpler than the complex structure of the device suggested; a rubber hose, five-eighths of an inch, would be attached to the hydrant, and two or three gallons of water was poured into the device. The gauges, with graduated scales, would indicate the height of the water column, "the result, varying as it does, so far as power is concerned."

Air was then forced in the upright left column with a pump. The pressure that was applied was "usually about five pounds" and sometimes "as much as ten pounds." By application of mere air, or by "bleeding of the gauges," the pressure was "regulated at will," and Keely's device was "what is technically called, set." The power was made from "hydrant water and ordinary air, no chemicals about it, and it will drive an engine and transmit power," as the vapor passed from the device into a steel shell and into a condensing apparatus. From there it went through a "small tube and thence to the engine."

During a private demonstration, the valves "were all opened to show the machine was clear, air introduced and the lever was lifted, the first move showing 1,750 pound pressure on the gauge to the square inch, and though the chamber for condensing was open the current did not blow out a match held over it. With 6,5 pounds air pressure, the gauge indicated 5,200 pound to the inch and then 6,700 on the third trial. On the fourth it lifted a large lever (weighted) registering 5,000 pounds dead weight. The vapor was turned into an expulsion chamber and the cap flew off with a report like a rifle, frightening half those present, and lastly a 5 horsepower engine with 3/4 inch stroke and twenty-four-inch flywheel, was driven at 680 revolutions to the minute." What else could "the skeptical engineers" who were present do but mutter that they "were convinced that the power was there, and that it could be applied"?[12]

In 1878, Keely was still busy constructing strange-looking devices, conducting extraordinary experiments and giving the occasional demonstration, but not producing anything of commercial value or ready use, and still no patent secured. Naturally, this lack of progress would be reflected in the press, and in 1878 the tone in the newspapers had changed markedly. Although most of these had been critical but objective at the start of Keely's remarkable career, now articles began to be published that were skeptical and sometimes even outright hostile in tone. In February, Keely wrote a letter to Edison, inviting him to visit his workshop and inspect his inventions: "Dear Sir, you have doubtless heard of my invention known exclusively through the newspapers as the 'Keely Motor.' My discovery consists in obtaining from air + water through vibratory action an elastic substance of great energy + capable of the same influences which produced absorption in water, giving a vacuum. Thinking that perhaps from your long experiments + extended knowledge in vibratory forces, be interested in seeing the operation of my invention I take the liberty of extending to you an invitation to visit my place at some time that may be convenient to you where I'd take pleasure in showing you my machine and operating it for you...."[13] Interestingly, Edison had started working on the first crude talking machine late in 1877, a forerunner of his phonograph that would be patented in 1878.[14] But around 1878 Edison was also experimenting with more exotic devices, one called a vocal engine that operated on sound waves, which he also patented in the same year.[15]

A person named Lynch offered to accompany Edison to Keely's workshop, but Edison's reply was curt. In a letter to Lynch, written on August 30, 1878, months after Keely's invitation, he wrote: "Collier just visited me. Keeley (sic) only willing to show pressure. I will not go until he works an engine."[16] In the end, Edison never came to Keely's workshop, giving rise to the legend that Keely refused to invite him.[17]

In March, a statement was published in which J.B. Knight, secretary of the Franklin Institute, announced, after examinations that had lasted for a period of five months, that he had reasons to suspect that the gauges on the motors which had registered 11,000 lb. to the square inch were altered. He also stated that he thought that the machinery was made "large and massive for the purpose of misleading those who care to view it and every one else...," and that "in operating the engine, no vacuum was produced or utilized as claimed." Knight therefore concluded that Keely's multiplicator with its attachments, including the reacting device, were not capable of producing the effects claimed. The enormous strength was, according to Knight, "entirely unnecessary in sustaining the pressure to which they are subjected, but are rather calculated to astonish and mislead those who witness his experiments." In his opinion, the extreme high pressure that Keely claimed was not produced, "probably in no case exceeding 500 pounds per square inch, and that the readings of the gauges and weighted lever apparatus were incorrect." Keely's vacuum was not produced by a condensation or absorption of the alleged vapor, but by mechanical means, "such as a previously exhausted chamber or its equivalents." Keely's vapor was to him "simply atmospheric air which had been previously compressed and stored up in the various hollow spheres and other chambers of the apparatus in the intervals between the experiments."[18]

Samples of Keely's vapor were submitted to Charles M. Cresson, a then well-known chemist, for a "careful analysis." Cresson's report stated that the vapor was "merely atmospheric air," and that the pressure in the tube was only 225 pounds, instead of the 1,200 pounds that Keely claimed.[19] Keely's reaction as might be expected was one of anger, and the reaction of the stockholders, who by now had expended millions of dollars was one of astonishment. Keely proposed to give "two public exhibitions a week with the motor in an endeavor to demonstrate that it is not a fraud."[20]

However, these public exhibitions did not do much to clarify things for the skeptics. Two weeks later, a Philadelphia newspaper remarked while publishing a letter by a stockholder of the Keely Motor Company, Oliver M. Babcock that, "...we would have been more gratified if the letter had thrown some light on what is admitted to be the very obscure nature of the moving power of the machine, and the mechanism through which it displays a transient force."

Babcock's letter was a defense of Keely against another attack, this time by professors Marks and Barker of the University of Pennsylvania. Just a week before, they had published in the same newspaper a rebuttal of Keely's claims, by stating that the Keely motor according to them worked by, not surprisingly, compressed air in combination with an ordinary hydraulic screw pump, which they saw standing "rusty and dusty in the corner of a remote room of Keely's shop." Babcock wrote: "...they can have $1,000 if they can, within three months by any means, direct or indirect, fill a twenty pint chamber with air at a pressure of 10,000 lbs." Of the hydraulic screw pump, Babcock had this to say: "...has not been in use since last December. It never was used for compressed air, and could not be so used with success on any considerable volume of air or other substance equally elastic. ...The compressed air theory, especially in connection with the hydraulic screw pump is simply absurd, and does not deserve a moment's notice."[21]

In 1878, Keely also began conducting several remarkable experiments involving the suspension of gravity, which he called "vibratory lift,"[22] and the construction of an instrument with the same name.[23] While experimenting with this device, he made his first discovery of the disintegration of mineral substances by means of the new force.[24]

But while these experiments filled his protagonists and those who claimed to have witnessed these experiments with awe, there were still the skeptical-minded scientists who thought quite different about the whole affair. Still no complete engine was forthcoming, and by 1879 this was to have its effects on the Keely Motor Company, all the more because the stockholders were jeered at by the press and others.

The company was now on the verge of bankruptcy, and Keely agreed to a scheme that the company proposed. He assigned the rights of two of his inventions, the automatic waterlift and the vapor gun, to the Keely Motor

Company. Stock was increased from 20,000 to 100,000 shares. The 80,000 new shares were to be divided equally: 40,000 to pay for the inventions, 20,000 to the treasury of the company, and 20,000 divided among stockholders. Of the 40,000 shares that should have gone to Keely, less than 5,000 reached him. Nearly 34,000 went to compensate fraudulent claims held by three separate persons against him. One of these claims involved the case of a man who, acting as Keely's agent or attorney, disposed of two-thirds of the vapor gun and the automatic waterlift, but failed to make the proper returns. The transferee came in for two-thirds of the 40,000 shares set apart to pay for the inventions. A number of shares of less amount went to other persons to whom Keely had made advance sales in order to carry on his work. The appearance of this stock on the market in 1880 broke the prices down to a nominal rate, and it discouraged many stockholders who had obtained their stock by fair purchases.[25]

During 1880 there still was not much hope or the prospect of an immediate success of Keely's inventions, and a newspaper even remarked that an office, kept in New York for the Keely motor, was "simply an enterprise for speculations in Keely motor stock—in fact, a stock gambling den."[26]

Suddenly in March, Keely announced that he would complete his engine in a short period of time, and not only that, he would feature an entire new line of exotic devices. A newspaper noted that the interest of the general public became "somewhat renewed in the Keely motor, caused by the statements of Mr. Keely himself that he will be ready in about six weeks to put his new vibratory engine to practical use. Instead of the great cumbersome machinery which he had heretofore, there are small, neat looking objects which he calls generators and engines. He claims now to have full control of the vapor which contains such great power, and can do with it as he sees fit. 'About two years ago,' says Mr. Keely, 'I abandoned the idea of applying my vaporic power to the ordinary piston engines, and by accident found that a new engine of a different sort was needed. It is not an invention, and I do not claim to be an inventor, but a discoverer. I am so confident now that I have succeeded, that I will stake all I have in the world on the results to be accomplished within the next three months.'"

Keely's new device, the vibratory engine that he had completed in his workshop, occupied a space of about four feet square. "All the machinery is contained in a cylinder which resembles an ordinary drum. Through this runs a double shaft, one revolving in a sleeve. It is upon this shaft that the difficulty at present exists. The negative and positive motions are nearly equal, and Mr. Keely is engaged in the graduation of these so as to cause them to harmonize. When he accomplishes this, which he says is a tedious operation, then the Keely motor will be completed."

A reporter was given the privilege of a demonstration: "Two small keys were turned, and immediately the shaft, containing an 18-inch wheel, began to revolve. There is no flywheel to the engine, only the one to which the pulley is attached direct. This moves at the rate of about 25 revolutions per minute. Mr. Keely claims that this is all that is necessary, as the shafting may be geared to run the machinery to any speed required...The new generator is also a curiosity. It occupies a space of about 6 feet in length, 10 in width, and a height of 5 feet. There are numerous small pipes, of mysterious appearance of the thickness of telegraph wire, bored to the fineness of a cambric needle. One of these leads from the generator to the engine, and it is claimed that all the power is secured through this medium, and the regularly motion secured by the vibratory apparatus contained inside the drum cylinder."

Naturally, Keely stated, as he had done on several previous occasions, that he was planning on obtaining a patent. But even then, Keely warned, "it will require at least a year of lecturing to demonstrate the secret of generator and engine. ...The apparatus will be in use some 20 years before the thing is fully understood." The reporter could state that, "There is one thing certain, Mr. Keely has succeeded in making the wheel go around. He has abandoned his idea of pressure. He has got hold of something which he says is the right thing, and has recently been creating some excitement in a private way among scientific men."[27]

The members of the Keely Motor Company could only declare, at their annual meeting which was held six months later on December 8th, that Keely had "discovered and developed a new motive power of extraordinary power and energy," and that they had every confidence that he would have "under his entire control" the "mechanical details connected with his engine."[28] Yet the next year was not to fulfill the expectations of the Keely Motor Company, and the interest of the public dissolved once again. A lecture in New York during which the motor was fully described and was illustrated by diagrams "of the most complicated and convincing character," drew no attendants. But it was also noted that "with the innumerable pipes and cocks of the generator, and the imposing simplicity of the engine, one could not help being convinced that the Keely motor is as genuine and satisfactory as the ablest perpetual motion machine ever patented."[29] A scientific journal published the lecture, together with the accompanying diagrams.[30]

What was Keely doing during these years? The overall picture that emerges is that, apart from the usual demonstrations in his workshop, he was building device after device, destroying and selling these as "old metal" after having constructed newer prototypes while trying to master what he had discovered. "Since the principle of the Keely motor was first discovered," a newspaper wrote at that time, "the inventor has made half a dozen different engines, each one of which has been simpler and better than its predecessors. In its present state the engine contains 'one hundred and fifty pints in a descending vibratory scale.'"

And in doing so, Keely would invent and introduce a new phraseology for his discovery and certain parts of his devices, unlike anything heard before. In a puzzled state, one was left to muse on terms and parts as "six tuning forks," though five would probably be sufficient, a "compound vitalizing medium," a "vibratory elliptic," a "positive wave plate," a "spiraphone box," together with several "positive and negative tubes," and as many sets of "triple vibratories" as are necessary for transmitting "sympathies."

And as if this wasn't enough, Keely at one time allegedly stated that he wanted "to add a 'compound deodorized vaporized shaft' to the generator and to enlarge the 'antinomian cylinder' of the engine by prolonging it at the end and inserting in its 'negative casing' a 'monophysite tube,' studded with thirty-six 'sabellian holes,' and terminating in a 'galvanic manichuan chamber.' With these improvements, Keely claimed, he would be able to obtain seven hundred additional revolutions per minute, and to reduce the supply of water needed in the generator to five-sixths of a pint."[31]

The months rolled on without Keely being able to meet the constant and pressing demands of the Keely Motor Company for a commercially exploitable engine. Instead it seems, he was more devoted to journeying over ever-new distances far beyond the horizons of known science, and following the new trails that his experiments showed him. The limitless possibilities! Who can say now with any certainty what went through his mind and what visions he had at night in the lonely hours in his workshop, with all Philadelphia around him at rest? There are some hints though, of what he must have thought during these years, for Keely later wrote: "There are moments in which I feel that I can measure the very stars, which shine like Edens in planetary space; fit abodes for beings who have made it the study of their lives on earth to create peace and happiness all around them."[32] Sometimes during his experiments, he would suffer accidents; once in a while explosions occurred, "sometimes harmless to him, at other times laying him up for weeks at a time," and it was said that for a decade, Keely made no progress.[33]

In 1881, he was in one of the gravest periods of his career. The stock of the Keely Motor Company had fallen very low, and as a consequence, a new exhibition had been deemed necessary. The demonstration would be given in the evening of April 22, in the presence of a large body of businessmen from New York. A few days before, Keely gave a private showing to several other important businessmen; among those present was a major of the United States Ordnance Department, Commander Gorringe of the United States Navy, the vice president of Erie Railway, the commodore of the New York Yacht Club, the president of the Lehigh Valley Railroad and 20 unidentified persons. Gorringe later said: "I am amazed at what I have seen. It is certainly one of the most remarkable curiosities I have ever looked upon, and appears bona fide."[34]

The demonstration of April 22 was attended by an equally impressive list of persons: the city Chamberlain of New York, a representative of the Continental Iron Works, the secretary of the American Wrecking Company, and other, unidentified persons. The demonstration was described as "a very extended one." When the visitors, who almost completely filled the front room of Keely's workshop, had been seated, "they saw before them an odd-looking machine built of steel, that shone like a mirror." A contemporary account described the device in a puzzled tone: "...it is wholly unlike any other collection of globes and tubes that has ever been exhibited."[35]

The visitors were given ample opportunity to inspect the device. Every cock and tube was removed to show that the apparatus was empty. Lights were placed underneath the engine, and those present were invited to look into and through the various chambers. All the plugs and attachments were replaced, and a member of the group "drew a glass of water from the hydrant" and poured the water into "half a dozen funnel-topped tubes." In "exactly 29 seconds a force was generated sufficient to raise a six-foot lever (one inch fulcrum) upon which were hung 700 pounds of iron." The pressure was asserted to be 15,000 pounds to the square inch. The vapor responsible for this pressure was then stored in a steel cylinder "about thirty inches long and five inches thick," through the center of which was stretched "a piece of piano wire."

The confined vapor was "vivified" by "external vibrations of great energy," obtained from a tuning fork of immense size. Then a long tube of "very constricted orifice" was attached to the steel cylinder, to form the connection with the engine in the rear room. When the visitors assembled in this room, they saw an engine that Keely called a compound generator. Keely explained this name by telling that "it can be worked with equal effect by positive or negative energy." After he pulled open a few cocks, a spirophone, contained in one of the drums, began "to roar," and the shaft that carried a belt-wheel began to revolve "with great velocity." The sound, a "whirring sound (much resembling the rising of a flock of quail)" gradually became regular and harmonious, and the engine settled down to a regular speed of about sixty revolutions per minute. Keely then made some "curious experiments" to exhibit what was named "vibratory energy." The revolutions of the engine were increased or diminished at will by Keely striking an iron disk or a gigantic tuning fork, or drawing a bow over a tightly stretched steel wire. When Keely changed from the negative to the positive energy, it resulted in an "almost instantaneous reversal of the engine." Keely declared that this reversal could be made "at the very highest velocity without breaking anything."[36]

A brake, specially made with wooden lining was then applied to the belt wheel with a leverage of five feet and the weight of two of the heaviest persons of the visiting party, but "no perceptible diminution in the speed resulted." This was not all that the visitors were shown: "many other strange experiments with the vapor gun and other appliances of the alleged invention were given, after which the party separated." The demonstration lasted for three unforgettable hours,[37] and one may now only speculate upon the nature of the conversations of the visitors, upon their return on the midnight train to New York.

Scientific American, however, noted cynically that the demonstration showed foremost that "the Keely managers still look to the New York men. It was from them that their first treasure was extracted after the original first exhibition; and the new show is doubtless expected to yield another yellow harvest."[38]

Unfortunately, the year 1881 would bring Keely no yellow harvest. Instead, towards the end of the year he would be on the brink of bankruptcy. He had continued his investigations, and from time to time gave out some new features of his discovery, and the new applications that could be made of it. But Keely had not reached a point where he deemed it safe to apply for patents, so difficulties arose between him and the Keely Motor Company. The company refused to pay his bills, and he was practically abandoned by the company.[39] Another effect of these unfortunate developments was that the Keely Motor Company once again came to be regarded as a fraudulent undertaking, Keely as a swindler and his allies as either "disreputable gamblers in stock or the dupes of his wizard artifices."

At this critical moment, John H. Lorimer—who was a member of the Board of Directors of the Keely Motor Company—and Babcock came to his aide. Lorimer published a pamphlet[40] in which he demonstrated that some of the company's directors were responsible for the existing sordid state of affairs. Already having published his favorable pamphlet on Keely, Babcock started a series of lectures with a similar message as Lorimer's.[41]

In this period, Keely was assisted financially by several persons whose names were kept secret at that time. One of these was a Dr. William Pepper, who was the provost of the University of Pennsylvania, of all places. He was deeply interested in Keely's inventions, and at one time he dontated $10,000 to his work.[42]

However generous these gestures were, bankruptcy stayed a constant threat and left the Keely household at the brink of starvation. At times, Keely was in such despair that he destroyed several of his "devices for research which had been the labor of years" in fits of frustration, and had to raise money from the sale of other devices as old iron. By pawning his watch and even by selling his costly scientific instruments, including a valuable microscope, he managed to earn enough money to pay the mechanics and to buy material in order to continue his work.[43]

When bankruptcy and starvation were imminent, Clara Bloomfield-Moore, a wealthy Philadelphian woman, came to his aide. She did this at a time when "...the public seemed to have become incredulous or indifferent, when a paragraph published in 1881-2 caught the eye of the widow of Bloomfield-Moore, the paper manufacturer. It related that the inventor, still working to perfect his apparatus, was on the verge of starvation and despair. Mrs. Moore, in speaking of the incident, said she had just been reading of the suicide of an inventor in New York who had been unsuccessful in getting any one to take an interest in his invention, which after his death was seen to have been a valuable one. Here, she thought, was an opportunity to save another inventor from a like fate. She made inquiries, called to see him, and supplied him with means to go on with his work."[44]

What caught her eye was Babcock's pamphlet on Keely that he privately published in Philadelphia in 1881.[45] Bloomfield-Moore would later write that Keely always spoke of that winter of 1881-1882 as "the darkest period of his life."[46] She found "his wife's roof mortgaged over her head" and Keely somberly pondering over the possibility of committing suicide. She took $10,000, with which she was originally planning to found a small public library, and gave this to Keely. He took half of the amount; more he would not need, he said.[47]

She would finance his experiments and research for almost 15 years, and already being a prolific writer, she would also write a number of articles about Keely and his work.[48]

Things seemed to have turned for the better, but then something unexpected happened.

Within the year, Keely announced a new discovery: the vibratory force of which the demonstration of April 22 was an example. In the meantime, Bloomfield-Moore had become a convert to his theories and revived the hopes of the Keely Motor stockholders. But Keely understood that the company had no interest in his new discoveries.[49] For the 10 years before 1882, Keely had limited his demonstrations to the liberation, at will, of the energy he had accidentally discovered while experimenting on vibrations in 1872. The ensuing years he tried to construct what was termed "the perfect engine" that he had promised the Keely Motor Company. According to Bloomfield-Moore, Keely "made the mistake of pursuing his researches on the line of invention instead of discovery. All his thoughts were concentrated in this direction up to the year 1882."[50]

So in 1882, the Keely Motor Company brought suit against the inventor.[51] Relations between Keely and the company officials were already strained because of his constant refusal or inability to obtain patent papers on his former discovery, the construction of a commercially successful device, and of the rift between him and the company a year before. But the reason for the Keely Motor Company to take him to court was of course Keely's announcement that he had abandoned his disintegration research and instead concentrated on vibratory energy, the new and different technology to which, Keely claimed, the Keely Motor Company had no rights or interest whatsoever.[52]

About this new line of research, Bloomfield-Moore wrote, "the two forms of force which he has been experimenting with, and the phenomena attending them, are the very antithesis of each other."[53] The genesis of this line of research may perhaps be traced back five years earlier; around 1875, Keely developed the notion that he would need a "new engine of a different sort."[54]

The Keely Motor Company tried to learn what the differences were between the two devices and what the nature of the secret was that they possessed. This situation brought on the suit against Keely by the Keely Motor Company. He was ordered by the court to explain his secret, which he absolutely refused to do. He claimed that to divulge any information would be to give away his secret, so the court ordered him to be committed to prison until he complied with the order. Bloomfield-Moore stated that "...had Keely obeyed the order of the court in 1882, and made his marvelous secret public, it would have collapsed."[55]

It was claimed that it was she who arranged a compromise. An eminent engineer was sworn to secrecy before interviewing Keely. What was said during that interrogation is unknown, but apparently Keely was able to convince the engineer by explaining the differences between the two devices. The engineer's report, while never disclosing the secrets of the devices, seems to have been satisfactory to both the judge and the Keely Motor Company, for the suit was dismissed.[56]

More details of this sad period in Keely's career were written down by Charles Fort, stating that the Keely Motor Company kept its faith until December 1882, when: "there was a meeting of disappointed stockholders of the Keely Motor Co. In the midst of protests and accusations, Keely announced that, though he would not publicly divulge the secret of his motor, he would tell everything to any representative of the dissatisfied ones. A stockholder named Boekel was agreed upon. Boekel's report was that it would be improper to describe the principle of the mechanism, but that 'Mr. Keely had discovered all that he had claimed.' There is no way of inquiring how Mr. Boekel was convinced. Considering the billions of human beings who have been 'convinced' by words and phrases beyond their comprehension, I think that Mr. Boekel was reduced to a state of mental helplessness by flows of a hydro-pneumatic-pulsating-vacuo terminology; and that faithfully he kept his promise not to explain, because he had not more than the slightest comprehension of what it was that had convinced him."[57]

The story uncovered from contemporary newspapers paints a different picture; in fact, the stockholders had lost their patience with, and trust in the inventor several months earlier.

Keely's invention was far from complete, and there was no immediate prospect of it being finished anytime soon. There were still no patents, and worse, only Keely seemed to understand the workings of his devices. He was "still groping for the evasive contrivance that will set everything working according to the original expectation." While the stockholders would have preferred that he do just that, Keely's mind was "scattered over so many inventions that this one cannot receive his constant attention." So the stockholders demanded that he apply for a patent, or at least "explain his invention to some other person," because in case he died, "all the beautiful machinery required in his experiments, and the well engraved certificates of stock will be turned into old iron and waste paper." And, what if Keely would become insane? "Mr. Keely's labors may be too much for him. His friends are afraid he will go crazy, and this would be just as bad for his backers as his death."[58]

On January 20, "John Keely filed a demurrer to the bill in equity presented against him by the stockholders of the Keely Motor Company. The demurrer is entirely technical and gives a number of reasons why the court should not afford the plaintiffs the relief they seek."[59]

Keely's legal troubles were far from over: In Philadelphia, on the morning of March 27, in the Court of Common Pleas No. 1, "argument was heard upon a demurrer by Jonathan Puzy, representing John W. Keely, to the bill in equity recently filed by the Keely Motor Company to compel Keely to divulge the secret of his motor. It was argued on behalf of the demurrer that the inventor could not be made to expose that which no one know but himself and which was hidden in his own brain."[60] But on April 1, Judge Pierce overruled Keely's demurrer and ordered him "to make known his process in the way indicated in the bill filed by the Keely Motor Company. This is to compel him to divulge his secret of the motor."[61]

Keely's reply came on the 24th of May: "The answer, which is sworn to, substantially admits the truth of the formal portion of the complainant's bill concerning the contract, & Keely adds that, although owing to certain abstruse difficulties by reason of the nature and qualities of the said force, he has thus far failed in his efforts to bring the said inventions and discoveries into any practicable use or to arrive at the utility required by the law, he believes he will ultimately succeed."[62]

In his defense, George H. Peabody, a stockholder of the Keely Motor Company, wrote that, "...The suit on the part of the directors and stockholders is not merely to compel Mr. Keely to give what exists in his mind only, but to give the company the present finished generator and the 'Secret' of working it, as he has done over 7,000 times during the last two years in the presence of hundreds of able men who have no doubt of its great value to day..." The letter also repeated the more sinister motive for the suit: "This is asked for so that the company may understand the power in case of Mr. Keely's death."[63]

On June 7, at a meeting of the committee that was appointed by the Board of Directors, an attempt was made to overcome the rift between Keely and certain stockholders. An agreement was signed by both President Randall of the company and Keely, which was then publicized: "...differences have been adjusted, and William Boekel, of Philadelphia, is agreed upon as the person to be instructed by Mr. Keely in the construction and operation of his inventions."[64] The honor of learning Keely's secret was bestowed upon Boekel, who was by all means a logical choice. Boekel devoted his life to "mechanical pursuits," and he knew Keely well before the legendary beginning on November 10, 1874. He stated that he was "intimately acquainted" with Keely and his inventions.[65]

So what became of Boekel's interrogation? It is suggested that Bloomfield-Moore was instrumental in forging the compromise even though contemporary sources never mention her, and while Fort, by absence of a reference to a time period, suggests that the interview by Boekel was quickly, and more or less satisfactorily done, quite the opposite seems to have taken place.

The answer came several months later; "Mr. William Boekel, the Philadelphia machinist who was selected by Mr. Keely as a proper person to study his mysterious motor and satisfy the doubts of the stockholders of the Keely Motor Company, has been pursuing his studies now for about three months without learning anything tangible in regard to the wonderful engine. ...As the weeks have passed and Boekel has learned nothing that he did not know before, the stockholders are again becoming suspicious...as Boekel had been employed to manufacture certain parts of the marvelous machine, some of the stockholders objected to him from the first, and declared that his selection was prearranged. Meanwhile, Boekel visits Keely's factory daily, and the people dwelling in the neighborhood are frequently startled by the sounds of terrific explosions in the building. The enemies of the inventor say that he explodes large quantities of gunpowder there for the sake of the effect, and they have determined, if Boekel has not told them something to satisfy them by Tuesday of next week, that they will console themselves with a monster mass-meeting for the purpose of expressing indignation." In a lighter vein, it was added that "Mr. Keely, however, does not appear to be alarmed, and complacently spends time with his new trotting horse over the drives in Fairmount Park on pleasant afternoons."[66]

Then, on December 13, at the annual meeting of the stockholders, Boekel's report was finally presented, supplemented by another report from Keely. His report stated that his engine would soon be completed and in full operation, and that it was his purpose "to bring the matter to the attention of the Pennsylvania Railroad Company with a view of having the engine first applied on their road between Philadelphia and New York."

Keely assured the stockholders that "he has passed the line of experiment and nothing is left to be done but the completion of the mechanical work now being done on his engine." Once again, there was talk of patents; Keely said in conclusion that, "At your request and with the aid of Mr. Boekel and of counsel I have prepared a caveat for my generator, which has been forwarded to the patent office."

Boekel presented his report; he painted the great possibilities of the natural forces, which, he said, Keely had been "assiduously studying until the inventor's investigations and experiments have carried him far beyond the laboratory experiments of such men as Prof. Tyndall and other scientific investigators." Keely had also succeeded in "exciting, harnessing and utilizing the subtle force which to them has been only a subject of scientific wonder." Boekel then went on to explain that what Keely claimed to have discovered, was "the fact that water in its natural state is capable of being, by vibratory action, disintegrated, so that its molecular structure is broken up, and there is evolved therefrom a permanent expansive gas or ether, which result is produced by mechanical action."

Unfortunately Boekel's report did not describe the mechanism that was used, as it would be "improper." He concluded his report by stating that Keely had "discovered all that he has claimed, and that the stockholders should abide in patience the success of the enterprise." Boekel also added that it had taken so long to perfect his invention, because although he built "three engines at different times, each being an improvement on its predecessor, but none of them equal to a fair test." But now, Keely was engaged in the construction of a vibratory engine that would deliver 500 horsepower.[67]

The construction of this engine would take longer than expected; halfway through 1883, word was out that he had completed his vibratory engine that would deliver an awesome amount of horsepower. "For several days past it has been rumored on the street that the much-talked about inventor, Keely of motor fame, has not only really completed his well-known and yet little-known invention, but has completed a wonderful new vibratory engine."

This new engine was giant-sized, but would deliver less than the promised 500 horsepower: "...and now Mr. Keely is completing a mammoth 'vibratory' engine which, he says, will develop at least 300 horsepower. It is rapidly approaching completion...by the end of the present month of July, or early in August, it will be complete. On its completion he will publicly exhibit it first at his workshop, testing its power with a dynamometer, and afterward at another location." Keely was also planning, according to Collier, to give an exhibition or a series of exhibitions that month "of most interesting character in vibratory mechanism."[68]

August came, and at a regular monthly meeting of the directors of the Keely Motor Company at the offices of the corporation on Walnut Street, the "monster engine was officially declared to be finished, and it was also declared that it would be ready for operation about the first week of September. It was announced that a final inspection of the machine by the trustees would be made tomorrow forenoon, at which the stockholders were invited to be present."

Optimism reigned; Treasurer Green proclaimed that he was "exceedingly hopeful, and declared that the day of Keely's vindication was close at hand."

In this high-spirited and triumphant atmosphere, it was almost overlooked that poor Boekel "had not yet been introduced to the mysteries of the motor by Mr. Keely; that the inventor kept delaying matters by telling the expert that he could explain to him the curious mechanism after its completion in less than two hours. This he had not done yet, because the engine had not reached that final and highly desirable stage." The rumor that Keely had completed his engine reverberated throughout the country: "Applications for passage on the first Keely train for New York have already begun to pour in from all parts of the country."[69]

The inspection came on August 29, but it would be a big disappointment. In the presence of the directors of the Keely Motor Company, headed by President Randall, Keely "played a tune on the vibrator, and remarked that it was quite ridiculous to expect any experiments on the big engine yet... One of the visitors, who has promised a number of friends a ride to New York on September 1 looked grieved, whereupon Mr. Keely called attention to the fine appearance which the black and white lining of the shell would be presented after it had been enameled." The response by the unfortunate visitor was unfortunately lost in history.

Amidst the disturbance, a spontaneous search of Keely's workshop began: "The directors then went upstairs and examined the wreckage of the 12 smaller engines used in past experiments. They concluded that the remains of the $5,000 one closely resembled the fragments of the $70,000 one downstairs. Then Mr. Keely and Mr. Becker, the foreman, went through a brief dialogue. 'You've worked for me 14 years, haven't you?' said Mr. Keely. 'More than 14, I guess' said Mr. Becker. 'And how much do you know about running the motor?' said the inventor. 'Nothing,' replied the foreman. 'If I did I wouldn't be here wearing a dirty shirt.' Afterward, Mr. Keely made some astute remarks about a 30 lb. vacuum, and Mr. Becker said that the stock would begin to rise again within a week. He refused, in a mysterious manner, to tell the reason of the expected 'boom.' The foreman then distributed among the visitors the card of a Walnut Street broker, of whom the company's stock could be bought, and the performance closed."[70]

The bizarre meeting with the unusual anticlimax stirred the smoldering resentment that still lingered in the Keely Motor Company. Two months later, on October 29, they were "up in arms again and are preparing for another suit against the inventor." This time, the complaints came from Keely's former friends, and the suit was to be brought "in the name of the company against Keely for the fulfillment of his many pledges."[71]

Yet that night, at a meeting of the Board of the Directors, Keely made a "statement explaining his progress, and stating that he was constructing a street chamber to contain the vapor, and that when this was completed an exhibition would be given. The directors voted Mr. Keely's explanation very satisfactory. And when the Keely Motor Company held a meeting on December 12, Keely was granted more time to complete his invention. It was also remarked that he 'had met with great obstacles, both mechanic and domestic.' For a year Mrs. Keely has been ill, her death occurring about one month ago."[72]

The year 1884 was to become another trouble-ridden year for John Keely. Unknown to him, in the same year a lean Yugoslavian immigrant arrived in New York with just a few dollars, a booklet with his poetry, some designs for a futuristic aircraft and an introduction letter to Thomas Alva Edison. His name was Nikola Tesla, and in the years to come, this genius would encounter several problems, not unlike those that Keely had to deal with. And in the end, Tesla would have something to say about Keely as well.

Meanwhile, new setbacks occurred. All the employees at Keely's workshop were fired, including the head mechanic, Albert Chance, who had worked on the motor for seven years. This measure was taken, Secretary Schuellerman explained to the reporter, "because their work has been accomplished and we have no further use for them. Today Mr. Keely will begin focalizing and adjusting the vibrators. He will henceforth operate entirely alone. This work of adjustment may take several days. A perfect adjustment of all the parts is necessary. This is a delicate operation, but for Mr. Keely it is not a difficult one, and as soon as he obtains one revolution, be it ever so slow, his task practically is finished." High spirited, like the optimism that had lived so shortly the year before, Schuellerman added: "I see no reason why we may not expect to hear almost any day now that the engine is running."[73]

And so, at the meeting of the Keely Motor Company held on March 25, the directors who left Schuellerman's office that evening were to the outside world, a very "hopeful-looking body of men." They had good reason. "The vibratory engine is finished," they said; "the work of adjusting and focalizing is progressing rapidly, and Mr. Keely has fixed the date for the actual exhibition of the motor on or before April 10." Keely himself was not present at the meeting. He told treasurer Green that "I am now so near done with my work that I don't want to appear before the directors again until I appear to exhibit to them in our final triumph." Green also read a statement by Keely, in which he was denying that he was delaying the completion of the motor in order to "apply his mysterious power to other mediums in the interest of other parties."[74]

However, around this time, when Keely once again was asked when he would patent his engine, he answered that, "I do not know how near success may be, nor yet how far off it is."[75]

September—and almost a decade after the first demonstration that had gained him his reputation—Keely would find himself for the first time giving a demonstration elsewhere than in his workshop, and with a new device called a "vaporic gun," with which he had given occasional demonstrations in his workshop as early as 1881. Until September 1884, all of Keely's demonstrations had taken place in his workshop. This was a severe point of criticism, since scoffers and skeptics doubted that his devices would work anywhere else. The Keely Motor Company had finally succeeded in arranging a demonstration at another location. The demonstration was an initiative of a Col. John Hamilton and Captain Van Reed, who had invested in Keely's inventions and who had visited him in Philadelphia, accompanied by A.R. Edey, at that time president of the Keely Motor Company.

The demonstration was given on September 20 at a government range at Fort Lafayette[76] at Sandy Hook. Keely arrived there by a special car, placed at his disposal by the Philadelphia and Reading Railroad.[77] This car was needed because Keely expressed grave doubts about transporting via the railroad the two containers with the etheric gas on which the gun worked, since the effect that would be produced upon the vapor by the sound of an express train were uncertain. Nevertheless, they arrived safely at the range.[78] Around 300 people, among them representatives of *Scientific American* and government officials, were present to witness his demonstrations with the "vaporic gun," sometimes also called "pneumatic gun," or "etheric force gun."[79]

Keely and his helpers also brought a small cannon to the area, the "vaporic gun," that worked, according to Keely, on this etheric gas. A lead bullet with a diameter of an inch was placed into the cannon, and Keely then obtained the force from the containers through a "flexible copper tube," attaching one end of it to the breech of the gun.[80]

The force was derived from an etheric vapor produced by his generator. The first chamber of the generator was filled with air that was "stimulated by vibrations" to "create a small disturbance." This was done by dropping water into it. When the air then reached the second chamber it would come in contact with the vibratory ether, which would "act upon the water and the air in such a way as to separate the particles of water and air. Expansion follows, and the force thus generated is irresistible."[81]

The gun had a spherical knob, secured to the breech, from which projected a round vibrator bar having a diameter about equal to that at the extremity of the muzzle. The breech was 4.5 inches in diameter and its length 3.5 feet. The biggest of two containers, made of wrought iron, that held this force, or vapor, was opened, but "nothing could be obtained from it." This had happened before the guests arrived. Keely feared that the vapor had become "negatized," so he "administered blows carefully" between the big and the little container. To this end, Keely used a wooden mallet. A stroke upon the small container "intensified" the vaporic quality in the larger container, and Keely kept the "vivification" up by delivering "a blow now and then."[82]

In loading the gun, the gas check, consisting of three disks having a diameter of almost two inches—two being of hard rubber and the third being of soft rubber—was first placed in position, and then the muzzle was screwed up tight. The bullet was placed in position in the gun and the valve was turned to admit the vapor to the breech. After waiting a few seconds, the end of the vibrator fixed at the breech of the gun would be struck, and the charge exploded. The time between the turning of the valve and the discharge was, on an average, about six seconds. Sometimes, he would need more blows on the vibrator to obtain the discharge. Keely also used his mallet on the gun itself, as it had certain "acoustic properties peculiarly its own," and "blows upon its exterior set a number of vibrators in action distributed through its breech." More vibrators were part of the interiors of the wrought iron containers.[83]

The operating of the vaporic gun seemed to have been a risky affair at times. Keely would later write that, "It has been impossible for me to write, my right hand and arm were so severely strained, but I have not been idle. I have had time for reflection, and I have been setting up a key to explain vibratory rotation. I have also a plan for a device to be attached to the Liberator as an indicator to show when the neutral center is free from its intensification while operating. In this way, the dangerous influences will be avoided which present themselves on the extension of the vibratory waves that operate the gun."[84]

At the demonstration at Sandy Hook, Keely fired 19 rounds at a target placed 500 yards away. There was no difficulty in sending the bullets that far with a five degree elevation. A conical steel bullet pierced four inches of pine plank placed a few feet from the gun. The noise made by the gun "closely resembled that caused by a common shotgun when loose powder, having no ramming on top of it, is exploded," and "a small cloud of white vapor, which almost instantly disappeared, followed the discharge."[85]

Opinions as to the effectiveness of Keely's gun varied considerably; a witness to its test firing would later say that "I saw the famous gun fired, and it did not amount to much. A good, healthy donkey could kick harder than the projectile struck,"[86] but "flattened missiles, spread out by the concussion to about three inches in diameter," convinced another witness of the "marvelous power and instantaneous action of this strange vapor, apparently equal to that of gunpowder itself."[87]

The day following the demonstration of the vaporic gun, Keely was visited by a reporter who found him in a lighthearted and talkative mood. Keely "sat in his dressing gown this evening in the second story of his residence on Oxford Street reading after demonstrating the new motor gun on Saturday in the presence of a party of government officials at Sandy Hook. A satisfied expression was on his face. A reporter...was cautiously examining a vaporic vibrator that lay on the table in the middle of the room. Laying down his paper the inventor wheeled around in his chair, and, after studying for a moment, broke the silence that he has so persistently maintained... 'My experimenting days are over,' he said. 'This will develop my active enterprise. Complete success is very near at hand. My experiments at Sandy Hook demonstrated that my vaporic force is a fact and not a mere creation of fancy, as many persons have persisted in declaring. I am now able to produce a power of projection thrice greater than that of gunpowder, and there is no limit to this force. My motor will be completed in less than two months and I will then make a public exhibition of its wonderful powers, which are already in a position to manifest themselves. The adaptation of my force to gunnery is positively assured. I can apply it with more effect than that of nitro-glycerine.'"

Being in the best of moods, Keely also explained the principle of his vaporic gun to the reporter. He took what looked like "a policeman's billy" out of a sachel, that, he claimed, was a vibrator. "It is a hollow coil of steel of the finest quality. In one end is an orifice, by which it is attached to the gun. It is the most peculiar piece of steel in the world." Tapping one end of the coil twice on the floor and holding it to the ear of the reporter, the reporter noticed that "the steel cone was humming in a very high key. The noise was like that of a tuning fork." When the reporter took the cone in his hand, he felt that the cone was quivering from one end to the other. "It hums, don't it? No other piece of metal in the world of similar shape will hum at all. ...That steel bar...was the beginning of my motor. By means of it I stumbled on my discoveries. For seven years I have kept flowing through that core a stream of etheric vapor. The action of the vapor has been to affect the relations of the molecules and to alter to a certain extent their conditions. For this reason it has become subject to these vibrations. ...There has been no apparent outward change in the steel. Its weight is the same as before, but it is in the process of silent dissolution. Were I to pass through it for 20 years longer this etheric vapor, it would crumble into nothingness. ...The steel cone is necessary for the promulgation of the projecting force of the etheric vapor when applied to gunnery."

The vaporic gun Keely described as "a breechloading rifle weighing 500 pounds. It was specially constructed for me. On Wednesday last I charged my tube, a five-gallon reservoir of wrought iron, one and a half inches thick—with etheric vapor. Then I boxed it up and did not ever test it, so I was certain of its powers. There, hours before the experiments, it remained untouched in my shop. The process of charging it consumed less than four seconds of time. You could not guess how much material was used in making the vapor. ...To project 20 leaden bullets, each weighing nearly five ounces, at a velocity of over 450 feet a second, there was required six drops of water and about a pint of air. From this combination I derived sufficient force to fire 20 bullets of like weight as those used." But during the demonstration of his vaporic gun, Keely noticed an odd effect: "The most curious thing about all is that I found at the end of my experiments, was that I had increased the power in my tube instead of diminishing it. ...The initial velocity of the last bullet was more than that of the first one."[88]

The day after his demonstration several daily papers contained favorable articles of what was termed as his "great success." As a consequence, the stock of the Keely Motor Company was sent up from 9 cents on a dollar to 15 cents, "the money of the deluded purchasers was thus successfully netted," as it was wryly commented.[89]

Three weeks after the experiment at Sandy Hook, *Scientific American,* which never saw much in Keely's inventions, published its opinion; to the editors compressed air was the real power behind the vaporic gun, which was referred to as "nothing more than a clumsy air gun."[90]

Scientific American found a staunch ally in Captain Zalinski of the U.S. Army, who was one of those present at the demonstration at Sandy Hook. He thought along similar lines. He would later declare to the press that he saw Keely "fire his gun by placing some mysterious appliance in the breech, sounding a tuning fork and then opening a brass cock on the reservoir." Zalinski stated that "it was evident to me that Keely had accumulated gas or air under very high pressure. Upon opening the cock spoken of, the air would rush through the copper tube into the air chamber of the gun. The several seconds that elapsed between the opening of the brass cock on the reservoir and the firing was requisite for the bursting of the diaphragms." Zalinski had his own reasons for denouncing Keely's demonstration, for at around the same time he promoted another invention, the "dynamite gun." This was the joint invention of a number of men and was undergoing tests ordered by the government, under the special direction of Zalinski.[91] So naturally Zalinski kept ridiculing Keely's vaporic gun in the press,[92] or downplayed the stranger effects that others had noted. He would even visit him in his workshop, but that did not change his opinion of the inventor and his works. He had, he said, taken a pressure gauge with him that would register 10,000 pounds, but Keely refused to use the tool. And although he witnessed him making "a globe revolve by a tuning fork," the demonstration didn't convince him. Returning to the topic of Keely's vaporic gun, Zalinski stated: "I question Mr. Keely's ability...to fire the gun continuously for 100 rounds, maintaining the same high velocity without discarding his flask."[93]

But elsewhere it was remarked, just as Keely had noted and told the reporter who had visited him at his home the day after the demonstration, that "Whatever the substance was that Keely carried in a steel tube, it was apparently inexhaustible, the projective force of the nineteenth and last shot being greater than any of the preceding ones, a circumstance that strangely combated the theory of compressed air."[94]

Zalinski offered no explanation of this strange detail. Instead he confidently assured that his "rather exceptional experience with air at high pressure enabled me to see possibilities that might not have occurred to others. The air chamber of the gun was so small and the reservoir so large, with a pressure of say, 3,000 pounds to the square inch, that he could continue firing for a number of shots."

Zalinski also claimed to have offered to produce a similar result "if the company would furnish an outfit which would look like Keely's. This offer was declined."[95]

It would be two years before the statement by *Scientific American* that Keely had used compressed air in the Sandy Hook experiment was vehemently countered: "That he uses compressed air or any known gas, as charged and insisted upon in the *Scientific American,* is absurd and totally impossible to conceive of, when we consider the available space for such compressed gas or air in all the cylinders put together which Mr. Keely employs. Besides, the phenomena accompanying the discharges of this gas or vapor after each experiment are entirely different from those of compressed air or ordinary gas. Mr. Keely justly complains that the *Scientific American* editors keep up the hue and cry of humbug and fraud against him, and at the same time have refused the most urgent invitations extended to them to come to Philadelphia and witness the operations of his discoveries before ridiculing them."[96]

Keely had always been pestered by persons who claimed that he was a fraud, and although enough was alleged usually nothing was proven beyond any doubt, and nothing substantial had been brought to light. In these uncertain times, another person joined the ranks of those who thought that Keely was a mere swindler, and he, too, claimed that he had a very good reason for stating so: "A story is current here," wrote a newspaper, "that a veteran machinist named Baker, an old resident of Bridgeport, has just returned from Philadelphia with a sensational story to tell. He is represented as having been for the past two years an employee at the workshops of the Keely Motor Company in Philadelphia, and as the representative of a New York capitalist, by whom he was to be paid for the discovery and exposure of Keely's much advertised secret. Now Baker returns to denounce Keely as a fraud, and outlines a book which he promises to write for the education of Wall Street and other parts of this too-confiding world."

Baker claims to have been drawing $300 a month from the New Yorker while pursuing his investigations and to have had the full confidence of Keely. Keely, he says, was very careful in engaging him, keeping him at the most unimportant employment until he felt that the man was trustworthy. Baker said it required a year and a half to discover the secret.

Baker also gave an elaborate description of the motor: "The motor proper consisted of a heavy outside covering of metal shaped to deceive the spectator in every way he may look at it. The outside looks as if the machine consisted of a large massive iron cylinder with valves, wheels, and outside pipes. These are supposed to assist in the act of generating the famous new force when in fact the outside shape has little to do with the working parts. Those parts are on the inside. The force is, pure and simple, air, the least bit tainted with a chemical to deceive, as everything else is made to do. The air is pumped from 7 to 21 steel tubes on the inside of the shell. The tubes are of sufficient strength to withstand a pressure of from 10 to 30,000 pounds. There is a mechanism inside the shell that permits the compressed air to pass from one chamber or cylinder at a time into a distinct and separate cylinder which contains the piston that operates the flywheel of the machine. By this method the machine can be kept running five minutes or perhaps longer, and yet show very little change on the pressure gauge. The plan is to allow only one third of the air to escape from one cylinder, and then that one is disconnected, and so on until but one cylinder has been used to that extent, when the machine is stopped and a great show is made, as, of course, the indicated pressure is exactly the same as it was before the wheel went around. Not a drop of water is used at any time. The water story is all bosh."

"Such in substance is a long story as it is told by Baker, who alleges that Keely is far from being a practical mechanic and never talks to one, though when a stockholder comes around Keely deluges him with a mixed fantastic jargon, using a hundred terms or more that no mechanic or scientist ever heard before. Baker avers that the idea of a motor was given to him in Newark, N.J., as long ago as 1867, when Dr. George A. Prindham, then of Newark, now of Philadelphia, constructed a machine in many respects like the Keely motor at the fire engine works of Gould Brothers, on Railroad Avenue in Newark. Keely, he says, captured the idea by haunting the shops. Baker omits to make public the name of the New York capitalist in whose interest he has been playing the detective on Mr. Keely."[97] The book that Baker promised to write was never published, and we may only guess at the identity of the New York capitalist.

Was there any truth in Baker's unsettling claims? To find that out, a reporter was sent to follow the trail, and dutifully went to the firm of George & Eberhardt, under which authority the firm of the Gould Brothers now was working. Eberhardt was interviewed, who admitted that "he certainly remembered a machine like the one which Baker claims was made in the shop of the Gould Brothers. We made the machine for Dr. Prindham, and he spent considerable money and time on it. I also recall that one part of the machinery called for a powerful screw of chilled steel and that we found considerable trouble in bringing it to perfection." A brother of Prindham, who was employed in the shop, had more to say: "Yes, I remember it all very well...The Baker you speak of is, I think, very likely A. Beckert, a German we had here. He was employed on work of the kind you mention. It was my brother who tried to get the machine first brought out. A man named Scarttergood one day introduced a tall, lean down Easterner to him and said that the Yankee had a wonderful patent, but he hadn't enough money to bring it out. He wanted to form a stock company, but my brother said he had money enough to bring the thing out, and together they started on it. My brother often went into ecstasies over the invention. ...He often told me that all he need do was to put a quart of cold water and the thing would go and be as powerful as a Cortiss engine, while only occupying one hundredth part of the space and costing only a few hundred dollars."

The device itself was described as "an elaborate affair, with a big cylinder, like the description of Keely's machine. It had a small engine attached to it. He used to put some cyanide of potassium into the pipes and make the thing go, which it would for a short time." The device, however, was not successful; after he paid $300 for the making of the machine lever, and at least $700 or $800 more for the other parts and the patent, Prindham finally gave up in despair and "would have no more to do with it." Unfortunately, Prindham's brother could not recall the name of the patentee, but thought it likely that "the inventor who caused the doctor to invest in it is now with Keely working the same racket in Philadelphia." When the reporter asked if Keely had visited the workshop at one time or another, Prindham's brother ambiguously muttered that yes, "I have seen him around the shop, I think. He was in Newark certainly, but where there are 175 men on one floor of a shop it is almost impossible to remember the names of those who have worked here and left. But I am of the opinion that Keely has been around the shop at one time or another."[98]

As Baker's claims foremost demonstrate, by now everybody was looking into the Keely affair and each uncovered something to their liking. Skeptics and cynics simply saw compressed air as the motive power; but Keely's supporters were sure they beheld the grandiose forces of the cosmos at work in his workshop. There were those who thought of Keely as a mere swindler and as a consequence saw swindlers everywhere connected in a sinister conspiracy of swindlership. There were others who saw a herald of a new and grandiose age for all mankind in the plagued inventor.

An interesting and amusing side-effect of this controversy was that the publicity surrounding Keely dragged many a curious episode of early American history in print that would otherwise have been totally forgotten today. Five years after Baker made his startling claims, but still not had written his promised book, it would be alleged far away in Chicago that, in fact, John Keely at one time had been a person named John Adam Huss. Of this the claimant was absolutely certain: "I knew John W. Keely, the motor man, twenty-two years ago, as John Adam Huss," as a certain Eustace Wyszynski confessed in a letter to a Chicago newspaper.

In 1856, Wyszinski had met a man with that name in Louisville. There, Huss unfolded his plans for what he called a "hydraulic air engine which would relegate steam as a motive power to the past." A number of prominent citizens became interested; a company called The Hydraulic Air Company was formed in three states. Huss was paid "several thousand cash down for the exclusive rights in those states and put on a large salary to superintend the erection of a factory."

After much delay Huss was finally forced to bring out his engine, but "It was a bald failure." A second demonstration was announced, but postponed from time to time. A year later, the date was fixed. The day arrived and the stockholders met. Who then could paint their disappointment when "the ugly rumor reached them that Mr. Huss had not been seen for a day or so?" Huss fled, and 22 years after the unfortunate incident, Wyszynski showed a reporter stock certificates, cuts of the engine, and records of the money expended.

"But how does that identify Huss with Keely?" the reporter understandably asked. Wyszynski explained how the Huss affair "broke up" the business of his son-in-law. Wyszynski went to work as a map engraver with a lithographic company. "I had been working for them about two years when an incident occurred which subsequent developments have kept as vivid in my memory as if it had occurred yesterday. All efforts to trace Huss had been fruitless, and he had almost passed out of mind. But one day, as I was bending over a lithographic stone, I was aroused by a voice in conversation with Charles W. German, the head of the firm. I looked up. Two men were consulting him as to whether he could make a print of a machine of which they had a small model. One of the men was Huss. As I recognized him he saw me. He and his companion made a bolt for the door and left Mr. German standing dumfounded. 'What in thunder's the matter?' exclaimed Mr. German, when he got his breath. 'That's Huss, the air machine man,' I said. 'It's John W. Keely, the machinist,' replied Mr. German, 'and his friend is C.M. Babcock, so they introduced themselves.'" That, according to Wyszynski, was the last time that he saw John Adam Huss, "alias John W. Keely, but it isn't the last I've heard of him by a great deal." When Keely began to attract attention in Philadelphia, Wyszynski obtained a description of his device "so far as it had been made public." Wyszynski claimed that he found out that "it was the same thing we had put our money in at Louisville."[99]

But while no comments of Keely or his supporters have survived on Baker's allegations, Keely did have something to say about Wyszynski's strange tale. In a letter to a Philadelphia newspaper, Keely explained that, "The whole of this story, so far as it relates to myself, is an utter fabrication. I do not know one of the parties named...and was never in the city of Louisville in my life. In 1866 I was residing and in business in this city (Philadelphia) at 817 Market Street, and was associated with Bennett C. Wilson. ...I repeat that the whole story, so far as it concerns myself, is a base falsehood."[100]

Considering all the publicity, a great deal would be written, alleged, claimed, speculated and pondered about Keely, both pro and con. Naturally, over the years, various details about him would appear. Several people came forward and told their remembrances of him, and how he appeared to them amidst the furor over his incredible engines.

One of these, when he met Keely, was "impressed by three things: the swarthiness of his complexion, the fact that he wore very large and brilliant diamonds in a very, very dirty shirt front, and the enormous size and malformations of his knuckles. He was playing checkers in his workshop, where his mysterious machine lay silent and grim. The checker-board was grimy, the draughts were grimy and his fingers were grimy."

The person admitted that these were "Trivialities to notice in the presence of a great discovery and a great discoverer," but still, "for some unaccountable reason those were the things which impressed the narrator most, and to this day he never thinks of Keely but what there comes before the mind's vision the spectacle of a man handling dirty checkers with still dirtier fingers. His knuckles as has been said, were enormously large."[101]

Another impression of Keely was that he had "a shrewd notion of the value of publicity so far as it whetted the appetite or interest in his mechanical affairs. ...He was affable, good-natured and hearty in his manner. ...When in his shirt sleeves dilating with robust energy of speech on the possibilities of the revolution in science which he would soon create, he seemed to have all the faith and sincerity of the typical inventor struggling with a great thought which those around him had not yet the intellect to grasp. He was...well dressed, well fed in appearance, and sometimes could be seen driving out into the Park with a sealskin cap on his head, and with all the outward indications of comfort. In those days, when out of his shop, he was something of a cross in semblance between a gentleman-like gambler and a sturdy mechanic...there were few householders in the Twenty-ninth Ward who seemed to live in more substantial ease. It was this prosperous condition, together with his personal enjoyment, which first caused the cynical to be suspicious of his purpose."[102] Yet it was also written that he was "a rough, rude, crude man, whose personal habits were not expensive."[103]

"I can never forget my first sight, and first impression of, and afterwards the introduction given me to this most remarkable man," wrote another. "At the time I speak of, in my daily ride east on a Chestnut Street car, on frequent occasions I was led to observe the entrance of a tall, gaunt, pale-faced man, which striking peculiarity of countenance was in strong contrast with his piercing black eyes and his well-oiled jet black hair. He always was dressed in a rather ill-fitting suit of black cloth, and wearing a very shiny and very conspicuous silk hat, and the bottom of his shirt front was magnificently bedecked with studs of glittering diamonds. I was curiously interested to notice that he invariably left the car at the Continental Hotel. The peculiar fascination of the man's appearance was so impressive upon my guileless heart that I was led to imagine that just such a creature must surely be a black-leg or a gambler, of whom I had been taught to have a holy horror, and who I was led to believe frequently haunted the doorways of that hotel at that time." Since a friend knew Keely, and since Keely was the talk of the town, he obtained an invitation to visit him in his workshop, so off they went: "...we made our way to the back door of an old shanty in the upper portion of the city, which, as I did not keep up an acquaintance with the locality, I am not sure of its being the same veritable workshop which is still, or was until recently, the scene of so much mystifying and wondering comment. Upon a gentle rap the door was opened, and to my surprise upon beholding was the working man of mystery, the diamond-bedecked creature whom I had so ingloriously connected with the Continental Hotel."[104]

Another attempt to put a frame around the personality of the man was made by a Dr. George Mays, who was one of Keely's neighbors in the northwestern part of Philadelphia. He too was impressed by the large diamonds upon a very seedy shirt front. "I met him often during the past twenty years and must say that he always seemed to me to be under the influence of some haunting Nemesis. Outside of his workshop he was always in a hurry, whether he had anything to do or not, and it was almost impossible to engage him in a social conversation for any length of time, so important did he try to make his work appear, at least such of his neighbors as he knew were skeptical concerning his motor." His neighbors though, appeared to "have been fascinated by the man as he posed amid the mysteries of his workshop."[105]

Keely's diamonds led to another minor controversy, but he told a friend that he merely purchased some diamonds "as an investment."[106]

Another sketch, although made 10 years after the time-period in this chapter, described him as "a large, powerfully built man, with a large head, square shaven jaw, with heavy, dark-side whiskers, tinged with gray, and dark eyes which move rapidly. His movements are nervously quick and his speech is extremely rapid, as though it could not catch up with his thought. He impresses one with the belief that he is absorbed in what he calls his life study, that time is short and that every nerve must be strained to accomplish practical results while life remains."[107]

Keely also possessed great physical strength when he was in the prime of his life, and "used to take pride" in exhibiting this: "One of his feats was to put two planks side by side, set on the top of them a barrel of plaster weighing 225 pounds, place a second and even a third barrel of plaster on top of the first, and then lift the whole from the ground."[108]

These were glimpses of the man who by now was about to enter the second, most dramatic decade of his life, and the final years of a most unusual and astonishing career.

3

Prophet of the New Force
The Third Decade

"I feel that the world is waiting for this force; that this advance in science is necessary to keep the proper equilibrium in our age of progress."
John Worrell Keely, 1885

The year 1885 would find Keely still toiling away in his workshop in Philadelphia, perfecting his remarkable engines. As he wrote at that time: "I am in a perfect sea of mental and physical strain, intensified in anticipation of the near approach of final and complete success, and bombarded from all points of the compass by demands and inquiries; yet, in my researches, months pass as minutes. The immense mental and physical strain of the past few weeks, the struggles and disappointments have almost broken me up. Until the reaction took place, which followed my success, I could never have conceived the possibility of my becoming so reduced in strength as I am now."

And, while contemplating to devote less time to his work in the future, and taking a few days rest since he was so absorbed in his research and driven to the brink of utter exhaustion, he was now paying "the penalty." He also intended to withdraw entirely from all contact with the press, "in view of the unjust comments in certain journals." Keely would "give no more exhibitions after the one which closes the series;" instead, Keely wrote in a letter to Bloomfield-Moore, he would devote all his "time and energies to bringing my models into a patentable condition."[1]

In June of that year, Keely invited a number of reporters to his workshop, but had taken care in the invitations to suggest that only those should come with a certain degree of technical knowledge. At his invitation, five scientific reporters from New York appeared at the workshop. With them were J.B. Waring, a mechanical engineer, and a number of stockholders of the Keely Motor Company. When a reporter who arrived early was let into the workshop, Keely remarked that certain "curious bits of apparatus on the ground floor were intended for the junkman and not for use." Except however, a "large iron globular object, swinging on axes like a school geographical globe." Keely stated that this device was to be a new engine, which he was engaged in building. "Keely has risen to new things in the past year and a half," the reporter wrote. "He has not only discarded his old apparatus, but in making a new one he has wholly avoided any resemblance in appearance to the one he formerly used. Even those parts of the old machine which he said were absolutely essential for developing his new force fail to materialize in the present one. The machines resemble one another in this particular, viz., that both are impossible to be described by reference to anything set down in mechanical treatises. Even Keely confessed...his inability to describe the parts of his machine, and although he has given names to these parts he insisted on using different names when speaking of the same parts at different times."

Nevertheless, an attempt was made to describe the devices that he and his workman then assembled. Keely "brought to one side of the room a big piece of iron casting, and on this put a sheet of thick glass, making a stand. On top of the stand was placed a metallic bed plate about one inch thick, with holes around its edge. Tubes were placed upright on top and around the plate, and in the tube were rods. At one side was a cylinder about 18 inches high and about 2 inches in diameter...On the top of the plate and surrounded by the uprights were put boxes, cylindrical in shape containing rims from whose inner circumference steel wire prongs jutted, converging towards the center. Then a round cap was put on and bolted to the bed plate by means of upright rods. On top of the whole was screwed a globe with several apertures, to which tubes were affixed. The tubes led to strong cylinders like so called water reservoirs, only not quite as broad." When Keely thrummed over the steel wire prongs of his device, a sound like "piano strings" was heard. Then he attached two tuning forks to the device by screwing them into the holes on the edge of the bed plate. A third tuning fork, "about twice the size of these and mounted in a wooden frame, was placed loosely on the stand beside the machine." After some tapping and thrumming of the tuning forks and the steel wires, Keely announced that the device was now ready.

He struck a violin bow against the largest tuning fork and "The force was then turned on by means of a tube into a little chamber or valve into which he had spit a mouthful of water. The chamber was part of an apparatus designed to show the pressure exerted. There was a steelyard arrangement with heavy iron weights at one end and a connection at the other with the water chamber." Keely then drew his violin bow across the large tuning fork, tapped the smaller tuning forks and opened some cocks in the device. Weights that were attached to the device went up.

The assembled reporters remained skeptical; it was, for instance, remarked that Keely had called a cylinder variously "molecular resonator, an etheric resonator, and an atomic presonator." The reporters were not allowed to touch his globular device or see the inside and noted a contradiction in statements concerning the device: "Keely said it was perfectly hollow and empty. His assistants said it contained some bits of mechanism."[2]

In the meantime Keely would claim to have made more discoveries while experimenting; "My researches teach me that electricity is but a certain condensed form of atomic vibration."[3] It appears that he was still pushing ahead into the very heart of an avant-garde science. He would also suffer more accidents in the process; "I have met with an accident to the Liberator. I was experimenting on the third order of intensification, when the rotation on the circuit was thrown down in the compound resonating chamber, which, by the instantaneous multiplication of the volume induced thereby, caused an explosion bursting the metal casing which enclosed the forty resonators, completely dismantling the Liberator. The shock took my senses from me for a few moments, but I was not even scratched this time. A part of the wall was torn away, and resonators and vibrators were thrown all over the room. The neighborhood was quite lively for a time, but I quieted all fears by telling the frightened ones that I was only experimenting."[4]

Keely constructed the Liberator after having entered a "new standard for research in an experiment often made by himself, but never before successful." The Liberator was the result of, and was much smaller than his generator of the year before.[5] The generator was left abandoned after unsuccessful attempts to construct an automatic arrangement, which would have enabled anyone to operate his devices.[6]

By the end of 1885, everything seemed to promise success for the following year. Around that time, Keely wrote, "...before many weeks have passed, a revelation will be unfolded that will startle the world; a revelation, so simple in its character, that the physicists will stand aghast, and perhaps feel humiliated by the nature of their efforts in the past to solve certain problems. Taking all matters into consideration...the month of January ought to find all completed."[7] News of all this reached the parties interested; the stockholders were jubilant, the stockbrokers were alerted and a great rise in the shares of the Keely Motor Company was expected.[8] A New York newspaper even wrote that Keely had "imprisoned the ether."[9]

In August, Collier wrote to Major Ricarde-Seaver who had convinced himself the year before that Keely "had grounds for his claims as a discoverer of an unknown force in nature." Ricarde-Seaver had done so while visiting Keely in Philadelphia and witnessing an incredible antigravity experiment; with the help of a belt and some devices, only vaguely described as "certain appliances which he wore upon his person," Keely was able to move on his own a 500 horsepower vibratory engine from one part of his shop to another. There was not a scratch on the floor and, later, astounded engineers declared that they "could not have moved it without a derrick, the operation of which would have required the removal of the roof of the shop." Ricarde-Seaver did not make everybody happy with his adherence to Keely's discoveries. He was elected as a member of the Athenaeum Club in London, but was politely informed by Sir William Thompson who had proposed him for membership that he would probably lose his election by supporting Keely. He was elected, however.[10] Collier wrote to Ricarde-Seaver that "The Bank of England is not more solid than our enterprise. My belief is that the present year will see us through, patents and all."[11]

Although Bloomfield-Moore remarked that, "The journals had ceased to ridicule, and some of them were giving serious attention to the possibilities lying hidden in the discovery of an unknown force,"[12] the next year would find Keely's invention not complete for commercial use, nor any of his engines patentable, except for the Liberator. Up to this time, his experiments had been conducted upon a principle of sympathetic vibration for the purpose of liberating a "vapoury or etheric product."[13] Keely once again was giving out demonstrations in his workshop, in spite of the fact that he had written the year before that "the loss of time and the interferences from exhibitions to which I have been subjected in the past" considerably hindered his research.[14]

Nevertheless, on one of these occasions, "there were some thirty or more invited guests present, including three ladies, all of whom took a deep interest in what they saw, the only drawback being the crowded condition of the room in which the exhibition took place."[15] The object of his demonstration was of course his magnificent Liberator, which he had demonstrated the previous year to the assembly of skeptical reporters and with which he also had suffered the accident that had startled the neighborhood. The Liberator was now completed.

With his Liberator, Keely liberated his "etheric vapor" or "interatomic force," which he then vitalized and stored for use. The apparatus was about three feet high and weighed about 150 lbs. The Liberator stood on a moveable wooden pedestal, between two and three feet high, and was "entirely disconnected from the floor, wall or ceiling by any rods, pipes or wires, through which power from a distance could, by any possibility, be conveyed into the apparatus."[16] The Liberator was "the producer of the force that, it is claimed, will furnish power to the extent of 10 tons to the square inch. It is composed of brass resonants, steel tuning forks, and two or three steel and brass dials. It is about as queer looking a piece of mechanism as could be found anywhere."

Like the year before, Keely first had to reassemble his Liberator before he could begin his demonstration, for "the shop was in disorder. Pieces of the Liberator lie about in every direction." Thus Keely showed, by having the Liberator apart, that there was no source of hidden power concealed in the device. The reassembly took him half an hour. Then, "Secretary Schuellerman went out and got a quarter's worth of lubricating oil, and Mr. Keely poured some of it on the piston of a big lever, then with a little copper tube he connected the Liberator with the lever."[17]

The bewilderment of the spectators was apparent in the poetic description of the device. It had a shape that was never seen before in any machine and an interesting comparison was made to the description of the previous year: "On the top of this pedestal are piled the various circular frames and other parts of the generating or liberating machinery in symmetrical order, consisting of scores of steel-wire rods about three inches long, secured at one end and free to vibrate when struck or snapped, somewhat resembling the tongues of a musical box.

Radiating from these metallic frames are also numerous tubes screwed to their sides, one set standing out like miniature cannons from port-holes of a circular fort. Over this fortification is another similar structure surrounded with two score (more or less) of resonant tubes six or eight inches in length, secured perpendicularly, resembling a colonnade surrounding some miniature ancient Greek palace. Surmounting these singular parts is a small metal box, called the liberator proper, of very singular form, and which would seem to hold about a pint or so of gas, water, or other material, if it were not for the resonators it is said to contain."[18]

Underneath all this was a large steel Chladni plate about twenty inches in diameter, which was fastened horizontally at the center by a "metal post running up through it." Above this plate were a number of tuning forks, and below the frame, and at one side of the pedestal, was suspended horizontally an oblong hollow cylinder made of metal, which Keely called a "receiver." In the cylinder the "etheric force" was stored after having been "vitalized." From the cylinder, this etheric force was led through a flexible, copper pipe "about three-eighths of an inch in diameter." The pipe could be bent in any direction so that it could be connected with different parts of the machine.[19]

With a violin bow he tested the vibrator by drawing the bow over the tuning forks. Then he let out the air in the two-pint tube under the Liberator, and said he was ready to charge the little tube with vibrating power to the extent of ten tons to the square inch. The visitors looked on in mystified silence as the inventor, with beads of perspiration on his forehead, explained that the piston of the lever was a half square inch in area, and that it took 1,600 pounds pressure on the half square inch of area to raise the bare lever. Keely also explained that he used no water with his Liberator, but instead "got an etheric force from the atmosphere by vibratory action, which is accomplished with the Liberator, and that there was no impingement or abutment or visible exhaust from the pressure, except a slight sound."[20]

Keely drew his fiddle-bow across three of the tuning forks, and as these sounded with what he called an "etheric chord," he struck the Chladni plate with a tiny hammer. The etheric force was thus liberated, the tubes vitalized and the receiver charged with some 10,000 pounds of pressure to the square inch that would, according to Keely, increase to no less than 25,000 pounds. A weight on a lever was lifted; "in order to assure ourselves of the full 25,000 pounds to the square inch claimed, we added most of our own weight to the arm of the lever without forcing the piston back again."[21]

Keely then demonstrated his vaporic gun by leading the force from the receiver "by the same kind of flexible copper tube, attaching one end of it to the breech of the gun."[22] Then, "He took some vulcanite and rubber wafers for packing, and then rammed a leaden bullet, one and one-half inches in diameter, into the cannon with a broom handle. An iron plate was passed outside of the back door. The cock from the given point tube was opened and the bullet went whizzing through the panel of the door and flattened itself on the iron plate. There was a report about as loud as the sound made by firing off a revolver when the bullet left the cannon. There was no recoil of the gun, and the barrel was about the same temperature as the atmosphere. Three bullets were fired in quick succession, and Mr. Keely said that there was sufficient power in the tube to shoot 500 more bullets..."[23]

The famous Keely motor of which the world had heard and read so much, "a smooth hollow sphere of metal about two feet in diameter," and which consisted in a 25 horsepower rotary engine, was also run with his etheric vapor. "What is most astonishing about the rotation of this sphere, by simply turning on the vapor," a witness remarked, "is the fact that there is no escape for the gas anywhere, after it has done its work, nor any outlet or exhaust-pipe for such escape, as is well known to be absolutely necessary in the use of any gas, liquid or vapor known to mechanics, and by which engines are readily driven."[24]

A description of Keely's "200 horsepower engine," on which he had been working "for some time," has also survived: "It is encased in copper and is full of brass resonants. It looks like a patent washing machine. Mr. Keely says it will be working in a machine shop on Vine Street below Sixth inside of 60 days, and that then he will be ready to take out patents. The machine can be put in any shop or factory, and will run machinery of 200 horsepower. With one expulsion of the Liberator of one-eight of a second the machinery will run all day. Mr. Keely claims that by simply charging the tubes daily with the vibratory power the machinery in a big factory can be run without even having a Liberator from which the mysterious power is originally produced."[25]

Towards the end of 1886, Keely was hard at work on his vibratory generator. He would enjoy his first vacation in years,[26] and would still be giving his regular demonstrations. On September 24, 12 persons witnessed his experiments "calculated to demonstrate the power of sympathetic vibration as applied to dynamics." "For the purposes of these experiments," a witness later wrote, "we were shown into an upper room about 12 by 14 feet in size, across one end of which extended a plain workbench. Upon this bench, extending more than half the length of it, were stretched two wires, tensioned to vibrate when agitated, the first wire giving forth a low note, the second a tone considerably higher." From this wire a small steel rod ran "really consisting of three sections, so arranged that the ends touched," and ended against a heavy plate of glass. At the end of the bench, a smooth, copper sphere of about a foot in diameter had been hung in a circular frame. The sphere was "expected to revolve by the force of sympathetic vibration."[27]

One axle of the sphere ended in a rubber bulb, "like a small syringe." The sphere was placed between the thick plate of glass against which the steel rod ended, and a similar plate resting against the wall. The four legs of the circular frame also rested on a glass plate. On the bench were also two large tuning forks, "fixed upright in their moveable spine resonant cases, standing some three feet apart, and the nearest one that distance from the sphere." Between the tuning forks was a small brass object resembling a snail shell mounted on a pedestal, which was called a resonator.[28]

At the opposite end of the room, "thrown carelessly upon the floor," was a flat ring about one inch wide, with a diameter of about forty inches. Fixed into this ring was a large tuning fork at right angles with its flat surface. Attached to the inner edge, and extending entirely around the ring was a brass tube half an inch in diameter. The two ends of the tube ended in a small sphere, "about the size of an apple." Resting on the floor, between the ring and the bench, was a small iron receiver that Keely called "double compressor." A small copper pipe led from this receiver to the little ball or sphere of the brass tube on the inner side of the flat ring. Another copper pipe was attached to the receiver and went into the next room to the Liberator. Through this long pipe, and through the double compressor and the short pipe went the vapor that charged the brass tube in the ring on the floor.[29]

The beginning of the actual experiment was marked by the strangeness of it all; Keely obtained what he called "the mass chord" of those willing to participate in the experiment. He determined this mass chord by putting a steel bolt in the hands of the volunteers. It, which he called a "sensitizer," resembled a "car coupling pin, but shorter, and having a one-quarter inch hole through its length." From this sensitizer ran a 10-foot "hair-like" wire that ended in a reed whistle. Then Keely dropped the whistle into the snail-shell resonator on the bench. This produced a certain sound by which he could determine whether or not the volunteer held the right sensitizer. If not, a smaller or bigger one was handed. No two persons used the same size. After this unusual ritual, the brass tube within the ring on the floor was "charged with the force. A sheet of vulcanized rubber in the top of the double compressor was blown out "with a report that indicated great power, and, as Keely had stated that this rubber sheet would only yield to a pressure of 2,000 pounds, it caused "some of the brave savants present to seek positions in the remote corners."[30]

Keely then went to the next room, where his Liberator stood, as well as another curious device called the "140 octave resonator." This was a brass tube of about four and a half inches in diameter and some eight inches deep. This mechanism was "supposed to be full of 'resonators,'" and the top "suggested an old-fashioned candle-mold." With this device, Keely claimed to be able to give 140 octaves.[31] One of the volunteers then went within the circle upon the floor, within his hand the steel pin of his mass chord. Other volunteers then bowed the tuning forks at the command of Keely. The sphere "away across the room," began to revolve. "Slowly at first, but with an increasing speed as the forks continued to vibrate. When the volunteer stepped out of the circle, the sphere stopped at once; stepping back in the circle, the sphere 'immediately responded.'" Collier, who was present, tried it. To him this experiment was also entirely new.[32]

The witnesses were greatly impressed; "The entire absence of careful preparation, for the gathering, as shown by Mr. Keely's repeated search for objects needed at various points, that should have been gathered and placed where wanted beforehand; the change made on the instant by the substitution of one article for another, that was found to be misfit or would not work, etc., all added force to the results shown."[33] This strange demonstration clearly showed that Keely had traveled a long way since the days that he wrested energy out of enormous engines.

In September 1886, an article appeared in Philadelphia's *Lippincott's Magazine* entitled "Keely's Etheric Force." It was the first article accepted by any Philadelphia editor, setting forth Keely's claims on the public "for the patience and protection which the discoverer of a force in nature needs, while researching the unknown laws that govern its operation. Up to this time Keely had been held responsible for the errors made in the premature organization of the Keely Motor Company, and the selling of stock before there was anything to give in return for the money paid by its investors," Bloomfield-Moore remarked.[34] A month later, two Philadelphia engineers, J.H. Linneville and W. Barnet LeVan, made a thorough examination of Keely's Liberator, and the lever by means of which the energy of the force that was generated by the Liberator was measured. Both devices were completely dismantled for that purpose. The outcome of the examination was that the engineers both certified that a weight of 550 pounds was raised on the end of the lever during the test, showing a force of 15,751 pounds to the square inch. It was noted that the highest pressure possible to obtain through compressed air was 5,000 pounds to the square inch.[35]

Elaborate etchings of his various devices were printed in a magazine the next year.[36] Reinforced by, or perhaps in spite of, eyewitness accounts of experiments, demonstrations and the occasional examination such as described above, the controversy became heated once more.

Another visitor wrote that, "The 'sympathetic etheric force' which Mr. Keely claims to have discovered may be best described as coming nearer to the primal force of willpower of nature than any force yet liberated from her storehouse. Its inventor seems to claim for it that it is that primal force itself; he speaks of the breath of life which God breathed into man's nostrils at the creation of the world. Whether Mr. Keely's force is itself elementary or not, who shall say? He claims that at least it is the last and greatest step in the analysis of matter."[37]

In the meantime, news of Keely's doings stretched out across the Atlantic. In 1887, in England a series of articles appeared in *The British Mercantile Gazette,* its June issue devoting more than eight columns to the progress and position of Keely and his discovery of the etheric force.[38]

Keely, who by now had totally abandoned water as the basis of obtaining the power, directed his attention solely to air as a basis. As a result, he announced in a circular dated June 9, 1887, that he would only use a wire instead of a tube as the connecting link between the sympathetic mediums to evolve the ether and operate his machinery.[39]

Apparently this would lead Keely into a whole new land of discoveries and possibilities, the exact nature of which we can now only speculate. A newspaper for instance wrote that Keely was involved in "making a flying machine,"[40] and on December 14 a new statement by Keely, a "voluminous report," was read by Collier during a meeting of the stockholders of the Keely Motor Company at

Sherer's Hall at Eight and Walnut Streets. In it, Keely reviewed his efforts and experiments since 1882, when he was engaged in the construction of a generator for the purpose of securing a vaporic or etheric force from water and air. When the device was completed, he soon found out that it was "impracticable owing to the impossibility of securing graduation." After a series of interesting but laborious experiments, Keely built a Liberator in March 1885, which he operated together with the generator. Although he considered this "a stride in advance of anything accomplished hitherto," he also claimed that, "Meanwhile new phenomena have unfolded to him, opening a new field of experiment."

As a result, Keely became "possessed of a new and important discovery," and he would not need his generator nor his Liberator anymore. "His operations will be conducted without either the vaporic or etheric force which heretofore played such an important part in his exhibitions. What name to give his new form of force he does not know, but the basis of it all, he says, is vibratory sympathy. It may be divided too, into negative and sympathetic attraction, these two forms of force being the antithesis of each other. As to the practical outcome of his work...Mr. Keely could make no promises. He had no doubt that he would sooner or later be able to produce engines of varying capacity, so small as to run a sewing machine and so large and powerful as to plow the sea as the motive force in great ships. Among the work yet to be done is the construction of a sympathetic machine of a very delicate character. While this will be a perfect vibratory structure itself, its function is to complete the work of graduation or governing the force, but as to what length of time it will take to complete the work he cannot say."

Keely's report was accepted, but a newspaper shrewdly remarked that "The most important fact contained in Mr. Keely's report was suppressed. The part that was not read to the meeting informed the stockholders that he had in contemplation the formation of a new company and that he had already sold a number of obligations for the new issue of stock in order to raise money to prosecute his experiments." And indeed, Keely told a reporter the same day that the obligations called for between 30,000 and 50,000 shares of stock, and that "the new capitalization would be on a basis of $15,000,000." He furthermore stated that his old shareholders would receive share for share of the new issue. He would retain about 40,000 shares.[41]

The year 1888 saw the publications of both a curious pamphlet and a remarkable book. In July, the pamphlet "Keely's Secrets," written by Bloomfield-Moore, came to light. The pamphlet, which had "a wide circulation,"[42] was both a theoretical exposé and a defense against Keely's critics. In its pages, she trusted Keely's visionary ideas of employing his discoveries for a system of airflight on which he had started working the year before, and of a possible application of his discoveries in the cure of disease. The pamphlet was published by the Theosophical Society in London, who also published *The Secret Doctrine,* written by Helena Blavatsky, Russian-born mysticist, occultist and founder of the Theosophical Society. In the huge tome, she devoted an entire chapter to Keely and his discoveries, and spoke out in favor of him. From that time on, theosophists and occultists who were not already doing so would direct their attention to the enigmatic inventor from Philadelphia.

In September of the same year, Blavatsky published an article on Keely in a French theosophical magazine.[43] The same month in Paris, *Le Figaro* printed the expectations of French inventor Colonel Le Mat saying that "the chain which holds the aerial ship to earth would be broken asunder by Keely's discovery. The nineteenth century holds in its strong arms the pledge, that sooner or later the aerial navy, so long waited for, will traverse the trackless high roads of space from continent to continent."[44]

Farther away in the Austrian city of Vienna, an Austrian nobleman, the Chevalier Griez de Ronse, printed a series of papers on Keely's discoveries in a Viennese journal called *The Vienna Weekly News,* of which he was the owner. One of these articles mentioned that the attention of English scientists had been drawn to Keely's claims, in regard to having imprisoned the ether, by Professor Henri Hertz's experiments in ether vibrations at the Bonn University. "Keely, like the late Dr. Schuster, claims on behalf of science the right to prosecute its investigations until a mechanical explanation of all things is attained." The Austrian nobleman was well informed; obviously he had read the pamphlet of Bloomfield-Moore, whom he might have known personally, and he was also aware of Keely's imprisonment in a jail in Philadelphia which had happened the same year.[45]

The year 1888 would also be overshadowed by legal matters. Keely would be sent to jail, this time because Bennet C. Wilson reappeared to claim what he thought was rightfully his. Wilson had a curious tale to tell. He claimed that he had sponsored Keely's first machines some 22 years ago.[46]

Wilson financially supported Keely's experiments from 1863 until 1872, and he provided him with a workshop on Market Street. But as the years past, and Keely's experiments met with little or no success, Wilson got tired of advancing money. He further claimed that, late in August 1869, Keely said that his device would soon be ready for sale, but he needed funds. Keely then made an assignment in writing of his whole right and title with all interest in the motor which, according to Wilson, was called a "reacting vibratory motor."

Wilson's patience was "exhausted when his fortune was exhausted, and Mr. Keely turned to new pastures, and with new names for the machine he had in mind, he found new patrons." From 1871 until 1878 Wilson had not been able to get "satisfaction out of Mr. Keely." In 1878, Wilson secured access to Keely's shop and, he claimed, there stood the machine "upon which Keely was experimenting for his newest and latest motor company," and it was the same with which he had "practiced on in the old Market Street machine shop." So now, Wilson's bill claimed that these assignments Keely made to him entitled him to all the patents that Keely had taken out for the perfection of his motor.[47]

The only thing that Keely could do was to prove that it was an entirely different engine that he had been working on. In all this legal confusion, it was somehow overlooked that Keely had not obtained any patent on any engine that had resulted from his line of research after 1871. Strangely, while always quick to bring that aspect under attention, the press fell silent over this important detail. This was unfortunately not the only legal case that Keely became entangled in. Sadly, the other case was once again directed by a faction of the Keely Motor Company, the very company that had originally been founded to support him. The case was a reflection of what had happened to Keely in 1882. In July, the same month that Bloomfield-Moore's pamphlet was published, four directors of the New York branch of the Keely Motor Company brought a suit against him in order to force him to "turn over his property to surrender patents, and to disclose his secret to some one appointed by them."

This suit was against the will of the three Philadelphian directors, who on September 8 organized a meeting at Keely's request. That afternoon, 60 stockholders met in the third-story room at Eighth and Walnut Street. Keely also issued a circular in which he stated that he had reason to believe the majority of the stockholders were opposed to the suit against him, "but he wanted to know definitely what they were going to do about it, as his own policy would thereby be shaped."

Not surprisingly, the New York directors were absent from the meeting, as was Keely, but he did sent a letter to the meeting, which was read. In it he wrote that he had received a number of replies from stockholders, "residents of New York and elsewhere, all of whom deprecate the revival of the suit against me and express themselves in favor of the proposition of reorganization of the company as submitted by me to the stockholders. I have also been requested, verbally and in writing, by many stockholders to take into my confidence Mr. J.H. Linnville and Mr. W. Barnet LeVan in connection with Mr. Boekel, and avail myself of their aid and advice in the matter of applying for my letters patent on my invention, and I shall exhibit myself of their aid and advice in the matter of applying for my letters patent on my invention, and I shall exhibit to them from time to time progressive experiments and explain the same to them."

The outcome of the meeting was that, of the 60 stockholders present, who held proxies for about 150 others, all were in favor of Keely. One dissenting vote though was heard when the reorganization was proposed. The proposition was to decrease the value of the shares from $50 to $20, but to increase the number of the shares to $250,000, so that the capital would remain the same.[48]

The incident also demonstrated that, as the previous years had witnessed, the Keely Motor Company had never been a unanimous organization. Some of its members were driven by the simple prospect of profit, and profit alone; but those who did so would become greatly disappointed over the years. Others felt more favorably of Keely and perhaps shared the same sense of wonder with him. A clear indication revealed itself at a meeting held three days after the meeting on September 8. Collier, Thomas and William Clark resigned from their positions as directors of the Keely Motor Company. In their place Boekel, New Yorker George Hastings and Henry N. Hooper of Brooklyn were appointed. New Yorker Guilian S. Hook was elected Treasurer.

Thomas vented his frustration of the whole affair in a long tirade: "We withdrew as a body...because the suit brought against Mr. Keely by the Board of Directors of the Keely Motor Company"—here Mr. Thomas spoke very sarcastically—"was brought by the New York directors without in any way giving the Philadelphia directors any intimation whatever of what they were going to do. They did not do it in the board, but acted as a board themselves without any authority from us. They not only acted without our previous knowledge in the matter, but as a Board of Directors of the Keely Motor Company they appropriated themselves money to push the suit. To get the money...they sold the stock of the company at a great sacrifice. Of this I am certain. It was roughshod all through. We had no say in running the machine at all and were treated disgracefully. They ignored our Treasurer to such an extent that he resigned. Whenever they had any money to pay they would pay it themselves, and would not allow it to come within 50 years of our Treasurer's hands. ...After we discovered that the suit had been brought we canvassed the matter thoroughly and withdrew. ...I see by the election that the New York directors gain practically two members and a Treasurer, leaving only one Philadelphian. Well, they are stronger now than they were before."[49]

When on September 25 the new Board of Directors met at the office of the Keely Motor Company at 911 Walnut Street, it was expected that the Board would take action upon the decision of the stockholders at their meeting of September 8 to reorganize and thus increase capital stock to "provide funds for the working out of the inventor's alleged new discovery."

But that was by far not the most important matter to discuss, for Wilson's bill of January 3 had reached alarming proportions and the court proceedings of Wilson against Keely took a large share of the meeting. The directors also stated that "it had been decided to investigate 'a point' in the matter which, it is believed, will place everything right between Keely and the directors." This precise point was not explained. Boekel, who knew more about Keely's devices than any other man, gave the directors "a description of several interesting experiments recently made by Keely on his new 'sympathetic attraction.'"

Keely stayed confident. The machine that Wilson claimed he had the rights to, and the machine that Keely was working on were "entirely different." Keely informed the experts that he would be ready to show them his machine and explain its working, according to the decree of the court. The decree did not demand that Keely would "put the machine to work."[50]

Wilson's case against Keely was brought to the Court of Common Pleas, and the committee of experts which was appointed to investigate the matter was, from Keely's side, an unfortunate assembly. One of the appointed was Professor William D. Marks, who already had his dealings with Keely,[51] and could hardly have been called impartial. Answering to a reporter on Collier's threat to have him arrested on a criminal charge and bring civil suit against him, he said, "Let me give you my statement of the affair. ...I think it was in 1878 that Mr. Collier approached me as secretary of the company and requested me to make an investigation of the Keely Motor. ...In the course of my investigation, after I made three visits to the shop of Mr. Keely, during which time I carefully watched all of his manipulations, I became convinced that the source of his alleged power was compressed air, located in a cylinder, which he called an 'expulsion tube.' I told Mr. Keely that I believed compressed air was concealed in the tube and asked to be allowed to take a monkey-wrench and unscrew a stop-cock in this tube, which would at once prove the correctness or incorrectness of my belief. I was refused both by Mr. Keely and Mr. Collier, under the absurd plea that it would 'desensitize' the machine. I then asked that they allow me to test a gauge on which there was an alleged pressure of 50,000 pounds. I was refused this opportunity also, on the ground that Mr. Keely desired to use this gauge on the following day. I asked this because I believed the gauge to have been tampered with. Then I said to Mr. Keely and Mr. Collier that both their machine and themselves were a swindle and a fraud."

Apparently Marks only joined the committee "at the request of Dr. Pepper," provost of the University of Pennsylvania, and in reply "to a very courteous note from Judge Reed."

Marks complained that the task was not easy: "Since this investigation has begun the committee has been shown a wreck which is apparently what is left of what Mr. Keely claimed to be the motor of 1878, and also a copper globe which has since been added. ...Messrs. Keely and Collier have refused to put together the machinery; to show its method of operation, or to furnish any explanation. ...A type-written copy of this alleged explanation, which is not any explanation, has been furnished to the committee. It is still my opinion that the Keely motor is a fraud, Mr. Keely a swindler and possibly also Mr. Collier."

Naturally Charles M. Cresson, who also was one of the experts appointed by the Court, and who also had dealings with Keely in the past as the chemist who analyzed his vaporic substance,[52] said that the charge that the committee was hostile towards Keely was not true: "I was chairman of that committee, and I told them that we were not to inquire into the merits or demerits of this affair. We were simply to find out whether two machines were alike. If so, then Wilson's claim was good; if not, his claim was not good." Was Cresson impartial? It was Cresson who claimed to have been present during the past fifteen years at several exhibitions of the Keely Motor and who had said a week before to a reporter that it was his opinion that "No exhibit shown to me has to my mind demonstrated the fact that any of the work performed (such as the lifting of a heavy weight, exerting an enormous pressure per square inch, or running a rotary or other engine, or projecting bullets from a gun) is of necessity the result of any unknown or mysterious new force."[53]

Naturally Judge Finletter, who presided over the case, denied allegations that he had remarked at a dinner party that Keely was a fraud and should be put in prison.[54] These were however, the ingredients for Keely's trial.

On November 17 at 10:00 a.m., the Court of Common Pleas was called to order. Rufus E. Shapley, Wilson's lawyer, addressed Judge Finletter and applied for a writ of attachment compelling Keely to appear in court at once. Finletter granted the request and an hour later. With the attachment, Deputy Sheriff Pattison headed for Keely's house and workshop. Keely was not in his house or his workshop, but in the vicinity of the Court House. Pattison searched the surroundings of the Court House, but could not find him. At 1:00 p.m. Keely was still not found, and Finletter adjourned court until half an hour later. Ten minutes before the court was again called to order, Keely walked in, appearing "to be laboring under suppressed excitement but walking erect." He was accompanied by his lawyer, Joseph J. Murphy. The two quietly took their seats, and at 1:30 p.m. Finletter called the court to order. Murphy told Finletter that Keely would like to defend himself, which he would do with a written statement. Keely then arose, took off his overcoat and kissed the book. Meanwhile Finletter told Keely that "You have been brought into court on an attachment for contempt in not obeying an order of the court. You have now an opportunity to purge yourself of the contempt." Keely answered that he had done everything that had lain in his power to obey the court. He then read a long statement, "which he declared was true in every particular, giving an account of his interviews with the experts, who, he said, were hostile to him and unable through prejudice to make a fair report regarding the motor."

After he finished, Finletter showed neither sympathy nor interest but instead, without for a moment considering or commenting upon Keely's statement, recited "in a low tone his decree, which he prepared during the reading of the statement. Finletter ordered that Keely 'shall be committed to the county prison to be kept there and confined in custody until he shall have purged himself of said contempt and until he shall have been legally discharged from said contempt.'" All the while Keely was standing in front of the Judge, "and listened attentively, his face bearing a look of suspense and anxiety." After Finletter was through, Keely "appeared to be dazed" and remained standing until Murphy asked him to sit down. Murphy immediately sent word to Wayne MacVeagh, Keely's senior lawyer.

Keely left the courtroom in the custody of Pattison, and they went to the county prison. Keely and Patison crossed Independence Square, "followed by a dozen pairs of eyes," with Keely leading the way, and Pattison in the rear. The two walked out Samson Street to Ninth, where they got a carriage and were driven to the prison.[55]

Notwithstanding Keely's typewritten explanation that he had given the committee, they too labeled his attitude as one of refusal; and it had been this refusal to disclose the exact nature of his discovery and to give information to the appointed committee that led Finletter to decide to imprison Keely for contempt of court. Cresson explained what that meant: "The commission was directed by the Court to ascertain such facts as would determine the similarity or dissimilarity of the machine and without any reference to their originality, economy or merit. The commission had no desire and has not made the slightest attempt to go beyond the narrow line of the duty imposed upon them by the Court. Four out of the five members which composed the commission reported simply the facts, which, briefly stated, were that Mr. Keely had not made it possible for them to do what they were directed to do, because he has not exhibited (except in dismantled condition) or operated the machine called the 'Keely Motor,' has not given them any intelligent description of it, and that he had obstructed rather than assisted them in their efforts to discharge the duty assigned them. They have done nothing to provoke or justify Mr. Collier's abuse."

Thus Keely was locked in a felon's cell in Moyamensing Prison.[56] The carpetless cell was 9 feet wide and 14 feet long. Prisoner #150, he was locked up and left to look "gloomily through the small cell window out upon that gray November day."[57]

He spent a "quiet Sunday" in Moyamensing Prison, and although the night was cold, he said that "he slept as comfortably as he could expect under the circumstances." When breakfast was brought in, Keely's cell was skipped, for he would have a "heavier breakfast" half an hour later. A reporter wrote how Keely "sat near the door and listened with deep interest" to a sermon during a religious service, and "While the inventor was trying to fill out his afternoon nap up in his cell, a number of persons were making anxious inquiries. ...A few friends turned away when they were told that nobody could get into the prison on Sunday."[58]

In the meantime, it was declared that the Court's commitment left Keely "to fix his own terms of imprisonment," meaning that he should remain committed "until he purges himself of contempt by complying with its order to explain his motor."[59] November 19 would find Collier, Murphy and MacVeagh in Harrisburg, where they were making applications to the Supreme Court of Pennsylvania for his release,[60] since "a very dexterous legal move had emanated from the brain of MacVeagh."[61] Keely's lawyers secured a writ of habeas corpus from the Supreme Court, and he was released on a $1,000 bail.[62]

A fortunate side-effect of Wilson's legal proceedings against Keely was that with this Keely was able to successfully defend himself against allegations of swindle, printed in a Philadelphia newspaper in January 1889. It was alleged that Keely at another time and place had been known as John Adam Huss, who had been involved in what appears to have been an elaborate swindle. Now, with Wilson's suit against him, he was able to prove that he was not.

Two months after his release, on January 28, 1889,[63] Chief Justice Paxton of the Supreme Court, upon hearing the case, reversed the decision of the Court of Common Pleas, and discharged Keely from the contempt of which he was adjudged guilty by Judge Finletter. The opinion of the Supreme Court was that "the order commanding Keely to exhibit, explain and operate his motor was premature, and that being the case the Court below had no right to enforce the attachment committing the defendant to jail for contempt."[64]

The methods of the committee, and even its installment, were also considered highly dubious: "After issue was joined an examiner could have been appointed and the proofs taken in an orderly manner. Instead of so proceeding a commission of experts was appointed to examine the defendant's machine." Keely was not only required to exhibit his device, but also to operate it and explain the mode of construction and operation, a measure considered to require "considerable expense to clean the machine, put it together and operate it."[65]

Even though Keely's attitude was labeled as one of refusal, the Supreme Court thought otherwise: "The defendant appears to have been willing to exhibit it and in point of fact did so. ...But to make an order not only to exhibit it but to operate it, the practical effect of which was to wring from him his defense in advance of any issue joined, was an improvident and excessive exercise of chancery powers. It is the more remarkable from the fact that the plaintiff's case, as shown by the exhibits and the drawings, was sealed up in an envelope and retained by the Court, access to the same being not only denied to the defendant, but even to the experts appointed by the Court."[66]

Court of Common Pleas Judges Finletter, Gordon and Read grudgingly said that the Supreme Court was "laboring under a misconception when they reversed the decision of the lower court in releasing inventor Keely from prison."[67] And so while the legal battle was over, and further legal details, pro and con, had become quite arcane, a newspaper concluded that "The right of imprisonment for contempt of Court is, and always will be, accepted; but when a vulnerable point can be found in the judicial mail, it will always be pierced to discharge the prisoner. The Judges don't just state it in that way, but that's about the way, all the same."[68]

It would be a year later, on February 25, 1890, that poor Wilson would finally give up. Again his patience seemed exhausted when his fortune was exhausted. Keely's release "practically ended all litigation," a friend of Wilson said, "there was no way of getting the evidence in, and without it the case had to end. Mr. Wilson is not a wealthy man and he could not afford to begin a new litigation. It was proposed some time ago to clear the records by making the suit disconnected, but Mr. Wilson hesitated to do that until yesterday, when he had his counsel discontinue the case."[69]

Keely's release was wryly noted in the press: "Just as the sympathetic world was beginning to despair of ever hearing from him more, Keely, the motor man, bobs up again as new and fresh as when he first came to the surface, and assures us that his mysterious force is really bottled at last and that he is going to let some loose in a few days. Some wicked people will doubt of course and say they were told this before."[70]

After his release, and during the winter of 1888-1889, Keely was turning his attention more and more to a series of experiments of a different nature. Perhaps his former experiments were done with a more or less commercial prospect in mind, but now he was fully concentrating himself upon the unknown force and its laws of which he still was ignorant. He admitted "that he cannot construct a patentable engine to use this force till he has mastered the principle."[71]

Nevertheless, and undoubtedly under pressure from the Keely Motor Company, he demonstrated a preliminary commercial engine in November 1889:

"before he had completed his graduation, he was induced…to apply a brake, to show what resistance the vibratory current could bear under powerful friction. A force sufficient to stop a train of cars, it was estimated, did not interfere with its running; but under additional strain a 'thud' was heard, and the shaft of the engine was twisted."[72]

In 1890, Keely was still amidst an astonishing range of new experiments based upon levitation. While its results are similar to those of antigravity, levitation is looked upon as the suspension or neutralizing of gravity, while antigravity is considered a force opposite of gravity. Although Keely had experimented with the suspension of gravity years before and built devices to do so, he was now turning his attention more and more to the navigation of the skies. This remarkable evolution in his ambitions was not that unusual. For years his experiments were limited to the production of his force, the raising of a lever, the firing of his vaporic gun and the demonstration of a vacuum greater than had ever been produced. Since 1887, he had been working on what a newspaper referred to as "a flying machine." A year later he pursued his research on a line that enabled him to show a certain progress year after year. This also meant that he never repeated his experiments. While discarding or improving upon his research equipment after having obtained the results that his theories would lead him to expect, he continued his investigations with the information thus obtained in ever-new directions;[73] his thoughts traveling along the same lines.

On April 6, Professor Leidy of the University of Pennsylvania witnessed certain levitation experiments. Leidy was greatly impressed, and declared to a reporter that he was convinced that Keely had discovered "a new force, distinct from magnetism or electricity."[74]

Another experiment that Leidy witnessed, and that resembled Keely's demonstration in 1886, was the application of the force through the atmosphere from one room to another "without any other medium of conveyance than a silk cord. The door into a little back shop, whose existence until then was unknown, was now open and a silk cord passed from the transmitter to a large bronze globe, which was mounted on an axis horizontally. The other end of the cord was not fastened to the globe, but to a slender bar of steel supported on an upright near it. A plate of glass an inch thick was between the end of the resonant steel bar and the globe. A similar piece of glass was put between the wall and the other end of the bar. Glass was put under the upright which supported the bar. Glass plates were also put under the upright which supported the axis of the globe. Keely then took a harmonica in his hands, and, allowing the silk cord from the transmitter to pass over the harmonica in contact with it, began to sound notes on it. When the sympathetic chord was struck the vibratory force, he declared, was conveyed along the silk cord. The bronze globe, which was about fourteen inches in diameter, began to revolve about its axis. The faster Keely played, the faster the globe whirled."

Having seen all this, Leidy was given to a futuristic extrapolation of his own: "Some day…I suppose a young lady will be able to play on the piano and set her father's mill to grinding. I see no possible source of deception. This demonstration is wonderful. There is no explanation of the effect thus produced, except by vibratory force such as Keely assigns as the cause."[75]

A month later, on May 8, Keely suddenly announced that he was finally ready. It was a rainy day in Philadelphia, with "nothing to break the silence of the street outside save the jingling bells of an occasional car. The air of the workshop, outer and inner sanctum, was as still as the murk of a tomb is supposed to be," a reporter with an obvious feeling for atmosphere wrote. "I have finished my work! I have discovered my force! I have accomplished my task!" Keely said to the startled reporter.[76] He also told him that "there is nothing for me to do but to wait until the mechanics can make me a perfect machine. When that is done I will at once demonstrate my discovery to the world. That is my great, and, indeed, my only difficulty now! The apparatus with which I am compelled to work is, and has been, mechanically defective. When it is correctly made I will challenge the world to deny what I affirm with and through it!"[77]

The reporter who dutifully wrote Keely's statements down was greatly impressed and drew an interesting description of him at that time: "John Ernest Worrell Keely stood with his elbow resting on one of his 'syrens,' the name he gives to a machine which sings, he says, and makes the atoms of universal ether dance. It was the first time I saw him alone, face to face," the reporter wrote, "Here he was, six feet one inch in height...heavy black eyebrows over deep-set, earnest brown eyes; high cheekbones over which rested the legs of the 'artificial eyes' his researches had long since reduced him to; a thick, black shock of hair, slightly streaked with gray; slightly bent across the broad shoulders and looking me straight in the eyes, with an expression at first guarded, almost crafty, soon opening into apparent confidence."[78]

Keely also confided to the reporter that his force was "not like steam, nor electricity, nor compressed air, nor galvanism—it is none of these, and it is not akin to any one of them." He also made a statement that would become quite familiar in the years ahead of him; "Now, if outside mechanics can make my instruments, I am all right. If they can't, I will have to wait until I find one who can."[79]

In 1890, the then world famous palmist Cheiro visited Keely's workshop,[80] a place that was the scene of so many wonders. Forty-five years later, Cheiro would publish his memoirs of this visit, along with other materials about him. Although sympathetic, these memoirs do not serve to clarify matters as they are untrustworthy, as are his other writings about him elsewhere, which we will see in chapter 9.

According to Cheiro, it was in Bloomfield-Moore's London house that he first heard of "the Keely-Motor."[81] At Keely's workshop, Cheiro amongst others witnessed his antigravity experiments and a "revolving globe of glass...It was of very simple construction, merely a large glass globe balanced on a pivot of platinum that, when spinning, kept its equilibrium by centrifugal action. ...This globe was also started by a vibration from the violin. When it had attained considerable velocity, Keely made me lift it off the table and carry the whole thing, wooden stand and all, several times round the room. As its revolutions became more and more rapid I grew alarmed, believing it might any moment fly to pieces. Again a discord from the violin and in a few minutes it stopped."[82] Unfortunately, as Cheiro's account is inaccurate, the question remains open if this was merely a distorted memory, or indeed a description of a device that Keely had actually built.

The year that Cheiro visited Keely, foreign publications again wrote about him and his discoveries. In March the *Anglo-Austria* contained two papers on the subject, and in October the London periodical *Invention* published its opinion that amongst others, reiterated Professor Leidy's visit.[83]

Yet in the autumn of that year, Keely would find himself again threatened with lawsuits and harassed by demands to give demonstrations in order to raise the price of stock. A subscription was even started to raise funds for the prosecution. Keely now found himself in a difficult position. He had to choose to either continue his research with, as the ultimate end, the completion of his system, or to divert his course and to resume his efforts on the construction and perfection of an engine that could be patented and made commercially profitable.

At this point, an attempt was made to hand out a written statement in which Keely explained that it would be far more profitable in the end if he could continue his research, since in the preceding years he had done "scarcely more than liberating the ether." The effort to circulate his statement failed, but instead an unfavorable exposé of the history of the company was circulated.[84] Around the same time news about him and his discoveries had not only been published across the United States, but also in foreign publications and theosophical magazines.[85]

Towards the end of January, 1893, he invited 30 "sanguine capitalists and promoters" from Philadelphia, New York, Chicago and Boston, to his shop to investigate the progress of his work on the development of the motor. The delegation assembled at Collier's office at 910 Walnut Street and held a meeting with closed doors. Those present were also pledged to maintain silence on the details of what they subsequently heard and saw. Collier read a report written by Keely on the progress he had made so far, explaining the difficulties that were delaying his success. Those that were present said that, although they could not give out any details concerning the report, Keely assured them that success was imminent and a "limited time and amount of money he would require to complete the greatest motor the world has ever known." After the meeting the assembly was driven to his workshop where for two hours he demonstrated his devices and explained the workings of the different parts of the machinery and the application of the new force.[86]

One who was present, a Jacob Bunn Jr. who was the vice president of the Illinois Watch Company, later told the press that he was not at liberty to give the contents of the report, or the names of those present, "except to say that they were men of large capital, who were accompanied in their visit of inspection by some of the best engineers and scientists of this country." Much of what he had seen was "wonderful," and he could only agree "with all the others, including the scientific men present, that Mr. Keely has discovered and utilized to a measure a seemingly new and powerful force which is understood alone by himself."[87]

Keely said that he had provided himself with "suitable furnaces and tools," and would begin immediately with "the difficult task of making these wheels" himself. "I can make them perfectly," Keely allegedly said, "and when this is accomplished I will put in motion a power that can run a train without fuel of any kind or any other force for thousands of years continuously if desired."[88]

Notwithstanding these claims, the public once again lost its interest, although the book *Keely and his Discoveries: Aerial Navigation,* written by Bloomfield-Moore, and published in London in October, fuelled the interest in England to a considerable extent. So much so that Prof. Dewar of the Royal Institution in England, who had "liquified oxygen" and was "the foremost chemist in England in all matters pertaining to gases" was asked if it was true that Keely would place all the facts relating to his motor in Dewar's hands. Two years before that, Dewar agreed to investigate Keely's engines if he went to America. However, by 1893 Dewar had still not visited America, but he remarked, 'Two days ago I received a letter from Mrs. Moore, written in Philadelphia, saying that Mr. Keely had completed his system and no longer needs sympathy or endorsement.[89]

But that grandiose millennium of power—evoked by Keely's completed system—had to wait for another year. The end, though, was now in sight, or so it was told at the annual meeting of the stockholders of the Keely Motor Company. Only a small representation of the stock "put in an appearance." A minimum of 50,001 shares were required to reach a quorum, therefore, no meeting was held. According to Secretary Schuellerman, this was not that much of a problem since there was "no business of importance to be transacted" anyway, and the stockholders were already informed that "in a few months all the arrangements of Mr. John W. Keely toward perfecting the machine which is to revolutionize power production will be completed, and patents will be taken out in all parts of the civilized world." A 200-horsepower commercial engine that had been "in construction for a long time" was delivered to Keely's laboratory the week before. This would not mean that the device would be ready for use immediately; a "large amount of adjustment and graduation" had to be done, but when this was completed, Schuellerman added, Keely would secure patents for all of his inventions.[90]

The year 1894 came, and as a consequence of Schuellerman's statement the year before, excitement filled the air in Philadelphia. A newspaper reported that, "It has been known for some time that Keely has his motor about ready for commercial work..."[91] a statement that was foreshadowed by a newspaper a year ago, which proclaimed: "It appears that Keely is still alive, and that his motor is once more just about to move."[92] But the years of hard labor had taken their toll on Keely; it was also reported that his eyesight was "somewhat impaired and he works under difficulties," a condition that was described by a visiting scientist of Boston as "half blind."[93]

The scientist was so taken with what he witnessed in Keely's workshop that he confided his experiences in a long letter to a newspaper, which show a marked contrast with other, more superficial reports on Keely's doings. This was not that surprising, as the scientist admitted that, "Impelled by a life long interest in the wonders of natural science and honored by the personal friendship of Mr. Keely and a few of his advisers, he has followed the course of this investigator for years."

And impressed the scientist was: "I have seen a spectacle I would have pronounced impossible, according to all the accepted theories of physics with which I am familiar. Without apparent exhibition of heat, electricity or any other form of energy hitherto operated by man, I have seen a strong metallic wheel, weighing some seventy pounds, in swift and steady revolution by the hour, and absolutely without cost...Long we stand around that flying wheel. The friend who photographed it at rest, again levels his camera upon it. In vain, its spokes cannot tarry long enough to be caught in its snare. It is still as death and almost as mysterious. We listen to long dissertations upon the reason for the relative positions of the eight discs on the wheel and the nine on the stationary rim, and how the adjustment can be so altered that instead of a revolution there will be a violent oscillation back and forth. We are shown the corresponding wheel and the rim of the large engine close by which is to bear the discs, not singly, but in groups, the steel resonating drum with the circles of tubes inside, the 35-inch Chladni plate underneath, the sympathetic transmitter on top, the extra wheel bearing on its spokes cylindrical cases, each filled solid with a hundred thin curved plates of steel, to get the utmost superficial areas, we are told."

And where Keely had complained the year before that two wheels of copper tubing were made imperfect, and had stated that he would make these himself, the scientist now wrote: "The engine you have been looking upon requires as part of itself...certain heavy tubular copper rings. Skillful artisans failed in various endeavors, by electrical deposition and otherwise, to make them right. The inventor contrived machinery for bending into semi-circles sections of copper tube one and one-half inch bore, three-eighths of an inch thick, forcing a steel ball through them to keep the tube in shape. To make a ring he placed two of these half circles together, and joined the ends in some way without heat, by what he calls 'sympathetic attraction' so that the resonant properties of the rings are satisfactory, and though you see the line of union, the two parts cannot be severed. You see one of these rings, some fifteen inches in diameter, hanging by a block and tackle from the ceiling, and lashed to the lower half swings a big iron ball weighing 550 pounds and there it has swung for weeks."

Now was as good as any time to remember some good tales of the past; "Nothing is said now," wrote the Bostonian scientist, "of other wonders of which other witnesses can speak, and which are said to have appeared in the slow progress this incomprehensible man has been making all these years; of a pressure obtained from the disintegration of water by vibration, of 20,000 pounds to the square inch; of a slowly revolving drum which went no slower when winding tightly upon itself a stout inch and one-half rope fastened to a beam, and no faster when the rope parted under the strain; of the disintegration of rock into impalpable powder; of raising heavy weights by aid of a 'vibratory lift,' recalling the 'negative gravity' of our modern storyteller."[94]

The year 1894 would also witness the appearance of two remarkable and futuristic novels, written in the vein of the "modern story tellers," that bore a direct relation with the visionary ideas of the enigmatic inventor. The first novel, titled *Dashed Against the Rock,* was written by theosophist, freemason and medium William Colville. His book was written after conversations with Keely, portrayed in the novel as the character "Aldebaran." The character of Professor Monteith was in real life Prof. Wentworth Lascelles-Scott, an acquaintance of Bloomfield-Moore. In his introduction, Colville made it very clear that, "In presenting to the public the following extraordinary romance, I wish to be distinctly understood that I am not in any sense the author of the scientific dissertations and tables which form a considerable portion of the volume. ...The unusual and distinctly technical terminology employed in some of the most important sections of this story may be considered out of place in a tale containing some amusing incidents and ostensibly published as a novel. ...I had no alternative but to what I have done or suppress this priceless knowledge altogether, for I have only received it on trust from a friend who is its custodian in a sense that I am not...I do not pose as a teacher; I am in these pages only a recorder."[95]

The most remarkable portions, aside from a wealth of materials pertaining to Keely's underlying philosophies,[96] were clear and definite descriptions of a proposed airship that would be built in such a way that with it one could also travel to the nearby planets and stars.[97] For these were the years that Keely was more and more contemplating not only a system of aerial flight, but from the text we may assume, even space flight. After all, Keely had been working on what was to become a complex engine for aerial navigation since 1887.

The second novel in certain aspects even outclassed the first; for whereas in Colville's novel we find rudimentary proposals for an airship, capable of traveling to the stars and nearby planets, in the second novel a ship did exactly this, all the way to Jupiter. This novel was *A Journey In Other Worlds* written by John Jacob Astor, one of the most wealthy men on earth and an acquaintance of Bloomfield-Moore.[98] Both books showed something of the far-reaching nature of Keely's visions.

It also demonstrated that these were to be Keely's most interesting years, as they show something of the spirit and the mind of the inventor who was forever pushing back the frontiers of known science and even groping for the stars. In the process, he would construct device after device and discard equally as many, becoming utterly incomprehensible to his contemporaries.

A reporter later remarked that he "found himself unable to follow Keely's explanations of the abstract principles on account of his unfamiliarity with the nomenclature of the new science...," and that he had to "abandon all preconceived notions of natural philosophy if he thought to make any progress in the study of the new power. The laws of gravity, as laid down by Newton, magnetism, as formulated by Faraday, Henry and others; light and heat, as outlined by Lord Kelvin, and the correlation of forces had to give way to this mysterious force, which practically creates something out of nothing, if we may unquestionably adopt the discoverer's theories."[99]

But no matter how far Keely's inner vision stretched, little understood by some and more fully appreciated by others, the troubles with his eyesight continued, necessitating a medical examination. On this, Bloomfield-Moore wrote: "Last week I received a letter from Schuellerman, telling me that when the final trial was made of Mr. Keely's eye, he wept for joy that his sight was restored..."

To her, Keely wrote: "My dearest and ever remembered friend, it is with the most exceeding pleasure that I can thus prove to you the perfection of my sight; and to be thus enabled to gratify my most ardent desire of sympathetic communication by letter to yourself."[100]

4

The Power Millennium
Keely's Last Years

"The dawn of a new revelation has already begun to shed its light which will show to the nations that pure science is true religion and the one sympathetic association with Deity."

John Keely in a letter to Clara Bloomfield-Moore
dated May 30, 1894

"Tesla has reached out almost to the crest of the harmonic wave, leaving all electrical explorers far behind him."

Clara Bloomfield-Moore
Keely and his Discoveries, Aerial Navigation, 1893

From 1895 until the end of his life, Keely would find himself in one of the strangest, most complex and most significant of late 19th and early 20th century history. While he quietly struggled to master his discoveries, many foundations were shaped around him that would influence the course of history. Science, in the exploits of Keely's contemporary Tesla, made enormous leaps. Occultism rose once again, and with it, numerous esoteric and hermetic societies appeared on the surface. Keely was steadfastly working on his engine that would enable an airship and possibly even a spaceship to fly as proposed in Colville's and Astor's books. And as a haunting echo of his dreams, the years 1896 and 1897 would witness an inexplicable wave of what was referred to as airships crossing the skies over large parts of the U.S

The year 1895, with Keely laboring day after day to perfect his inventions and to master his discovery, was the year that public interest would be renewed in his work. The year began uncertainly enough. In January, the U.S. government was on the very brink of bankruptcy. Through the financial scheming of J. Pierpont Morgan, one of Tesla's financiers, and with the help of August Delmont, a wealthy businessman, Morgan was able to secure $60 million in foreign gold reserves, thus saving the country from total bankruptcy. This incident also created a significant shift of power; it marked the rise of Morgan, also known as "the octopus" since he had a hand in almost every American industrial branch, as "the king of Wall Street."[1] The year 1895 would also see the smoldering beginnings of what would eventually become the Spanish-American War.

In February, Keely submitted two circulars amongst the stockholders. In the first circular, he explained that, "It is well known that during the period of research and discovery, which has consumed about twenty-two years, I have on many occasions imagined I had neared the end of my work, and under the influence of this thought, or for the purpose of helping me over some unforeseen obstacles, I have entered into so many obligations that at the present time, even assuming that I am right now in thinking I am at the end or near enough the end of my researches to take action looking to a practical development of my system, it would be impossible for me to deliver to the world this system, in the shape of property of any commercial value, until I have positive knowledge of the character and amount of all the outstanding obligations against me. To obtain this knowledge is now the first step to be taken, before I can even offer my work for the inspection of those who may be selected to determine the soundness of my present belief in my ability to produce and deliver a commercially valuable system of Vibratory Physics."

Possibly the compelling episode a month before, in which the U.S. found itself on the brink of bankruptcy, reverberated shockwaves among those who had financial interests in the Keely Motor Company, and thereby Keely was approached with questions and feelings of uncertainty, which prompted him to write his circulars. He continued by stating that, "...for until such disclosure of interests, claims or obligations is complete, further development of my enterprise under them is absolutely impossible."

He therefore decided to have all obligations registered by the Citizen's Trust and Surety Company of Philadelphia. The registration should take the form of "an exchange of the obligation for the Trust Company's receipt for the same; this receipt to contain a detailed description of the obligation deposited, and to carry with it a contract to return the original obligation to the owner or his assigns, at the expiration of a time to be set forth; or such equivalent obligation of a new organization as may be acceptable in full cancellation of the original claim." The circular ended by stating that persons could obtain blanks for filling out and attaching to the original obligation by writing to Keely.[2]

It is possible that Keely was also approached by the directors of the Keely Motor Company, who must have felt something of the threatening U.S. bankruptcy, and who wanted assurances from him. But his dissatisfaction with how business affairs were handled by the Keely Motor Company, plaguing him with court proceedings and harassing him with unreasonable demands, prompted him to establish a new company. In the second circular, Keely wrote: "...I also find the obligations entered into while I was yet groping after light in my work are impossible of execution, because they have no intelligent bearing on as it now stands; and so, in order to establish a commercial value to my former work, these old speculative obligations must be avoided entirely. I believe I am now able to produce a thoroughly practical commercial engine for the use of the force in all railways, and I wish to make such engine the basis for the organization of an entirely new company, capitalized in such a way as to enable me to accomplish something practical and of commercial value.... I propose to sell to the new company all rights for the whole world (exclusive of the territory covered by the 'Pacific Slope Concession' and the Philadelphia Manufacturing Company's contracts) in the use of the engine above mentioned. The price for this is to be the entire capital stock of the new company."

Furthermore, after the sale, Keely would return to the treasury of the newly formed company three-tenths of the capital stock, which would be divided in two-tenths "to be used to cancel all outstanding obligations," the other one-tenth "to be used to raise the working capital necessary to develop the property of the company." Seven-tenths would be kept by Keely. He would also set aside the equivalent of 51 one-hundredths of the entire capital stock "for my personal account and voting power, for a period of five years, so that it cannot be sold or given away by me during that period."

This was all necessary, explained Keely in the second circular, "to give me voting control of the company until it shall have been thoroughly established in the development of its rights under control of the engine before mentioned, and at the expiring of the trust I am to have absolute control of the capital stock covered by the same."[3]

Obviously Keely had discussed his plans for the formation of a new company with his supporter, Bloomfield-Moore, for under the second version of his circular she submitted a short notice, written under her nom-de-plume, H.O. Ward. In it she approved of his proposed scheme, writing that "it in no way violates the contract made with me by Mr. Keely on the 12th of April, 1890." She also urged the members of the Board to "accept Mr. Keely's terms as set down in circulars nos. I and II...as there then need be no further delay from want of that cooperation which now not only jeopardizes all commercial interests in Mr. Keely's inventions, but the interests and advancement of science in the long delayed announcements of the value of his great discoveries."[4] While this gave the Board of Directors certainly something to ponder upon, another unexpected occurrence took place several months later.

On November 6, 1895, a newspaper wrote how interest in what was by now called the "Keely motor mystery" had again "been aroused," the reason being a published statement that a group of "New York capitalists" had become interested in Keely's inventions. But somehow, perhaps as an aftermath of his circular, or as a herald of the following year's coming events, a large part of the article was devoted to Bloomfield-Moore's dealings with Keely. And from it, we learn more about the nature of the agreement. While she had helped him from 1881 on, she apparently made an agreement with him in 1890.[5] This agreement to which Keely was bound was that he must not attempt to build or construct an engine to "demonstrate in a practical way the value of his discovery, until he had entirely mastered the principles which lay at the bottom of it." Perhaps this would somehow explain the installments that Keely had made over the years.

But the term of this agreement was now drawing to a close, and for the past few months Keely "has been preparing to make a practical demonstration of his vibratory system, which is different from the one which he first discovered and abandoned because he could find no means of controlling the force."[6]

This did not mean that Bloomfield-Moore disposed of her interest, instead her "chief anxiety" was that someone other than Keely would possess his secret, and that it might not "die with him before the apparatus for utilizing the newly-discovered force shall have been patented." In her view, until such patents were issued, the stock of the Keely Motor Company would only have "speculative value." Bloomfield-Moore also "removed the restriction" from Keely, who would "soon commence the construction of a railway traction engine," the plan of which he had set forth in his February circular. "The researches being concluded, he says he can now devote his energies to the utilization of this hitherto unknown force."[7]

The New York capitalists who so expressed their interest in Keely visited and talked with him and were "favorably impressed with what they saw," so they would "probably become financially interested as well in the Keely enterprise." And a mighty group this was: the puissant rich John Jacob Astor; William K. Vanderbilt; William Cullen Brewster, president of the Fifth Avenue Trust Company; and other unidentified New York millionaires.[8] Astor had already purchased a large interest in the Keely motor "from a person who for some years had been an enthusiastic advocate of Mr. Keely."[9]

Never before was Keely approached by such an extremely wealthy group, and the news of the group traveled far and wide. One of Tesla's acquaintances wrote an alarming letter to Tesla, expressing his "astonishment at Astor's gullibility" in investing money in Keely's discoveries.[10] The reason for this alarm was that Tesla himself was trying to obtain funds from Astor.

The news also reached the directors of the Keely Motor Company. They responded by "proposing some sort of settlement of the business questions existing between the company on the one hand and Mr. Keely and his patroness, Mrs. Bloomfield-Moore, on the other," very likely also meaning the announcement in February of Keely's plans to establish a new company. Bloomfield-Moore rejected their propositions and stayed "in negotiations with Mr. Astor and other New York capitalists."

Nothing came of it however. "The negotiations that have been pending for some time between John Jacob Astor, William K. Vanderbilt, William Cullen Brewster and other New York millionaires, Mrs. Bloomfield-Moore and Keely for the control of Mrs. Moore's interest in the Keely motor have come to an end. At a conference in this city on Thursday, at which were present Mrs. Bloomfield-Moore, John Jacob Astor, William Cullen Brewster, J.K. Lorrimer, H.O. Ward and others, the question of organizing a Keely Power Company, to bring out Keely's invention when complete, was discussed."

One of the conditions of the wealthy New Yorkers was that Keely would move to New York. There the millionaires would "erect for him a new building and back him to the extent of several millions." The proposition was rejected, because neither Bloomfield-Moore nor Keely "would consent to the latter going to New York." Furthermore, both rejected the idea of forming a new company until a demonstration had been given of "certain operations of the invention." Another reason was that Bloomfield-Moore and Keely wanted to prevent speculation in Keely stock until the inventor had completed a patentable machine.[11]

One of the reasons for Keely and Bloomfield-Moore's rejection of the proposition of Keely moving to New York might have been that they had the chilling thought fresh in mind of Tesla's New York laboratory that had mysteriously burned to the ground in March of the same year.[12] The incident was widely noted in the press. Probably Bloomfield-Moore learned more of this disastrous incident first hand. She had been aware of Tesla's work since at least 1893. She also corresponded with Tesla on several occasions and met with him at least once, although in Tesla's letters to her that have survived there is no mention of the fire that destroyed his laboratory and many years of work. But from his letters we learn that he studied Keely's theories and was aware of the inventor from Philadelphia, although largely from a distance.

Thus towards the end of the year before, Tesla wrote to her in a letter that, "I sincerely wish that I would share your opinion about Keely and his works. That he has ingenuity and skill in experimenting I will readily believe. But his method is an unscientific one and his exposition is wanting in the extreme. It is painful to read his theories. Can he have recognized something and yet be utterly incapable of expressing it? This seems impossible for there is no truth which cannot be told in simple language. The best I can do after all I have heard of him from Mr. Theodore Puskas,[13] Mr. Andrews and others who have been thrown in close contact with Keely is to say that I pity him. I devotedly wish that you may be spared the pain of deception. The solution will eventually come but from other quarters. I myself am only one of many, who believe to have found a way and who strive and hope to accomplish...of your esteem. Please take care of your health and guard yourself against influences which can only result in a disadvantage to you."[14]

Tesla's harsh criticism on Keely's methodology did not cool Bloomfield-Moore's relationship with Tesla; why not see for himself instead of forming an opinion based on second-hand comments and the reading of Keely's visionary prose? So she invited him to come to Keely's workshop. A month after the fire that destroyed his laboratory, with Tesla obviously having other things on his mind, he replied to her in writing concerning her invitation: "I could not have accepted your invitation (to witness Keely's demonstrations) nor express myself in regard to Keely, but I wish to assure you sincerely that I would be very pleased to comply with your wishes in any other way. You are evidently following a noble and generous impulse in defending Keely, and I only wish that he would be worthy about your duties to yourself. You must not sacrifice your own welfare for others whomever they may be."[15]

And like Edison who had rejected an invitation some 17 years before him, Tesla would also never meet Keely. Notwithstanding Tesla's evasive maneuvers towards Keely, he felt different for Bloomfield-Moore, as his correspondence continued in the warmest of tones: "You are a noble-hearted woman and your evident anxiety makes me feel keener the disappointment of not being able to respond to your invitation. For many years I have been at work on great ideas. I have used scientific methods in the investigations of the problems which I proposed myself to solve and my progress in these fields though very slow was nevertheless sure. Now, after all these efforts in overcoming great obstacles when I have partially reached my dreams I cannot afford to dissipate my energies which many scientific men say are precious for my fellow men."[16]

In a subsequent letter, Tesla further expressed his concern over Bloomfield-Moore's health: "I hope that you will soon free yourself from your obligations here and go abroad as the affairs which you are devoting at present your attention to are certainly telling on your nerves. I regret this the more as I fear that your energies are wasted on a task which is not worthy to mankind as a single hour of your life."[17]

While Tesla never visited Keely, he did come to Philadelphia on a number of occasions - in 1893 when he lectured at the Franklin Institute during which he told the audiences about his visions of the precursor to television,[18] and again in May 1895, when he, Edison and Alexander Graham Bell met during the National Electrical Exposition.[19] It is not known if Keely attended any of these occasions. Based on the documentation that is at hand, there is no indication that Keely ever mentioned Tesla, or referred to him in either a negative or positive sense.

There is documentary evidence that the same year Tesla visited Bloomfield-Moore in Philadelphia at least on one occasion; on a letter, she scribbled, "After Tesla's visit to me in Philadelphia he wrote." In the letter, Tesla expressed his fascination with Bloomfield-Moore: "I need not assure that I have been deeply touched by your evident kindness for me. Yet, I would wish you were not so, for I feel that I am unfair as my whole soul is wrapped up in the investigation I am pursuing. It has grieved me to be a helpless onlooker on the last occasion for I would have much desired to help you, at least in sound advice. Please remember what I told you about your duties to yourself. You must not sacrifice your own welfare for others whomever they may be. Please take care of your health. You know that at your age one has not a very great receptive power. Your weakness at the moment of my leaving you Sunday evening alarmed me and I reproached myself for having remained so long depriving you of the much needed rest. But I was so interested in you that I forgot my duty. To be frank with you anything concerning Keely has not the faintest interest for me."[20]

The reason for the fact that Tesla so shunned away from a first-hand investigation of Keely's discoveries, while at the same time remaining on friendly terms with Bloomfield-Moore, is not so much found in Tesla's opinion of Keely's work. It was probably tainted by the frustrating fact that he hoped to procure funds from Astor, a seemingly impossible task, whereas with Keely, Astor apparently displayed no such hesitations. Nor is it found, in fact, in the qualities of the discoveries or the visionary and highly original scope of the ideas of either Keely or Tesla.[21] The reason is to be found in the enigmatic John Jacob Astor himself, the multimillionaire who eagerly negotiated with Keely and Bloomfield-Moore, while at the same time curiously evading Tesla's requests for financial funding.

John Jacob Astor (1864-1912) was one of the wealthiest men on earth, with assets somewhere around $100 million. By comparison, powerful Morgan, "the king of Wall Street" and one of Tesla's financiers, had amassed a fortune of $30 million.[22] John Jacob Astor was not only a multimillionaire and a distant relative of Theodore Roosevelt, but also an inventor with a visionary mind. He is characterized as a somewhat estranged man, a playboy who was dominated by his mother and his wife. He showed early on a talent for things mechanical, and he owned a workshop at Ferncliff, where his vivid imagination played on the possibilities of copper tubing, wires and electric currents. Later he built his own laboratory, in which he produced inventions such as a bicycle brake, a new kind of marine turbine engine, and a machine for removing surface dirt from roads. He was one of the first Americans to own a motor car.[23] Another of his inventions, a pneumatic walkway, won a prize at the 1893 Chicago World Fair. Other inventions included a storage battery, an internal combustion engine and a flying machine. One of his dreams was to find a way to create rain by pumping warm air from the surface of the earth into the upper atmosphere, an original idea, but one that the U.S. Patent Office turned down.[24] Complex and furtive as Astor was, he also had a reputation for eccentricity; he started writing his cryptic novel *A Journey In Other Worlds* in 1892, when he was only 28 years old.

The same month that Tesla visited Bloomfield-Moore in Philadelphia, he also met with Astor. He did so, writes Tesla's latest biographer, in his "progress of netting more millionaires,"[25] hoping to amass enough funds for his experiments. This uncovers yet another layer of Tesla's disinterest in Keely's work while at the same time keeping a fascination with Bloomfield-Moore, Keely's loyal and ardent supporter. For Bloomfield-Moore not only knew Astor intimately since at least 1882 when they mingled in the highest diplomatic and noble circles in Italy, but she writes that "...the mother of Mr. Astor (had) been a connection of my relatives, the Wollcotts."[26]

One time, Astor visited her to offer her "a paper for the *New Scientific Review,*" Bloomfield-Moore writes, "before he went over to New York for the yacht races...I corrected and sent back my proof—Spiro Vortex Action."[27]

She also mentions one instance when she had been on board the Nourmahal, Astor's steam-driven three-masted schooner that measured an impressive hundred yards and was equipped with four machine-guns to frighten off potential pirates. Bloomfield-Moore wrote about this occasion: "...on Sunday, August 4th, while on board the Nourmahal I agreed to go with Mr. Astor and the electrician who invented the dynamo plates a patent for which he has taken out.... Now I can only say that Mr. Samuels has given up electricity and is going to study Keely's system under Keely's instruction, thoroughly convinced after witnessing the operation of drawing forces from space, the action of the test mediums and the levitation of the weights in the jar."[28]

What, if any, advice Bloomfield-Moore gave Tesla in his hopes of obtaining a funding from Astor, if he ever sought her help in this matter, is not known. Bloomfield-Moore was a highly intelligent and kind-hearted woman, and her conversations with the genius Tesla must have been stimulating and interesting for both of them. In the end, Astor remained elusive and kept his ambiguous attitude towards Tesla. Tesla wrote Astor several letters after Astor met with Keely in Philadelphia, asking him for financial funding[29] and inviting him and his wife to visit his laboratory.[30]

In 1899 Astor finally agreed to fund Tesla, but not for his grandiose ideas, nor for a large or generous amount of money. During a meeting the year before the agreement, Tesla painted to Astor the possibilities of his teleautomaton. Astoundingly Astor, who four years before had published his book that also showed no limits in visionary concepts, replied coolly that, "You are taking too many leaps for me. Let us stick to oscillators and cold lights. Let me see some success in the marketplace with these two enterprises, before you go off saving the world with an invention of an entirely different order, and then I will commit more than my good wishes. Stop in again when you have a sound proposal or call me on the telephone."[31]

So when the agreement was made, Tesla was forced to limit himself to oscillators and fluorescent light bulbs, and on January 10, 1899, papers were signed whereby Astor gave Tesla $100,000 for 500 shares of the Tesla Electric Company, a rather insignificant sum compared with the "several millions" with which Astor and others were originally planning to support Keely. Astor was also elected director of the board of the Tesla Electric Company. At the same time, Tesla moved into Astor's hotel, the Waldorf-Astoria.[32] Tesla would not exploit his oscillators and fluorescent lights, but went to Colorado Springs instead, to embark upon another of his grandiose projects. After that, their relationship became strained. "Tesla had deceived him," writes Seiffer in his biography. "As wealthy as he was, Astor wanted to invest in a sure thing."[33]

History is left with an uncomfortable mystery. Astor's reluctance to invest in Tesla's most brilliant ideas, and in the end only giving a mere fraction of what he had at his disposal for one of Tesla's lesser inventions, is interpreted as shrewd and keen businessmanship, bent on investing only in what is sure. At the same time Astor was described as being very gullible. His willingness to support Keely with a far greater sum is not understood, since the weight of argument is shifted towards that fraction of Keely's contemporaries who were merely interested in depicting him as a fraud.

But this contradiction in the descriptions of Astor's character and motives fails to explain why he appeared so disinterested in Tesla's visionary ideas, while he himself had published a book overflowing with fantastic technological extrapolations. Or why he, not an entirely unsuccessful inventor himself and having at least some specialized knowledge, at the same time preferred to erect a Keely Power Company and together with other millionaires finance it with several millions, based on Keely's own fantastic discoveries.

Possibly Astor, having invented a flying machine, was more impressed with the prospects of Keely's proposed concepts of aerial navigation than Tesla's futuristic plans, for aerial navigation was a field of invention that concerned Tesla only at random. Aerial navigation also fascinated Astor's acquaintance Bloomfield-Moore. It very likely may have been she who pointed Astor's attention towards Keely's discoveries.

Perhaps Astor was involved in an intricate financial plan, design or intrigue based on division and domination. Whatever his reasons, Astor took these with him to a watery and ice cold grave. After he died in 1912 during the tragic Titanic disaster, only those 500 shares of the Tesla Electric Company were found. In all the years, Astor had never even taken the trouble to increase the stock.[34]

As a side effect of the negotations between Keely, Bloomfield-Moore and Astor and his rich allies, interest in Keely and his discoveries renewed itself somewhat, and the usual interview appeared occasionally in the Philadelphia press. In one of these, a puzzled reporter tried to describe one of Keely's avant-garde devices. "A disk-shaped apparatus was shown, about 18 inches in diameter and two inches thick, covered with brass, and divided into segments. Upon its face were three tuning forks, with heavy tines, two of the same pitch, and the third a half note below. On the lower edge was a small detachable piece of brass, with two small narrow strips of steel fastened to it. A brass cap, like the top of a fruit jar, filled even full with an amalgam paste, which Mr. Keely said he had studied three years to produce, was attracted to these steel strips, and hanging to the cap (or disk as it might be called) was an iron weight of 140 pounds." The reporter also witnessed a levitation experiment, similar to the one Leidy saw, involving disks and metal balls "the size of an orange," held in glass jars filled with water. Keely also confided to the reporter that he expected "to build his railway engine at once and have it ready for operation some time during the next year."[35]

Yet again, towards the end of 1895, notwithstanding earlier statements that Keely would commence the construction of the railway engine, nothing of any substance was reached. In November, a newspaper wrote that the rumors that Keely would exhibit an engine to "New York capitalists," were not true. "It was learned that no engine has yet been built; that the New York parties will not be here, and that, by particular request Mr. Keely had decided to give no exhibition of any kind at present."[36]

On December 11, the usual annual meeting of the stockholders of the Keely Motor Company was held. A newspaper ironically remarked that this time the meeting would likely "be one of more than usual interest, on account of the reported efforts of well-known capitalists to obtain an interest in Keely's inventions." But the most important reason for this more than unusual interest was that it somehow was discovered that the Keely Motor Company actually had no real basis of existence; "the New York parties who are said to have been looking into the matter have discovered that the Motor Company has no legal claim upon the results of the work that Keely has been engaged on for the last five years." This shattering conclusion was based upon the ground that "the terms of the contract binding Keely to turn over to the Motor Company all the inventions of whatever nature he may make during his life are illegal and cannot be enforced."[37]

If that was the case, the Keely Motor Company was without assets, since no patents were taken out, and no specifications of Keely's devices were ever drawn. "In fact," it continued, "Keely gave up the attempt to harness the power, and directed his investigations to the discovery of a means for utilizing the force which he believes to exist in the vibrations of the universal ether." Keely, however, was "willing to give the old company a minority interest in the new enterprise when formed, not as a matter of legal right," it was pointed out, "but as an equitable adjustment of all matters between him and the company."[38]

In the meantime, Bloomfield-Moore answered a letter to Collier in which she made it clear that no effort was made to buy her share in "the invention of the airship propeller and the railway traction engine."

She furthermore stated that, "had the Keely Motor Company made such a proposition to her, after she had refused theirs, she was "prepared to place her share under their control." But now, she valued her shares "as I never valued them before," although she admitted that she would have given her shares away if "they would have held the gift for the good of the masses, for the progress of humanity, for research in new fields, and not for selfish aggrandizement." An indication of what she must have thought of the true intentions of the Keely Motor Company also lay in her following statement: "Now there is nothing that could induce me to place these privileges in the hands of financiers," and she made a strange remark: "Had the negotiations been completed," she said, "within two weeks the scientific world would be set ablaze by the announcement that their syndicate was to employ apergy instead of electricity as the motive power."[39]

Nothing came of the negotiations between Keely, Bloomfield-Moore and Astor and his allies. In the meantime, Tesla was still in pursuit of the millionaire's support, even though a slight connection between Astor and Keely still remained. The next month, Keely announced that he would impart with some of his learning. He would instruct Zak Samuels, "Astor's expert," who had met Bloomfield-Moore on Astor's imposing ship, "in the taking up of dead lines and in the sensitizing of metal so that it is acted upon by radiation on the same principle as in the human brain." This apparently could not be done immediately; when Keely was ready, he declared, he would "notify Mr. Samuels (who is the only man I will have time to instruct) and this at the earliest possible moment."[40]

On the same day, a far more important statement was issued to the press: a dark herald of the dramatic event that would unfold the following year. But the statement was not only a herald; it was also an indication of what the true cause of that event was. It all began on December 14, when the Philadelphia press got a glimpse of the sordid family affairs of Keely's most loyal supporter; Clara Bloomfield-Moore. Bloomfield-Moore had accused her son, Clarence B. Moore, of having wasted the money of his father's estate of which he had been the sole acting executor since 1878. Of course, Moore denied this: "The answer of Clarence B. Moore, the son of Mrs. Bloomfield-Moore, to her petition to have him dismissed as executor of his father's estate, was filed today by his counsel, C. Berkely Taylor and John G. Johnson." Moore vehemently added: "He denies mismanagement of the estate, tells a detailed story, alleging his mother's persecution of him, and asserts that she is insane."[41]

But this sudden, poisonous outburst withered away as quickly as it sprung up. Far more urgent affairs were taking place; Bloomfield-Moore, under her nom-de-plume of H.O. Ward, was giving out long statements to the press regarding what she held was the nature of Keely's discoveries, which somehow hinted to a less passive role for her in her relation with Keely than was generally assumed: "In 1894, when I made the suggestion to Mr. Keely, he began with his first attempts to vibrate hydrogen, with the result that he is now able to show the correctness of my random conjecture."[42]

On Monday, December 11, The Keely Motor Company had its usual chaotic assembly during its first stockholder meeting in five years. During the meeting the "harmonizing of conflicting interests" was attempted and a reorganization was announced. All in all, some fifty people were present in the hall at the southeast corner of Eight and Walnut Streets.

After the election and re-election of company members of the Board, a special three-member committee was appointed, consisting of Collier, treasurer Sylvester Snyder and Lancaster Thomas, vice president of the Keely Motor Company. This team conferred and formulated a plan of reorganization of the company, which was then submitted to Keely. Naturally a part of the plan was that Keely should apply for patents. Ackerman presented his annual report, which in part related that "more than a year ago Keely undertook to instruct his patent attorney, Mr. Charles B. Collier, relative to his inventions, including the sensitizing process, to enable Collier to take out patents. One of the members of the Board, Mr. Lancaster Thomas, was present during the sensitizing of several disks and assisted in its performance, and both gentlemen stated that it was absolutely new." Collier, the report stated, had told the Board that "this as well as Keely's other structures were capable of being protected by letters having the broadest scope known to our patent system."

But one year ago, Keely, "for some reason not known to the Board nor understood by Mr. Collier," ceased all communication with Collier and gave him no further information regarding his inventions. Instead, Keely sent out his two circulars in regard to reorganization of which the Board did not approve. The Board announced that "reorganization would be effected as soon as Keely operates a large commercial engine and enters again into the matter of procuring patents to cover his inventions."[43]

On December 24, almost two weeks later, Bloomfield-Moore made a statement to the press that Keely had abandoned all ideas of patenting his inventions, but that he would adopt a system of royalties for the commercial use of his discoveries. Keely decided upon this course because he was fed up with the delay of the stockholders of the old Keely Motor Company in accepting his terms for a reorganization of interests that he had announced almost a year before.

According to her there were "twenty or thirty instruments to be patented, covering the various branches of the system." Among these she enumerates the "spiro-vibrophonic" which would require eight or ten patents; the "resonating" four or five more; the "vitalizing" six, and that before these could be prepared, or commenced even, a "vibratory dynamo" must be set up, which would require weeks to set up and graduate. The "sympathetic transmission" system also would require months to put into a patentable shape, also the "sympathetic governor."[44]

With Keely's plans to form a new company, the ensuing reorganization of the old company, and Astor's attention towards him, the year 1895 ended undecided. In February 1896, the committee members of the Keely Motor Company were studying a counter proposition that Keely submitted.

Keely met with an accident on February 11, the day after he demonstrated his Vibratory Dynamo to the members of the Board. He was run over by either a car or a runaway horse on Chestnut Street.[45] The accident shook him up severely and confined him to his house, so the whole reorganization plans were halted until he had time to recover. The stockholders met again on March 2, following the adjournment of their last meeting on December 11 of the previous year, an adjournment that was effected to "allow time for the preparation of a satisfactory plan of reorganization which should harmonize all conflicting interests." At the meeting of March 2, nothing of any substance was decided. Due to Keely's accident another adjournment was taken. A new date was set and the new meeting was planned for April 2.[46]

On the day before his accident, Keely demonstrated his Vibratory Dynamo Machine, that he also called a First-circuit Engine.[47] The demonstration was successful; power was transmitted from an engine to a dynamo by means of a belt. The members of the Board who witnessed this test run thought that he had "obtained such control over his discovery as to entitle him to letters of patent broadly covering his process, and the machinery employed conducting it." The reorganization plans of the committee would be based upon his application for patents "at the earliest practicable period," wrote Ackermann, president of the Keely Motor Company.[48]

This letter, written some three weeks after Keely's accident, and the reorganization plans that were put forward with a sudden elan, were partly a response to the bitter statement made the previous year quoting that Keely had decided to "take out no patents."

On April 2, during a somewhat stormy meeting, the report of the appointed committee was read. The plan was to secure to the Keely Motor Company the exclusive rights in the territory covered by former agreements, Keely's two systems of producing power. The plan also stipulated that no reorganization should be made until Keely had produced a practicable machine and had applied for patents. A letter from Keely, dated March 31, was also read. He agreed to this plan, while remarking that he believed that he was safe in assuming that those present at his demonstration of the First-circuit Engine were satisfied that he had indeed attained "substantial control of my power, and that complete success in a commercial form of engine is clearly in sight and a question of but a few months at furthest."[49]

The report of the committee recited the agreements made between Keely and the company, and acknowledged the ownership of his new system that he discovered in 1882 and of which he claimed that the Keely Motor Company had no rights whatsoever. The report stated that Keely's claim "seems to be a valid one, and the directors evidently took that view in 1887, as also did the stockholders, as shown by the minutes of the annual meetings." The proposition was therefore to reorganize the company for the purpose of acquiring the rights to his other discovery "within the territorial limits of the company."[50]

There were still some loose ends. The report noticed that one of the outcomes of the suit against him made in 1882, namely a decree of the Court that specifically directed him to file with the Court the necessary specifications for taking out patents on the inventions owned by the company, had not been done. Keely insisted that another clause in the decree, namely to provide him with "a proper amount of money for his maintenance and for perfecting his inventions," also had not been done. He then explained that, in order to procure means, he had been obliged to make agreements, called "assignments," that could be converted into stock of "any new company that may be formed." He also claimed that the agreement called for the use of 100,000 shares of the Keely Motor Company's stock as a working capital, to "which use it had not been put." The committee concluded that "there was matter of difference enough between the parties to litigate upon for a generation," and that the best thing to do was to "wipe out the past and start with a fresh deal."[51]

And as always, there were differences. A number of wealthy stockholders from New York expressed their "considerable dissatisfaction" with the way things were going. Motions were presented and defeated, and the plan was finally adopted "after considerable argument and some bitter words." To mend things, Collier, Ackermann, who claimed to be an expert mechanical engineer, and another member of the committee who "was looked upon as an expert electrician and chemist, all cried in chorus that 'They saw enough of Keely's new etheric vibratory system...to satisfy them that it was the greatest invention on earth.'" So Keely was to be given one-half of the capital stock, one-fourth to the stockholders and one-fourth to be held by the company. Of the 50,000 shares present, about 45,000 shares voted in favor of the new plan.[52]

But what about the obtaining of patents? After all, without patents on Keely's devices there would be no reorganization. When pressed for an answer, Collier stated that "he did not know, and he then explained that the question was one of 'much difficulty.'" Collier further said that, "If it were an ordinary invention, I have no doubt that suitable drawings could be prepared, and an application made for a patent in four or five days. In some respects, the machinery is simple, but as it includes the workings of electricity and chemistry the matter of preparing a lucid explanation of it is difficult. In order to get a patent it is necessary to have specifications so plain that when the patent expires anyone can construct a machine from them and the drawings. That is what the drawings and specifications are for. It is right here that the difficulty begins, and I can't say whether the patent can be secured in three weeks or in six months. In dealing with this invention, which I regard as the most important ever made, ordinary plans cannot be used."[53] It was announced that the meeting would be adjourned until June.

With matters as inconclusive as always, opinions concerning Keely once again widely varied. The press certainly grew wary, and two weeks after the meeting, one Philadelphia newspaper cynically reported that, while feeling obliged "in common with other papers" to publish the "vague items of news respecting the Keely motor," that "Much of the apparatus which he exhibits is trivial in character, developing little power, or in its present form is of no apparent mechanical value." And while it declared that "it has never joined in the ridicule to which Mr. Keely's invention or discovery has been subjected," and recognized that "there was a field for such discoveries," it now had lost its patience. "Mr. Keely can at any time, if he possesses a new force, have his discovery generally acknowledged by taking his apparatus to the Franklin Institute or the Philosophical Society and exhibiting its manifestations of power under test conditions excluding the use of known forces, and until he is ready to do this it will be a waste of time for sensible men to discuss the subject." It concluded that the world was not required to prove anything; that was Keely's business, for "he is the challenger."[54]

A week later President Ackermann of the Keely Motor Company picked up the glove and stated that "The commercial engine is now practically complete, and will probably have its first public test on a street car..," and as a reaction to the negotiations with Astor et al. that had failed, he also declared that "It is true that we no longer need or desire outside capital..."[55] No public test occurred in April, but the next month a new and shattering event took place. Because its real causes have never been properly examined, the event has always wrongfully been interpreted by antagonists as another indication that Keely was a fraud.[56]

On May 4, Bloomfield-Moore, Keely's staunch supporter, notified both Keely and the Keely Motor Company that she would no longer continue her support, and five days later she issued a statement to the press: "The announcement that the Keely Motor Company no longer needs or desires assistance has terminated the contract between Mrs. Bloomfield-Moore and Mr. John W. Keely, entered into between them verbally in 1888 and signed by them on April 5, 1890." She further stated that the Keely Motor Company would resume its responsibilities and would provide him with all the needed funds for his "costly researching instruments and for the construction of an engine which can be operated by others as well as by himself." According to her, until that happened there was no commercial value that warranted the proposed reorganization, and the putting on the market of an additional 500,000 shares of stock.[57]

To examine this event, we must first look into Bloomfield-Moore's private life, and discover her motivation for the support of, and then withdrawal of support of Keely. While it is already established that her motivation laid in her thinking that she would save another inventor whose invention would possibly hold great merit for mankind,[58] other important reasons are to be found in her private life, which had a sad history attached to it.

To begin with, while she is always referred to as a rich woman, this was a rather exaggerated viewpoint. As she herself made plain when her husband died in 1878, he left her no money other than his life insurance. From this money she gave "every dollar" to institutions that he had "promised to endow during his life time." This was about $50,000. Her husband's estate was locked up in mills and machinery. What was worked out of this, she gave to her children and grandchildren. In January 1892, she also renounced all dividend claims of her late husband's business, the Jessup & Moore Paper Company, to her children and grandchildren. Her second daughter Lilian, suffering from a mental illness, was taken away from her. "Had she recovered, I could not have helped Mr. Keely in any way with my Puritan strictness of ideas as to doing the duty that lies nearest to me before extending the field," wrote Bloomfield-Moore.[59]

In 1888, her daughter was again taken from her and conveyed to Sweden. In 1874, her daughter had married the influential Baron Carl Nils Daniel de Bildt, the first Lord Chamberlain of Sweden and Swedish Minister to Italy, who was also considered "an eminent historian."

Rumor had it that her daughter was taken away to secure control of her money in Sweden, although Bloomfield-Moore denied this.[60] The reason that her daughter was removed for the second time, as she wrote to a friend several years later, "was because of my having employed Dr. Hartmann who was the first to tell me of the curative powers of light in mental disorders."[61]

Possibly those that decided that her daughter should move to Sweden were abhorred by Hartmann's reputation, who met with Keely on two occasions and whose character was steeped in occultism, esoteric societies, neo-templarism and theosophy.[62] He had studied medicine in Munich, but in 1865, without having qualified, sailed to the U.S. as a ship's doctor. There he would obtain "some sort of medical qualification" in St. Louis, probably at the Eclectic Medical College, an institution that was "notorious for its low standards." Hartmann practiced medicine in several different states.[63]

In Sweden, Bloomfield-Moore's daughter "fell back into a state worse than the first." Bloomfield-Moore was summoned at "the Police Direction" in Vienna to answer questions about Keely, it having been stated that she was under "a delusion in furnishing with money a man known to be a fraud."[64]

We will not digress here any further in the tragic family affairs of a woman who by every contemporary source is being pictured as intelligent, unselfish and working for the benefit of mankind. What it illustrates, however, is the entangled and confused state of Keely's beneficiary, and it provides the background of Clarence B. Moore's statement made the year before that his mother was "insane." In fact, there seems to have been a bitter struggle between son and mother over the will of her late husband, as illustrated in a letter of Bloomfield-Moore: "If my eldest daughter is the next victim of my children's rebellion against their father's will, my son will be solely responsible for it..." Elsewhere she writes that "...futile would be his efforts against the invincible armor which I wear."[65]

We need not speculate over this point here, for a confirmation is to be found elsewhere: "By 1888, however, the courts again threatened, and this time Mrs. Bloomfield-Moore's son was able to deprive his mother of all legal and material rights, thus stopping the flow of his 'inheritance' toward Keely's endeavors."[66] And about this, Bloomfield-Moore herself wrote that "In 1889 efforts were made to prove that I was incompetent to take charge of the estate and of my invalid daughter."[67]

Bloomfield-Moore's letters on this sad affair that have survived reflect her anguish and kindness to all those concerned to a great degree, but also the concern of others who knew of her pains. Strangely, in one of her letters, written in 1895, she writes of having received a letter "from an Englishman who thinks that my son may be as well as I the victim of a conspiracy...,"[68] a topic that she only mentioned in her letter, but upon which she digressed no further.

Suffice to say that, aside from the earlier mentioned reason of her wanting to help a struggling inventor and through him, hoping to do something for mankind, she may have wanted to find some consolation for her own much-troubled soul: "Had it not been for the new interest in Keely's life which his discoveries brought to me at a time when the ploughers ploughed upon my heart and made long furrows, I should have been, years ago, in my grave or the inmate of a madhouse," she wrote.[69] Furthermore she also wrote that she could help Keely only because she could not properly extend her help to her children and grandchildren: "I would have prayed that the work might be given to another; feeling that I could best serve my Creator in serving those whom he had given to me to lead in the ways of righteousness.... I now have seven grandchildren, who are as children to me; two of whom are married. As each one marries I wish in consideration of the uncertainty of all kinds of security in America as is thought in Sweden, to have his or her share invested in their own country, and this desire has led me into some injustice to myself; having advanced to them all of my income for this year and, my agent writes, well into the next year."[70]

That was the reason that she could not provide Keely with any more funding. Since they both agreed that her funding would stop as soon as he reached a commercially exploitable stage, and Keely had proclaimed this, all came together. She wanted to mend things between her and her family members, and it relieved her of the responsibilities of which she had written some six years before that she "so long carried...at the cost of placing a barrier between myself and all those members of my family, who do not approve my course in assuming obligations which belong to the company to fulfill under its contract with Keely."[71]

Naturally the Keely Motor Company protested and "made overtures to Mrs. Moore with a view to her continued support," but to no effect.[72] Keely's reactions are not known, but he seems to have held no objection to her stopping her funding, as the terms of settlement offered by the company were unsatisfactory not only to Bloomfield-Moore, but also to Keely.[73]

On June 4, a meeting of the company stockholders was held, but nothing of any substance was reached, since no quorum was present. Schuellerman stated inter alia that the plan, adopted on April 2, was not "in accordance with the by-laws," since no quorum was present when the plan was adopted. Since now, too, no quorum was present, the meeting was again adjourned, but without setting a date. A special meeting would be held to consider the plan.[74] The annual meeting on December 10 also showed the same total lack of interest; only five stockholders were present and so the meeting was again adjourned, but no date set. Collier remarked that he was "about to draw up the patent specifications covering Keely's inventions, but that it would take considerable time for him to study the apparatus and comprehend the essential features sufficiently to draw up the specifications." It furthermore was remarked that, "Until they have been filed and the claims are allowed the company has no tangible property."7[75]

The year 1897 came, and after a quarter of a century, still no patents based on Keely's discoveries existed. In January, newspapers in America and England told of a scientist who had witnessed a test-run of the Keely motor and was greatly impressed with what he saw. The unnamed scientist declared that Keely "had certainly discovered a new force, which ought eventually be of inestimable value." Several other scientists who were present also found no evidence of the working of any power known to them and they felt "in the presence of something wonderful and beyond their comprehension." But it was also noted that Keely's explanations did nothing to help matters, as these were "couched in hopelessly obscure language." The most that the visiting scientist could make out was that the wheel was kept in motion "by some occult sympathy, allied in a certain way to the elements of music."[76]

Keely must have felt quite lonely at times; after all, Bloomfield-Moore had been much more than a funder. Her directions had helped steer his research, and the books she gave him provided him with a theoretical foundation. Perhaps she was one of those rare persons who understood his explanations, or at least grasped the new horizons that lay concealed in his dense prose.

Now he worked solely for the Keely Motor Company. He was becoming an old man suffering from Bright's disease. From photographs taken at that time, his eyes burned defiantly from an emaciated face. There is no way that we may establish with any certainty what he was thinking of the years that lay behind him with his grandiose plans of unlimited free-energy and a sky filled with immense airships propelled by his aerial propeller that lay still and quiet in a corner of his workshop. Keely stayed in contact with Bloomfield-Moore though, a contact that must have given him great comfort and consolation at times.

There were the occasional rumors in the press; in May, W.K. Fransioli, general manager of the Manhattan Elevated Railroad—who went to Keely's workshop to see the legendary engine on the invitation of the Keely Motor Company —gave rise to the story that the New York Railroad Company would try his engine. This was denied. The man went to the workshop because he was "always interested in anything appertaining to motive power," although he did admit that he went to Philadelphia "entirely unbiased and I returned impressed."

Fransioli had been accompanied by M. McNally, the master mechanic of the railroad company. They not only saw the motor work, but Keely also took the motor apart for them. They saw all its parts, but Fransioli would not say what he saw in the dismantled motor, as he did not think that "would be just for Mr. Keely."

When pressed for more details, Fransioli described the motor as being a "round covering, as big as the diameter of a stove pipe hat. An ordinary non-insulated wire was run from the motor to an engine that was run with high speed." Fransioli was certain that no electric current was used. The wire was removed with bare hands and no shock was felt. The engine stopped immediately. A member of the Keely Motor Company declared that, in his opinion, "John Keely has...found the foundation source of electricity. He has the aurora borealis chained down, sir."[77]

Bloomfield-Moore was informed in writing of the visit, and from it we learn of another, underlying motive of the visit that was not mentioned in the press. She wrote to the correspondent: "I should be, as you write, 'gratified with the renewed attention paid to Mr. Keely by the New York railroad people,' were it not that, in my opinion, this is another attempt to 'boom' the stock of the old Keely Motor Company, before the time is ripe for science to announce the discovery by Keely of an unknown force in nature. Not until this announcement is made, and the exact position of Keely's present mechanical progress is given to the public, will speculation become legitimate.... Until this has been accomplished these efforts of stock-jobbers do more harm to the cause than good; although personally I am benefited by their premature announcements."

She also announced her plans to visit Keely's workshop with Professor G.F. Fitzgerald. She told Fitzgerald that "the days of experiment and investigation are ended. He does not wish to investigate, but will go to Philadelphia for the purpose of meeting Mr. Keely and to witness the operation of the Vibrodyne." Professor Fitzgerald writes: "I look forward to the time when Mr. Keely will be able to describe his methods that his experiments may be repeated by every scientific man who desires to do so, and be then satisfied by his own observation of how they are produced." Fitgerald, a member of the British Association, would then travel to Toronto. "It was with this hope in view," continued Bloomfield-Moore, "(of having the announcement made in Toronto) that I was working when the action of those commercially interested brought Mr. Franzioli (sic) into premature communication with Mr. Keely. But this time they have rendered me a positive service personally, and should this 'renewed interest' in Keely's work last, what I most desire may yet be accomplished in Toronto; viz., that science shall hold the helm, instead of commerce, until the mechanical combination is as obedient to the law of nature, which governs all planetary masses throughout the universe."[78]

In June, Keely gave another demonstration, and "As usual in the Keely seances, the spectators were much impressed with the queer looking engine moving in repines to musical vibrations." One of the reporters remarked that, "It did not revolve rapidly, for the reason that the inventor has just begun work on it, more to determine its strength than anything else. Next week he will give an exhibition of its power, and later of its velocity." A newspaper scathingly remarked that "this is the true Keely mark, next week and next year...Mr. Keely is now an old man, and has devoted his working years to a motor that nobody can understand after it is explained, and nobody can say whether it is deception, delusion, or an elusive idea founded on something real."

The press and the public were growing weary of an inventor who, as a newspaper wrote, "for thirty-five years, at intervals...has been calling in a select company to witness the first practical operation of his etheric engine. To a certain extent it moves, but something in the machinery is found to be not quite right, and a few more weeks for rectification are necessary. Perhaps a year or two elapses, and the performance is then repeated. A generation has passed since the inventor announced that he had discovered and applied a new force in nature. He has never had much trouble in making converts.... Yet it is always next month or next year that the final practical results are to be attained and the mechanical world revolutionized. The request for a little more delay wound up last week's exhibition, as usual."[79]

In January 1898, the event of a paper[80] read to the Engineer's Club, drew headlines.[81] Its writer, E.A. Scott, electrical engineer by profession, declared that Keely had discovered "no new force" and had shown "no apparatus which demonstrated, or could demonstrate, such a force, but that many of the things done by Keely could be readily performed by the operation of well-known laws of nature." He had visited Keely's workshop on a number of occasions. About Keely's famous levitation experiments, involving glass jars filled with water and metal balls and discs he saw no mysterious etheric principle at work. To Scott it was merely the principle of the "Cartesian Diver," and the weights in the jar were so constructed as to just sink in the normal condition of the jar, but to rise when a slight increase of air pressure was put upon the surface through a fine tube carrying, not surprisingly, compressed air. Of this alleged tube, Collier emphatically stated that this was a platinum wire. Collier also offered to give a lecture to the Engineers Club, but when asked if he would bring Keely along with the jar in question, he said that Keely had repeatedly refused to do anything of the kind.

Collier was asked if he knew the real facts regarding Keely's discovery, and if he could continue Keely's work in case of the death of the inventor. He replied that he and Thomas knew "all that Keely had discovered, and that in the event of Keely's death, he would be willing to give the desired explanation."[82]

Collier would not let matters rest there. His countermove was quick and definitive, and toward the end of March he wrote to the press: "Since the delivery of Mr. Scott's address, upon my invitation not less than fifty gentlemen, including engineers, electricians, chemists and mechanics, have visited Keely's laboratory, all of whom have seen his machinery operated and some of whom have seen it as well in its wholly dismantled condition. To each and all I have called specific attention to Mr. Scott's assertion that said wires are tubes and invited them to investigate for themselves and they have done so, and all found them to be what I assert, wires."[83]

And what was Keely doing? While there are no direct references as to his activities during his last years, he once again stated that he at last solved the problem of harnessing the ether, and proudly declared: "My work is now complete."[84]

But was it? A reporter wrote that he announced that "after a labor extending through three decades he has discovered a new cosmic force. For years Mr. Keely claims to have been conscious of a power, but was unable to harness it for commercial purposes. Now, he says, he has solved the problem and that by this force steam and electricity will be at a discount, as his discovery costs practically nothing to operate it."

Keely unfolded grandiose visions, everything was possible through the medium of his new cosmic force: "With the new power railroad trains can be run at the rate of five miles a minute. I am aware that this is a broad assertion, but I have thoroughly studied the power of my new force, and know what I am talking about. Under the present condition of railroads, it will be practically impossible to attain this rate of speed, as it would tear the rails up and shake the cars too." Keely also expected to be able to place a machine on the market "in complete form inside of two months which will fully demonstrate his claims."[85] The months passed by, but this year too would see no engine that pushed trains on reinforced rails at tremendous speeds, propelled by the forces of the cosmos.[86]

Keely died on November 18, 1898, the same day that a Philadelphia newspaper reported that he was seriously ill with pneumonia at his home located at 1632 Oxford Street: "Mr. Keely contracted a severe cold last week, which necessitated his taking to bed on Saturday. He was resting quietly today, and his recovery is expected."[87]

A day later, another newspaper reported the sad news in a long article, containing a number of errors, such as that Keely never applied for, or obtained a patent.

"John Worrell Keely, the inventor of the 'Keely Motor' died yesterday afternoon at 3:00 at his late residence, 1632 Oxford Street, aged 71 years," the article ran. "Mr. Keely had been sick only a week. He first complained of being ill on Thursday of last week to Dr. Chase, of Boston, who called on him at his office while on a visit to this city. He contracted a severe cold, which developed into pneumonia, and for the last three days his condition was considered critical. He rallied somewhat yesterday morning, but soon after grew rapidly weaker, and it was evident he could not recover. He had always enjoyed robust health until three years ago, when his friends noticed that he had lost much of his vigor, and about this time he had an operation performed on his eyes, prior to which during a critical physical examination by his medical attendant it was discovered that he was a sufferer from Bright's disease, although Mr. Keely would never admit this. About two years ago he was run over by a team on Broad Street,[88] and very much shaken up, and this seemed to have affected his health to a considerable degree. Mr. Keely was twice married, and leaves a widow, but no children."[89]

Four days after his death, a newspaper published a list of pallbearers for the funeral services that reflected the wide and intriguing scope of the Keely history. Among others this list included John W. Keely, with whom was probably meant his brother J.A. Keely, Charles Hill, Barton Kincaid, with whom was meant Burton Kinraide and who we shall meet again, Ackermann and Peabody, John Jacob Astor, Professor Lascelles Scott, Professors Joseph Dewar and William B. Crooks, Wayne MacVeagh, Daniel Brinton, Collier, Zentmayer, Boekel and Schuellerman. Singing during the funeral services would be by the choir of the Twentieth Street Methodist Episcopal Church, Twentieth and Jefferson Streets, and by Rev. and Mrs. John G. Wilson and D. and G. Conquest Anthony.[90]

The funeral services were held at 10:00 a.m. on November 23 at Keely's late residence at 1632 Oxford Street. Long before 10:00 a.m., a large crowd of those who had known him, and those who were curious to take a last look at the inventor, assembled into the parlor of the first floor of the residence. There his body was laid in a mahogany coffin with oxidized trimmings. A silver plate on the lid bore the inscription, "John Worrall (sic) Keely, in his sixty-third year."

Keely rested among flowers, and a huge bouquet was presented by the directors of the company, and another by the trustees of the Twentieth Street Methodist Episcopal Church, of which he was a member. There were also bouquets from his friends in New York, Boston and Philadelphia. Anna Keely had placed near the coffin "a touching tribute of her love. It was a heart of red immortals. This design was so arranged that one-half of it might be buried with Mr. Keely and the other half kept by the bereaved woman."

The funeral services were largely attended. Among those present were the directors of the company, the trustees of the Twentieth Street Methodist Episcopal Church and friends from New York and Boston. Not all appointed pallbearers were present; notably Lascelles Scott, Dewar, Crooks and Astor were absent, as their names do not appear in the contemporary press accounts. The choir sang, and several speeches were made, among others by his friend William Colville. After the services, his body was taken to West Laurel Hill Cemetery, where it was put in the receiving vault to await the preparation of his tomb. When the coffin was placed in the vault Anna Keely broke the red heart in two, "leaving one part with the dead and taking the other home with her."[91]

Immediately after his death, strange rumors and legends sprang up. It was written for instance that "It is not known whether the secret of his motor...died with him. It is said that a few of his intimate friends know the secret, but it is not known whether they will be able to make any practicable use of it."[92]

And a little later, Collier said that "when Mr. Keely was taken sick he was just about to take out patents on his last machine, which he confidently asserted was to be the successful finale of his 25 years of experiments. He further asserted that he and Lancaster Thomas, a noted local chemist, were in possession of sufficient practicable information to enable them to complete the work and take out the letters patent. "The machine will be patented," said Mr. Collier. "When, I cannot just say now."[93]

Keely's one-time generous patroness Bloomfield-Moore died only a few months later on January 5, 1899, in London, England.[94] Five days later and a continent away, Tesla would sign his so-desired contract with elusive Astor, the former with whom Bloomfield-Moore had undoubtedly exchanged mutual hopes and ideas for the elevation of mankind, the latter with whom she had sailed on the Nourmahal across the boundless seas of expectation.

Science writer Henry Dam, who was also Bloomfield-Moore's literary executor, said: "I knew that when Mr. Keely died she would not live long. Her whole life was centered in his work, to the exclusion of all other interests and hopes. She had the most profound faith that neither Mr. Keely nor herself could die until the invention had succeeded. After receiving the cabled announcement of Mr. Keely's death she began to sink rapidly. Her ailment seemed more mental than physical."[95]

John Worrell Keely

Keely's workshop, the place of many incredible demonstrations for over a quarter of a century. In 1874, Keely settled into this building, which would be his workshop for life.

Stock certificate issued by the Keely Motor Company. The company would ultimately raise $5,000,000.00 in capital.

Musical Dynasphere

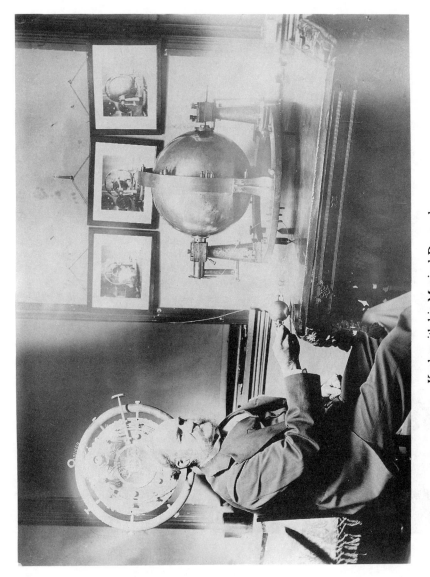

Keely with his Musical Dynasphere

Interior view (shell and drum cover removed) of a Musical Dynasphere

Hydro-Vacuo Motor or Engine

Hydro-Vacuo Motor or Engine

Liberator of the sound force from atoms of the atmosphere. Another attempt to get at the source of all force, but was eventually scrapped as unworkable.

Aerial Propeller

Airship Propeller

Vibrodyne and Sympathetic Negative Transmitter (1896)

"First Circuit" for showing polar and de-polar action. Operated by connection to the Sympathetic Negative Transmitter.

Combination for controlling and transmitting vibratory forces by sympathetic negative attraction (incomplete).

Vibratory Switch-Disk

The Generator was used to generate etheric vapor from pulsating
pressure and vacuum working on water and air (front view).

Generator (left side view)

Generator (right side view)

59

Generator (back view)

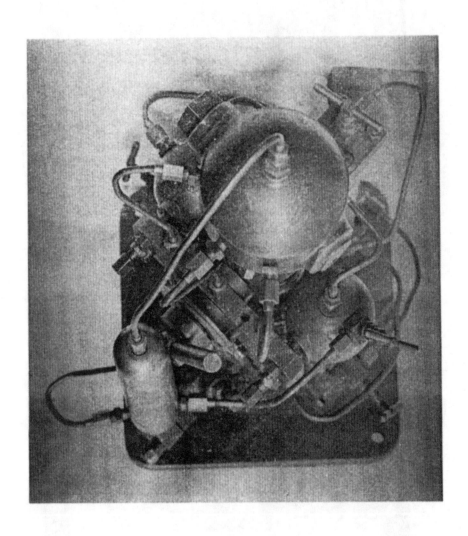

Generator (top view)

5

Into the Void
The Final Stage of the Keely Mystery

*"Not long since, Canon Wilberforce asked Keely what would become of
his discovery and his inventions in case of his death before they became
of commercial value to the public. Keely replied that he had written
thousands of pages, which he hoped would, in such an event, be mastered
by some mind capable of pursuing his researches to practical ends."*

Clara Bloomfield-Moore,
Keely's Discoveries, Aerial Navigation, 1893

The final mystery in the Keely story that gave rise to much speculation in
the years to come, was the disappearance of his papers, drawings and devices
after he died.

In his lifetime, Bloomfield-Moore also wrote about manuscripts in which his
philosophy was explained.[1] On one occasion, Keely told a reporter that he had
"two books on hand...one of which was to come out within a year, possibly
sooner," and the reporter was then shown some "proof sheets prepared by a
friend from Keely's manuscript books." During the interview Keely confided to
the reporter that, "I am at work at present on two books; the one which will
announce my force to the scientific world will contain from 1,500 to 2,000 pages
with 50 or 60 charts, describing the whole thing symbolically. It will appear in
1891."[2]

Notwithstanding Keely's statement, 1891 did not witness his book in print.
Nor would another projected book be printed later. Collier, always ready to come
forward with one or another boisterous claim, obtained from Keely "elaborate
descriptions of his various machines and devices, and much time was spent in
preparing drawings and specifications," another hint that drawings of Keely's
devices did exist. Collier also had "in preparation a book entitled *Two Decades
with Keely,* that would 'prove intensely interested reading.'"[3] What its contents

were we will probably never learn, since as far as is known, Collier's book too was never printed.

Even this facet of the Keely mystery is shrouded in uncertainty; we do not know if and exactly how many manuscripts or papers Keely wrote in his productive life.

In her book, Bloomfield-Moore quotes extensively from what she terms "Keely's writings," unfortunately without identifying this source.[4] From a statement by the attorney of Keely's widow, this source of written material may perhaps be identified: "A week or so after his death access to his papers was secured for the first time. Bundle after bundle was unrolled and minutely examined, but no secret appeared. Fragmentary MSS, in the shape of a diary, unfinished letters, proof sheets of matter ultimately appearing in Mrs. Bloomfield-Moore's book *Keely and His Discoveries,* were found in abundance, but no revelation."[5] And some months afterward, during a second investigation of his workshop, an unfinished letter from Keely to Bloomfield-Moore was found.[6] The whereabouts of these documents are not known.

According to Bloomfield-Moore, Keely also wrote "three treatises to explain his system" and she lists the titles of these papers: "Theoretical Expose or Philosophical Analysis of Vibro-Molecular, Vibro-Atomic, and Sympathetic Vibro-Etheric Forces, as Applied to Induce Mechanical Rotation by Negative Sympathetic Attraction." The second treatise is titled, equally verbose "Explanatory Analysis of Vibro-Acoustic Mechanism in all its Different Groupings or Combinations to Induce Propulsion and Attraction (Sympathetically) by the Power of Sound-Force; as also the Different Conditions of Intensity, both Positive and Negative, on the Progressive Octaves to Ozonic Liberation and Luminosity."

The final paper has the equally long title, "The Determining Principle of Matter, or the Connective Link Between the Finite and the Infinite, Progressively Considered from the Crude Molecular to the Compound Inter-Etheric; showing the Control of Spirit over Matter in all the Variations of Mass Chords and Molecular Groupings, both Physical and Mechanical."[7]

Unfortunately she does not elaborate any further on these treatises and nowhere in her book do we find a description of these writings, such as the number of pages, their contents or the locations. It is doubtful that she read or even saw these papers at all. It may be that Keely wrote to her about the papers in one of his letters, and that she took his word for the existence of these treatises. Most probably the papers were never published and remained private. Contemporary newspapers, which usually devoted much space to various details of his life and his inventions never mentioned their publication.

There is also the hint in *The Snell Manuscript,* a document that was originally written around 1934 and kept in private hands until it was recently published. Its author, C.W. Snell of Detroit, Michigan, writes that, "The secret of Keely's instruments died with him. These notes were made from books written by Keely himself, found in an old library. The books have since disappeared...."[8] This intriguing allegation is unfortunately ill-fitted for closer examination, since the name of the library is not mentioned and no bibliographical details of these books or papers, allegedly written by Keely himself, are given.

Some parts of Keely's philosophies did appear in print, however. There are, for instance, the portions in Colville's *Dashed Against the Rock,* published in 1894. And recently the title page of what appears to be a book, written by Keely, was published, called *Keely's Acoustic Theoretical Charts* with a long subtitle,[9] rivaling those as mentioned by Bloomfield-Moore. The book was printed in Philadelphia in 1887 and consists of a series of remarkable drawings as far as can be determined, since it is by no means certain that the charts were the sole contents of the book. Unfortunately this book is not included in the collections of the two most extensive libraries in the Western world, and thereby cannot be studied. A routine search that I conducted in the Library of Congress showed that only four titles about Keely, and not written by him, are in the possession of the American libraries today. A consultation of the British libraries delivered only a handful of pamphlets written by Bloomfield-Moore.[10]

So, while we are by no means certain that the charts were the only contents of the book, this is at least an indication that Keely did in fact trust some of his philosophy to paper and took the effort of publishing it. Bloomfield-Moore never refers to this 1887 book or any other work in her writings, and the contemporary newspapers make no mention of this particular book whatsoever, and all in all this is suspicious. What the press did write about was a large manuscript written by Keely. There is the tale by a reporter who was shown "a large book filled with heavy cardboard plates, each of which was a treatise in itself, all drawn with a pen by Mr. Keely with beautiful exactness. Nearly every one of these plates had one or more bars of music, showing the dominant chords of vibrations which governed certain combinations."[11] Probably these plates were the same as those which Keely published in 1887. It is claimed that he handed out handsomely framed reproductions of such a chart to various backers.[12]

It was also written that when Keely died, he left "a large number of drawings illustrating the principles of vibratory philosophy. Each plate had one or more charts of music, showing the dominating chords or vibrations which govern certain combinations," and sometimes a newspaper would print a number of these striking charts as a curiosity.[13]

Two days after Keely's death, a newspaper wrote that Keely "left a great mass of manuscript relating to the progress of his experiments, which is the property of his estate, and can only be disposed of in the administration of his effects."[14]

More allusions to the existence of his papers kept cropping up in the days after his death. A newspaper stated that "we did find Keely's voluminous diary covering his experiments, his successes and failures, from hour to hour and day to day through long, torturing years, from 1872-73 up to the time of his last illness. It is a complete record and exceedingly interesting. In view of this daily detailed and minute history of his experiments with his motors there is every reason for believing that Keely was honest and told the truth about his inventions."[15]

At the time of Keely's death, there were other documents in existence as well, such as those that were kept in the corporate office of the Keely Motor Company in the attic of a brick building on Walnut Street above Ninth in Philadelphia. The place was tenanted by Schuellerman. "The office is uncarpeted and almost destitute of furniture. There are no shades on the windows, and the light that enters them throws its rays upon a strange variety of photographic blueprints, working drawings and queer mechanical designs which illustrate the early and later stages of Keely's researches in the mysteries of molecular and atomic forces," a reporter wrote.[16] The whereabouts of these documents are not known.

Three days after Keely's death, B.L. Ackermann, president of the Keely Motor Company, visited Anna Keely upon her request by telegram. He told a reporter that "On my arrival at Mrs. Keely's house...I found her so prostrated that she was unable to more than state that Mr. Keely had left in her possession a manuscript of 2,000 pages, which explains the whole system and the work he has done."[17]

Two days later, a Philadelphia newspaper wrote that, "It is not at all impossible...that litigation may result over possession of Mr. Keely's papers and the machine, which he left uncompleted. Mrs. Keely is unable to be seen and has not in any way signified as yet her intentions with respect to the disposition of her husband's papers." Perhaps the Keely Motor Company had some dark misgivings, for the article continued that, "In case she should elect to refuse any members of the company access to the papers, and that she may adopt this course has already begun to be hinted at, the matter will be taken into court." Collier seized the opportunity to repeat that he was "in possession of much information on the subject of the motor," and that he was "confident that he knows enough about it to construct a machine similar to Mr. Keely's." Collier was in fact "now engaged in preparing for the press a work bearing on Keely's invention, which he expects, will occasion a great revival of scientific interest in Keely's work." The article ended with Collier's optimistic statement that neither he nor Thomas expected "any refusal on Mrs. Keely's part to hand over all her husband's papers and the contents of the laboratory."[18]

On November 24, only two days after the first mention of Keely's manuscript in the press, and a day after his funeral, the first strange contradiction occurred. That day Mr. Schuellerman said that, "he did not know anything one way or the other about the alleged manuscript which it is said Mr. Keely left with his wife. That would be known after the conference with Mrs. Keely."

It would become even stranger. President Ackermann, who mentioned two days earlier that he had heard from Keely's widow about the existence of the manuscript, now told an entirely different story when he was asked how he knew that Mr. Keely left in his wife's custody a manuscript describing his discoveries in detail: "...he said that Mr. Keely had told him he had written such a paper several times in the last twenty-five years. Mr. Ackermann said he had not read it, but Mr. Keely had pulled out a drawer of his desk and pointed out papers as the manuscript, so he was sure it existed."[19]

Another newspaper meanwhile remarked on the same day that, "The future of the Keely motor depends on the manuscript of 2,000 pages which, it is said, the dead inventor left his widow."[20]

On November 25, the word was out, and intrigue was to be expected: "A lively legal contest is expected between Mrs. John W. Keely, widow of the inventor...and the Keely Motor Company, over the possession of all papers left by the dead man bearing on the mysterious apparatus, the alleged secret of which he alone knew."

Apparently his widow had foreseen all this, for she hired an attorney named John G. Johnson. She had "surprised" the directors of the Keely Motor Company on the day of the funeral. What the surprise was, we may never know, but the directors went straight to Collier. Collier, we learn, was no longer a member of the company, and had made a sudden turn of loyalty: "It is said that Dr. Thomas and Mr. Collier were both friendly to the interests of Mrs. Keely, and that should a legal fight come up they will be on her side....There are other interests involved, and it is asserted that Mrs. Keely will resist any attempt by the company to deprive her of what she considers to be her lawful inheritance."[21]

A day later, Keely's grief-stricken widow refused to see anyone, and reporters were not able to communicate with the directors of the Keely Motor Company, therefore "no formal request has yet been preferred by those interested for permission to examine any papers Mr. Keely may have left."[22]

By November 29, the situation had not changed, and "So far as known no will has been found, but certain manuscript is said to be in the possession of Mrs. Keely. She is said to have made a search of the inventor's workshop for valuable papers, but the result of this is not known. All the machinery, tools, etc. in the shop are the property of the company, Mr. Collier stated...The keys of the workshop are still held by Mrs. Keely."[23]

On the same day, Mr. Schuellerman also said that, "Until Mrs. Keely's position and intentions are known, and until something definite is known about the manuscript, which, it is said, Mr. Keely left, describing his secret and his discoveries in detail, no action...can be taken by the directors of the company."[24]

Indeed, the Keely Motor Company did not fare well under these setbacks. The monthly meeting on November 29 was canceled, and "neither the secretary nor Charles B. Collier...were able to state just when the meeting would take place. Inquiry at Mrs. Keely's house at 17th and Oxford Sts. failed to elicit any information."[25] In reply to a reporter, Johnson said that "complications have arisen, which make it impossible for him to represent Mrs. Keely." Those complications were not specified.[26]

Doubts among those interested in Keely's legacy arose once more: "The late inventor Keely's mysterious secret is just as securely locked up today as when he was alive, and none of the directors of the motor company or the many stockholders know what Keely has left in the way of papers or models."[27] And some days later *Scientific American* wondered, "whether the mass of the manuscript which he left will be of any value or not, remains to be seen."[28]

On December 1, Keely's will did turn up; perhaps it was found during Mrs. Keely's search of his workshop. Keely's widow was made sole legatee and testatrix of an estate of $10,000. The will made no mention whatsoever of his motor.[29]

By December 5, the Board of Directors of the Keely Motor Company were in high hopes again. Much was expected from a conference to be held between Ackermann, Schuellerman who represented the company, and Charles S. Hill, attorney for Mrs. Keely.[30]

Nothing substantial was reached; Ackermann had his "short conference" with Hill. The discussion turned to the rights to the manuscript which Keely left, "and will be continued today." Ackermann did say that Mrs. Keely would "surrender the manuscript for an interest in the company."[31] On the same day, an anonymous stockholder confided to a reporter that according to him, "The talk about that manuscript is all nonsense. Keely made some copious notes for his own benefit, but how could he commit to paper that which he could not understand himself?"[32]

Precisely a week later, on December 14, the annual meeting of the Keely Motor Company was planned, but it was not expected that a quorum would be present to transact any business: "The by-laws of the company require that 50,000 shares of stock shall be represented at the meeting and it is not expected that this requirement will be met....Another reason for the postponement of the meeting is due to the fact that the representatives of the company have reached no conclusion with Mrs. Keely." It also appeared that Ackermann and the attorney of Keely's widow held several meetings, but were unable to "come to an understanding."[33]

The meeting was postponed due to the "small attendance." Of the 50,000 shares, only 3,500 were present. Ackermann was not present, but he drew up a five point letter that Schuellerman read. In it, Ackermann again changed his statements concerning Keely's manuscript. Now he alleged that Keely had informed him that the manuscripts were in the possession of Mrs. Keely. Point five related to the Keely manuscript and projected a gloomy view of the future of the Keely Motor Company in case of an everlasting absence of his paper: "Knowing, or, at least presuming, that with such manuscripts in the possession of the company, in addition to the devices, machinery, and so forth, there would exist a very favorable chance in attaining the same results, and that without such, or the same papers in possession of unfriendly or hostile persons unknown to us, the condition under such hypothesis, appeared to me very problematical of ever getting at results, or if so at any time, only in the far future."[34]

And now, a month after Keely's death, the conclusion was reached, but it would not be in the most harmonious of circumstances. Amidst the fights, the scandal, the accusations, and the further dissolution of the Keely Motor Company, we also learn more about the fate of Keely's alleged 2,000 page manuscript.

On December 20, the last remnants of solidarity between members of the Keely Motor Company now completely crumbled, and a newspaper ironically mentioned that, "Round one of the Keely Motor Company stockholders fight ended at 2:00 by a sudden adjournment which just saved Charles Collier, Jr., from an impending knockout at the hands of B.L. Ackermann and C.L. Schuellerman." Ackermann accused Collier of pocketing large sums of money, "and the mere mention of money aroused the thirty stockholders present." Young Mr. Collier...shouted to Mr. Ackermann, "That's a lie! That's a lie!" Ackermann himself faced accusations of having secured large quantities of stock by unsigned letters.[35] All in all, the Keely Motor Company was in a disgraceful state and its members a sorry lot - but the worst was yet to come.

A month had now passed since Ackermann alleged that he had heard of Keely's manuscript. At the meeting, Hill, Anna Keely's attorney, demonstrated that he had studied the financial proceedings of the company very well, and he said that the history of the relations between the late John W. Keely and the Keely Motor Company was "a tangled web of unfulfilled promises, broken contracts, grievous misunderstandings and mutual recriminations."

Hill pointed out that, "if any business operations are to be carried on, having the Keely discoveries as their basis, the present Keely Motor Company is not in an advantageous position to do so. A body of men attempting to exercise corporate powers where there is a question as to the validity of their existence as a corporation per se, would subject itself to unlimited litigation which otherwise might be avoided."

Hill could therefore easily enforce his position by stating: "If you force me to the point where I disclose certain things, I don't think the Keely Motor Company will ever do much more business."

So it seemed that the rumors about financial fraud that had always surrounded the Keely Motor Company were devastatingly true. But what about Keely's mysterious manuscript? Did it really exist? If so, where was it, and what happened to it?

Hill stated: "The reports concerning immense quantities of manuscript left behind him are so far as I am aware untrue." A little later Hill said, "...I do not think that Mr. Keely's secret is written down,"[36] which, of course, was a strange statement from someone who according to Ackermann had previously said that Keely's widow would turn in the manuscript in exchange for a substantial share in the Keely Motor Company.[37] Hill also stated that "he had made a careful examination of the effects of Mr. Keely and that no detailed manuscript had been found."[38]

According to Hill, "A few scraps of memoranda in the form of a diary, possibly 500 or 600 words, giving no clue to anything whatsoever, half completed drafts and letters written by him on business, and other such unimportant papers comprise the bulk of all that I have been able to find." Then Hill said ambiguously: "The whole thing is as uncertain a state as ever."[39]

All this of course did not end the controversy surrounding the manuscript, which by now had attained an almost mythical status. Collier maintained a day after the meeting, that Keely "did prepare a manuscript pertaining to his experiments." Collier said that "some time previous to Keely's death he spoke to him regarding the preservation of the results of his work. Keely told him that he had manuscript, and showed it to Mr. Collier. Mr. Collier read part of it at the time."[40]

Later, Hill would claim that not only manuscripts had been found and that the bulk of these had been published in Bloomfield-Moore's book, but furthermore that, "Previous to his death Mr. Keely repeatedly asserted that all of his secret formulas by which his wonderful experiments were performed had been committed to writing so fully and completely that were he to be taken away the work could go on uninterruptedly. He had given tantalizing glimpses of great piles of MSS to some of the officials of the company. ...Such things as these naturally led me to hope for the existence of some valuable papers."[41]

The strange story of Keely's manuscripts and papers relating to his inventions does not end here. There was also another alleged storage of documents pertaining to him and his discoveries. Bloomfield-Moore left "immense files of writings concerning the motor, including a hundred letters to herself from Mr. Keely."[42] But the story of what happened with these "immense files" is as strange, confusing and inconclusive as what happened with Keely's alleged 2,000 page manuscript.

More than half a century after his death and the ensuing tumult over his legacy, his nephew was located and interviewed. He was an old man at that time, and professed that he never had much confidence in Keely's experiments, but he recalled how Bloomfield-Moore sent "many of Keely's secrets" to Count von Rosen, Commander-Captain in the Swedish Navy, then living in Scotland. He apparently sent this material to Stockholm in 1912, and nothing more was heard of these "secrets," although in 1972 a ring-bound volume was published there in which the foreword suggested that the book consisted of these materials, to which claim there is some circumstantial evidence.[43]

What these secrets originally consisted of, Keely's nephew did not care to explain. We are by no means sure that, given that Count von Rosen did indeed send the secret materials via Scotland to Sweden, that these were the "immense files" purportedly in possession of Bloomfield-Moore. Count von Rosen was a grandson of Bloomfield-Moore; his father married Ella Carlton, the daughter of Bloomfield-Moore in 1864.[44] While he was the executor of her estate, he was but one of the executors of her will. She left her manuscripts to "another grandson," his cousin Harold de Bildt, who acted as her literary executor.[45]

Von Rosen told a reporter that Bloomfield-Moore "had no papers of which I know giving any idea of Keely's secret. Her papers are all here in the Fidelity Trust Company's care, and include the stock that she held in the motor company." He furthermore said that, when she learned of Keely's death, she "expressed a hope that he had imparted his secret to some one before he died."[46] Based on his statements, although made long before he allegedly sent the secret materials to Sweden, we are left to assume that her files contained nothing of importance concerning, and directly relating to his inventions.

Nevertheless, this may have given rise to the claims of Keely's nephew. The Count, while staying in Philadelphia, was "making some inquiries in regard to the motor, and says he will probably visit the old laboratory where Keely did his work before leaving the city."[47]

It is also alleged that a Philadelphia occultist, Isaac Meyer, who wrote a learned treatise on the cabala may have held some files on Keely. Meyer apparently willed the New York Public Library 36 boxes of notes, books, translations and bibliographies. Most of this material concerns the research he did on the cabala. Although a large portion is said to have included collections on Keely and his inventions, nothing was found during a search.[48]

And there the matter rests. Keely's mysterious manuscript disappeared into nothingness. The fate of the manuscript became shrouded in such an impenetrable mist so soon after the announcement was made that he left such materials, that one may even doubt if it ever existed at all. For even this is not entirely certain, or conclusively proven, although as we have seen, it was suggested many times that it did indeed exist. His 1887 book is nowhere to be found; the ultimate fates of Bloomfield-Moore's "immense files" and Keely's "hundred letters" are equally shrouded in the same impenetrable mist.[49]

There were also Keely's engines, which by this time had also disappeared. It was stated on several occasions that Keely constructed and discarded 129 different models of his wonderful engines,[50] while Keely himself once said that he had made "2,000 of them!"[51]

Many of the engines were no longer in existence because they were sold as "scrap iron" as Schuellerman stated. Babcock wrote: "many tons of these have been sold from time to time as old iron, brass and copper. One apparatus thus disposed of weighed twenty-two tons. Several similar ones, though somewhat lighter, have likewise gone to the scrap-heap. From this source money has often been raised by the inventor."[52] With these funds, he was able to continue his researches at a time when the Keely Motor Company left him without any financial support. Not that he earned much money from the sale of his discarded devices. At one time Collier lamented that, "Our generator, which cost us $60,000, we were forced to sell for $200 as old form."[53] Keely, or the companies that built his devices destroyed his older prototypes whenever a new, smaller and better one was built.[54] This was customary from the beginning, and in 1881 Babcock remarked that, "had they all been preserved, (they) would make an interesting museum of mechanical curiosities."[55]

When stockholders obtained court orders against Keely in 1882, he destroyed many of his machines and drawings in fear of confiscation. When the courts threatened again in 1888, Keely destroyed several devices and valuable papers.[56] When he was threatened with imprisonment by the Keely Motor Company, he not only destroyed a lot of his instruments "in a fit of despair," but also announced that "they may take his corpse to jail."[57]

Which devices he destroyed and which escaped destruction is impossible to say. Bloomfield-Moore confirms the destruction on those occasions, but from her statement we learn a little: "In the winter of 1881-82, when threatened with imprisonment by the managers of the Keely Motor Company for not disclosing his secret to them, which then would have been like pricking a bubble, he destroyed his vibratory lift and other instruments that he had been years in perfecting. At this time so hopeless was Keely, that his plans were made to destroy himself, after destroying his devices. ...Again, in 1888...he broke up his vibratory microscope, his sympathetic transmitter, and some devices, which have taken much of his time to reconstruct."[58] She too confirms that in 1882, "engine after engine was abandoned and sold as old metal in his repeated failures to construct one that would keep up the rotary motion of the ether that was necessary to hold it in any structure."[59]

But what became of the devices that Keely did not destroy? Whether he constructed 129 or as many as 2,000, is a mystery. Various writers and researchers have put forward their own theories about the fates of those other wonderful engines. Most theories agree that the devices were taken out of Philadelphia to another location. It is alleged that Dr. Chase, who visited him when he was ill, took some fundamental pieces of machinery from his workshop to Boston for unknown reasons. A variant of this strange story had it that Dr. Chase returned some of those devices during an investigation of Keely's workshop by Clarence Moore and some scientists.[60]

It is also claimed that the machines were shipped to his workshop in Boston by a man named Konrade, but the engines didn't work there. Because Konrade wouldn't admit that he had failed, he claimed that he had discovered hidden levers and cogwheels, which were propelled through a water pump that he discovered in a nearby building.[61] When we once again turn to the contemporary sources, we not only find that Dr. Chase was referred to only once,[62] but that there never was a Konrade. One who did exist however, was T. Burton Kinraide of Boston.[63] It was Kinraide who paid for the lot on the West Laurel Hill Cemetery where Keely found his final resting place.[64]

The subsequent fates of some of the engines are entangled with the strange story of Kinraide, which is embedded in that distressing period with its scandals, frustrations, hopes, accusations, and battles between the Keely Motor Company, Collier, Hill and Mrs. Keely over Keely's legacy.

The first direct reference to Kinraide without actually naming him—apart from his misspelled name in the list of pallbearers at the funeral—appeared in the press a week and a day after Keely's death. Kinraide "came on to Philadelphia after Mr. Keely died, was at the house every day between the death and the funeral; rode with Mrs. Keely in the same carriage to the cemetery and returned with her to the house. This alarming stranger is really not a stranger to the directors of the company. Several years ago, it is said, he appeared at the Keely establishment and has been a constant visitor and intimate of the family ever since. He remained on some of his visits as long as four or five weeks, during which time he devoted day after day to the study of the Keely motor, remaining in the room with it for hours at a time."

It seemed that the directors did not approve of this, but Keely was quoted as saying: "I can guard my secret all right." Nevertheless, a "full inquiry" was launched, to determine "the character and standing of Keely's mysterious friend."

The inquiry yielded some intriguing details: "It was discovered that he was a man of great wealth and lived in princely fashion in the vicinity of Boston. It was also learned that he was a scientist with theories of his own almost as remarkable as those advanced by Keely. Like Keely, he pursued discovery on the tones of vibration, and in order to conduct some special experiments, he had a huge cave hewn out of the solid rock on his estate, which he fitted up as a laboratory."[65] While experimenting, he succeeded in obtaining rotary motion on the compass needle from vibrations.[66]

It appears that stockholders in the Boston area already noticed this man: "On several occasions Boston stockholders in the Keely Motor Company wrote that Keely's ideas were being absorbed by a man who had already made a device similar in physical construction and operation. The apparatus was duplicated and forwarded to the directors, who handed it over to Mr. Keely. Keely laughed at it before them, but promised that he would take up the question. It was remarked after this that the friendship between the two men became closer than ever. The Boston scientist and man of means was at the Keely house every week, and it is alleged, is in the city at the present time..."[67]

A few days later, Collier revealed the name of this mysterious Bostonian, who had a huge grotto on his estate filled to the brim with weird and wonderful devices: "...The statement that a man who knew the inventor in life and who possessed the valuable secret was now in town was not denied by Mr. Collier. He said that this man was a Mr. Kincaid (sic) of Boston, and that he was an inventor himself. He frequently had interviews with Mr. Keely, Mr. Collier said, but it is unlikely that the inventor told him the secret of the famous motor."[68]

On December 20, when the fighting over Keely's legacy had reached its peak and accusations flew high and wide, another reference to the mysterious man from Boston appeared during the annual meeting of the stockholders. It was also stated that he had known Keely since 1895: "Before the battle began, the statement spread that Keely had confided in T.B. Kinraid, (sic) of Boston, who had been summoned to the dying bed of the inventor. President Ackermann said he had been told by Charles S. Hill, attorney of Mrs. Keely, that Mr. Keely said that Kinraid was the one man who could carry on the work successfully. At this statement one stockholder stood up to say that three years ago he heard Mr. Keely say that Kinraid might know enough to be a little inventor, but that he was of no great ability."[69]

Hill, the attorney of Keely's widow, outlined the plan of the estate, which was to place the machines and data in the hands of T.B. Kinraide. Apparently this was Mrs. Keely's plan: "It is Mrs. Keely's intention to have Mr. Kinraide give his entire time and his best abilities toward the completion of a practical engine, and he, until he has produced a machine capable of being patented..."

To all this, Hill also mentioned that this would happen within a year, at the end of which "he will present a report, supplemented by the allowance of an inspection of what he has done." Yet Hill was unclear of the time period: "Mr. Kinraide will work out the invention as the invention of John W. Keely, and will sign a contract to that effect. ...Mr. Kinraide thinks he can determine in six weeks time whether he can bring the experiments to a successful issue."

Hill not only stated that he thought it unwise to have one group of people to have the sole control—"The Bloomfield-Moore and other interests have to be considered"—but also said that Kinraide "will make no promises that any machine will be completed...I can say that if this matter should be opposed the company is practically signing its own death warrant."

Interestingly, this part of the final mystery holds a smaller mystery as well. A repetition of what happened to Boekel some fifteen years ago seems to have taken place. Somehow, Hill deemed it necessary to state at the meeting that "he had private information that would influence their decision, and that, in order to secure the cooperation of the members, he would disclose it in secret to one of their number. John J. Smith, of New York, was appointed to receive this secret, and held a conference with Mr. Hill."

While this is strange, considering the thunderous threats that Hill rained down upon the Board (why should he need the cooperation anyway) the outcome of the conference would be even more surprising.

At the evening session, Mr. Smith made his report. He said that if he were the sole owner of the interests of the Keely Motor Company, he would make the same report, being that the suggestions of Mr. Hill be followed. "I think I have the Keely motor secret," said he, "but I shall use it no more for my own interests than for the interests of the company. In a short time, Mr. Kinraide will be able to tell whether he can work it out or not. From what I've seen today, I am thoroughly convinced that Mr. Kinraide can take those machines and operate them as Mr. Keely did, and produce the same results."[70]

What was it that Smith learned and saw that totally convinced him, and even made him think that he now possessed Keely's secret? Smith did not disclose what he knew, and he kept his mouth shut tightly. For now!

When Smith was selected, Hill "at first demurred to any selection..." and then made the following very strange remark: "You will rue the day you let the thing get out."[71]

Collier tried his best shot and asked Hill if it were possible for Kinraide to "take those machines before a scientific body and operate them." Smith curtly stated that "it would be not advisable to do so, at this time." When Collier insisted, by stating that "for years and years we asked John W. Keely to make a demonstration before the Franklin Institute. Why can't it be done, now that we have somebody else at the head of the affairs?" Hill's reply was short: "As you have shown your confidence in me I shall have to ask you to consider that part of the matter closed for the present."[72]

The takeover by the men from Boston was now complete. Hill had eliminated any possible control of the Keely Motor Company over Keely's legacy. The fate of Keely's engines now lay in the hands of Kinraide, and no engine was to be exhibited. Meanwhile, Keely's engines were taken from the workshop and were placed in the vaults of the Land, Title and Trust Company, "fifteen of the more delicate pieces of machinery have been packed in flannel and removed for safety."[73]

All the material and data would be placed in the hands of T.B. Kinraide, which "was the dying request of Keely." Kinraide was to have no compensation until he came up with some tangible results, and he was to have a year in which to produce these. Kinraide would also remove all appliances to his own laboratory in Boston at his own expense. According to Hill, Kinraide would be able to tell "in a very few weeks...whether or not he can successfully complete the Keely inventions."[74]

What could the stockholders do but to "have resolved to allow T.B. Kinraide, of Boston, to continue the work left unfinished by Mr. Keely...," and, puzzled enough by all this, declare that, "The man who is to take up work on the mysterious motor is said to have enjoyed the fullest confidence of the deceased inventor, and had a knowledge of the experiments."[75]

The directors of the company thought otherwise. They were extremely skeptical, "to say the least, of Mr. Kinraide's ability to make any progress, inasmuch as they believe that he does not know Keely's secret... Keely was eccentric. He refused to divulge the one secret idea, which, he said, was the basis of all his experiments, even to his closest associates. Charles B. Collier and Lancaster Thomas, who for many years stood by Keely, while they were privy to many of his plans and had free access to his workshop, were never acquainted with the mysterious 'vitalizing force'—Keely's secret, and they are firm in their belief that Mr. Kinraide or no other person knows it."[76]

A dry conjecture of the situation was published a week later in a short editorial in a Philadelphia newspaper: "It appears that we must wait another year to learn the secret of the Keely motor, and we may not learn it even then. A Boston man who is supposed to have penetrated farther into the mystery than anybody else has been entrusted with the task of completing Keely's great life work—if he can. He is not entirely confident of himself, but it is agreed that if he cannot make the motor mote, nobody can, and he is to have a year for the trial."[77]

More specifically, Kinraide said that if at the end of one year he was convinced that there was absolutely nothing that would lead to a practical machine he would abandon his attempts. "If it is shown that there are clues enough to lead to the construction of such an engine it will be pushed rapidly forward to the patentable stage and further agreements made. Mr. Kinraide will make no agreement that any machine or engine will be completed at a certain time, or even that one will be completed at all. He will simply devote one year to try to solve the secret possessed by Keely. The work will be done at Mr. Kinraide's laboratory here, and during his working hours he will be inaccessible to all callers."[78]

On December 28, Keely's engines were shipped to Boston. But, contrary to what researchers suggest, not all of Keely's engines were to go to Kinraide's laboratory: "All the devices of the late John W. Keely that are held to be of importance in connection with the experiments to be made by T.B. Kinraide of Boston, will be shipped today, to be placed in his laboratory. The more delicate machines and parts have been in safe keeping in the vaults of the Land Title and Trust Company, having been placed there shortly after the death of Mr. Keely. Charles S. Hill, of Boston, attorney for Mrs. Keely, was in the city yesterday, and with Mr. Kinraide supervised the work of moving the machines. ...Mr. Hill says that the older machines and those which Mr. Kinraide believes will be of no practical value in his work will be stored by the Keely Motor Company in this city, and that a room will be rented for that purpose."[79] Hill personally escorted Keely's engines to Boston, where he arrived a day later.[80]

Keely's inventory, filed on January 3, the same day that his devices arrived in Boston, made a reference to "Fifteen pieces of experimental apparatus."[81] A Boston newspaper wrote that "Twenty large packing cases have arrived here, containing the material part of the famous Keely motor. T.B. Kinraide, an inventor, ordered the boxes removed to his laboratory. There he will experiment in trying to supply whatever is lacking in mechanism."

Kinraide also told a reporter that, "he had often talked with Mr. Keely on the principles of his invention. He never fully explained the secret of his perpetual motion to me, said Mr. Kinraide, but I feel that I know more of the motor than any other man. Mr. Keely, after being taken ill, expressed the wish that I be allowed to carry out his inventions. Before the hour set for the interview had arrived the inventor was past recovery. It was, however, at Mr. Keely's request and that of Mrs. Keely that I have consented to conduct these experiments. I am not a director of the company, neither am I an employee of the directors in any sense. I am not hired to do the work which I have undertaken. So far, all I have done has been at my expense."[82]

So Keely's engines, which were already crated on December 20,[83] were stored in two places. The contents of 20 large crates were on Kinraide's estate on Jamaica Plain in the vicinity of Boston, where his laboratory was situated,[84] and in Philadelphia, where Keely's "mechanical property" was removed and stored in a building on North Broad Street.[85] Apart from the machinery that was moved to Boston, the "heavier apparatus" was relocated to a storehouse at Broad and Vine Streets, possibly the same storehouse on Broad Street.[86] Elsewhere it is written that the storehouse was located at Broad Street and Fairmont Avenue.[87] In a safe deposit vault at the Land Title and Trust Company, Keely's disintegrator and other "fine pieces of machinery," his sensitized disks, wires, etc. were placed.[88] Before being placed in the vault, the disintegrator made a short stop at the Hotel Stratford, were it was examined by Smith, Kinraide and Hill.[89]

Perhaps a detailed inventory was made of the remaining devices at the time of his death, showing which devices Kinraide obtained, but if so, it did not survive. We are therefore left with superficial descriptions that Kinraide obtained the "remarkable machines, vibrators, lever machine and others used by Keely."[90] It was suggested that Kinraide also obtained his latest engine, the work on which was almost finished by William F. Rudolf at the time of Keely's death, and that resembled the Globe Motor but only larger, with a two feet copper globe weighing 75 pounds and the mechanism made out of decarbonized steel weighing 600 pounds.[91]

Behind the scene, the battle over the rights of ownership of the engines continued. Hill stated to a reporter that, "Whatever the property rights may be, that question will remain in abeyance. The company has simply agreed to turn over the machines to Mr. Kinraide for his experiments."[92]

On January 3, 1899, an inventory of Keely's personal estate was filed with Register of Wills Hackett: "An inventory of the personal estate of the late inventor John W. Keely, the appraisal being placed at $1,536." The only reference to the famous motor was the following unspecified clause: "Fifteen pieces of experimental apparatus, a certain right of unknown valuation in certain uncompleted and unpatented inventions."[93]

On January 29, 1899, precisely a month after some of the devices reached Jamaica Plain, Hill issued a remarkable statement to the press. Hill timed his statement well. At the same time the newspaper ran a huge article, together with detailed drawings and elaborate explanations of Keely's exposure as a fraud. In the article, Hill's statement received considerable space. "...It was arranged between the president of the Keely Motor Company and myself that on December 20, 1898, I should address the stockholders, giving my views, as Mrs. Keely's counsel, as to the best course to pursue. Shortly after this agreement, while examining the laboratory, Mr. Kinraide discovered the first evidence of fraud. Till that moment our confidence in Mr. Keely's integrity and honor was as firm as any of his friends who had invested thousands. Here was a new element, that of self-evident fraud, affecting, however, only one machine, and not vitiating, so far as we knew, any other of the numerous machines Mr. Keely employed."[94] Surprisingly, J.J. Smith, who claimed to know "Keely's Secret," accompanied Hill and Kinraide and also delivered a damning statement. Was this the secret that he came to know?[95]

Another character, J. Ransom Bridge from Philadelphia, came forward and said that, "When Mr. T. Burton Kinraide took charge of Keely's laboratory one of the first discoveries was how Mr. Keely did his experiments. He could vary the initial performance in a dozen ways, but the principle was always the same."[96]

The statements of Hill, Smith and Bridge should have settled the matter, but it didn't. Strangely, Kinraide vehemently disagreed with Hill's statement. Kinraide was "greatly surprised that Mr. Hill had written anything against the Keely motor, and refused to say anything" until he saw Mr. Hill, as he doubted the validity of the report. He was unable to find Mr. Hill, but consulted his lawyer instead. On the advice of his counsel he declined to give out anything further than this statement, which was carefully written out: "If, as you inform me, Mr. Hill has made a pretended exposé of the Keely motor in the New York Journal, and used my name in connection therewith, he has done so without my knowledge or assent, and I consider it a gross outrage and a dastardly breach of faith."[97]

Bridge answered that the knowledge of the fraud was "imparted to him without solicitation or request," and that he made no promise of secrecy and was under no obligation to keep silent. He also claimed that, "In addition to the evidences of fraud which I saw in Mr. Keely's laboratory in Philadelphia I also have seen since Mr. Keely's death the most important of the experiments performed by Mr. Kinraid (sic) himself. Mr. Kinraid told both Mr. Hill and myself that plain evidences of fraud covered every experiment done by Mr. Keely and as we had ocular demonstrations of the tricks we could not doubt the statement."[98]

Be that as it may, puzzled and alarmed stockholders told a reporter that they wanted to determine whether or not Mr. Kinraide believed that Keely's work was fraudulent. If he thought so, they wanted the machines returned to Philadelphia. And indeed they were puzzled by a certain ambiguity: "If it was known that the Keely machines were fraudulent, why were they taken on to Boston, and why was it held out that it would take Mr. Kinraide time to probe into the matter, and that it might take a year?"[99]

The same day, Thomas Collier, who the year before had so boldly claimed that he could build a machine similar to Keely's devices, and Rudolf, the man who built Keely's latest engine, visited Kinraide's laboratory in order to "satisfy themselves that the Keely machines are in the same condition as when they left the laboratory in this city."[100]

Rudolph, on his return from Boston, said that he would be going back "to continue the investigation set on foot by the directors." More could not be learned from him, as he did "not intend to give out any information, as it was the decision of the whole committee that we should say nothing in advance."[101]

On February 29, at a meeting of the Board held at the company's office at 913 Walnut St., Collier told the assembly that "the motor is now at Jamaica Plain near Boston, in the keeping of T.B. Kinraide: ...we will follow the events at Jamaica Plain with interest."[102] So at that time, the part of Keely's equipment that was shipped to Kinraide was still in the cave in Jamaica Plain near Boston, and we must assume that the rest of Keely's devices were still located in the vaults in Philadelphia.[103]

For four months, everybody, including the newspapers, was silent about Kinraide and his experiments with Keely's engines, and the status of the devices in Philadelphia. On July 16, 1899, the cloak of silence would be lifted, but under very unusual circumstances. The event announced its coming in a short notice in *Scientific American:* "Mr. Kinraide, of Jamaica Plain, Mass., has abandoned all work on the Keely motor, and will ship back to the Keely Motor Company all the machines and manuscripts left by Keely. Mr. Kinraide was on terms of some intimacy with Keely, and it was thought that he might discover, if possible, some virtue in the motor. The exposure of the frauds which Keely perpetrated in his Philadelphia laboratory, which we have already illustrated, has helped to induce Mr. Kinraide to abandon the whole matter."[104]

This statement offers more puzzlement than clarification. Just when one is about to frustratingly dismiss the whole manuscript-affair out of pure lack of further information, it appears from the article that Kinraide did have manuscripts left by Keely. Were these Keely's mysterious diaries which were, as a newspaper had written, given to Kinraide by his widow? And why would Kinraide abandon the work due to the news of the alleged exposure by Moore and his team? Kinraide disagreed with Hill's version of the confusing tale, even going so far as to consult his lawyer. And how about his experiments with Keely's engines? The eagerly awaited report finally came on July 16, almost two months after the allegation that Kinraide would ship the engines back to Philadelphia.

The circumstances in which Kinraide delivered his report were unusual. First, his report was not delivered by himself as one would have expected, but by Clarence B. Moore, the son of Bloomfield-Moore. And second, he did not obtain the report directly from Kinraide. Moore's informant was "a well-known gentleman" from Philadelphia who had invested "heavily in the stock of the alleged motor, and who recently paid Mr. Kinraide a visit at his laboratory." Kinraide told this anonymous informant that Keely had not discovered a new force, and demonstrated his claim "to the satisfaction of his visitor." Kinraide made this informant promise not to make public the information given to him, but he did, "however, receive permission to report the result of the investigation to Mr. Moore."[105] This is strange, since Kinraide could have known that Moore, whose antagonistic views about Keely were well known at that time, would head straight for the press.

And so Moore did. He told a reporter that "Mr. T.B. Kinraide, of Jamaica Plain, Boston, on whom the mantle of Keely was supposed to have fallen, and who actually did receive the Keely motor mechanism early in January, admits that the motor was a fraud, that the machinery was moved by well-known forces, and that the duplication of Keely's 'demonstrations' is a simple matter. When Mr. Kinraide was reported, a short time ago, to have returned the Keely motor machinery to Philadelphia, the inevitable conclusion was forced on any one conversant with the history of the Keely motor that Mr. Kinraide, having discovered the nature of the fraud, washed his hands of the whole affair. It seems however, that Mr. Kinraide did not send the apparatus back, but got to the heart of the delusion by a careful study of the motors in hand."[106]

According to the unnamed informant, during this "careful study" Kinraide set up the machine room in his cave at Jamaica Plain "almost exactly after the manner of Keely's, suggesting that he had begun on the premise that he must reproduce the whole thing if desirous of success." Then, Kinraide walked around the machines for some time, and "at last he turned and smilingly remarked that he might excel Keely at his own trickery if he had the same fluency of words as the latter."

Kinraide applied compressed air, hydraulic pressure and a powerful spring, and even a magnet, concealed in the wall. The hydraulic pressure and compressed air came from "hidden sources." With the help of all this, Kinraide obtained the same results. "One of the most unique pieces of mechanism I found in Philadelphia," Kinraide reportedly said, "was a spring to wind using a key as big as a crowbar." With the proper winding the spring would be able to "run for three or four days," and produces enormous energy. Kinraide also showed how he could start the hydraulic and compressed air pressure by picking up a violin, after which the "instrument wheels began to revolve," because he touched "a bulb," hidden under the floor at the same time.

It took the unnamed informant nearly eight hours in Kinraide's laboratory, "and when the Philadelphian emerged he was convinced that he had lost his investment." Kinraide repeated that those who would produce evidence of being "victimized," could obtain an invitation to visit Kinraide's laboratory at Jamaica Plain and "see the whole thing disclosed." Kinraide refused entrance to reporters, "for it has been decreed that, as far as the public is concerned, the Keely affair can rest in peace and the Keely victims have their sorrows to themselves."[107]

This, of course, was a strange thing as well, since this effectively barred all further verification of Kinraide's claims. It also made no sense, since everything was published in the newspapers anyway. A newspaper remarked that, "The announcement of Mr. Kinraide's present conviction is all well enough so far as it goes, but in the premises Mr. Kinraide certainly owes it to the world of science to make a full and complete report on his examination of the Keely machinery and his reasons for his belief that the motor is to be classed among the great frauds of the present century. In this issue Mr. Kinraide has not only a responsibility to himself but to the public. So long as he kept quiet after he discovered the fraudulent character of the motor so long was he, doubtless unintentionally, aiding in maintaining a delusion. ....So it seems he has spoken his mind freely but privately. This is where he errs. He should go on public record."

It was also remarked that, although others, scientists and the like, had already made their minds up that Keely was a fraud, Kinraide had one advantage; "he and not they had the machinery to work over. ...For this reason a brochure entitled 'Kinraide on Keely' ought to be issued. To keep quiet is not scientific and savors of disingenousness."

It was also suggested that Keely's machinery should, if possible, be sent to a museum, "in Philadelphia preferably."[108] And since this suggestion was published almost two months after the statement in *Scientific American* that Kinraide would ship Keely's devices back to the Keely Motor Company, we may conclude that no such thing occurred until that time. It seems that in the end, nothing came of this. The newspapers fell silent once more, and after a while, lost all interest. And although *Scientific American* had announced the return of Keely's engines to Philadelphia, no Philadelphia newspaper in the months to come mentioned their delivery to that city.

Several months before, an intriguing notice appeared in a Philadelphia newspaper in which Keely's brother, J.A. Keely of New York, told a reporter that the secret was known to "an inventor and electrician of Asbury Park," a Richard Croker, to whom it was shown five years previously upon an oath that he would not divulge it "as long as Keely was alive." Keely's brother was now "negotiating with this man for the manufacture of the motor."[109] Nothing more was published in the coming months and years. Keely's brother once again wandered off into the pages of history, and we hear no more of him or the electrician of Asbury Park.

The whereabouts of Keely's engines which were shipped to a cave somewhere on an estate on Jamaica Plain and the heavier ones that were stored in Philadelphia were lost in the mist of time, and there the mystery remains. Conventional contemporary sources offer no further explanation.

It is not known what happened to devices such as the huge aerial propeller, on which Keely worked as late as 1895. His aerial propeller, for instance (a description of this huge construction will follow in chapter 8) is strangely absent in contemporary accounts relating to the removal of his engines. No mention is made of either the destruction or the removal of the aerial propeller. Taking the sudden loss of interest of the conventional contemporary sources into account, we might speculate that Keely's devices were dispersed as time went by, and subsequent generations lost all curiosity in his incomprehensible inventions when even the memory of Keely himself became a thing of the past.

But as we will see in the chapter on Keely's connections with the occult undercurrents—and there were many—it is from those quarters that at least one, although an altogether very different, solution to this mystery is offered. For while conventional sources such as the contemporary newspapers were quite satisfied with repeating Moore's allegations that Keely was a fraud, and investigative journalism does not seem to have been a virtue in the Keely history, it is in the occult quarters that a very different story was told. There it is alluded to that interested parties, who went to a great deal of effort to remain anonymous, did see the enormous significance of Keely's discoveries and his engines, and took great care of collecting and hiding his devices from public view.

There is a suspicious incident that may give the viewpoint of these occult quarters some weight. In 1910, Astor—who fifteen years before had wanted to finance Keely to the extent of several millions—was reported to have docked a "mysterious craft" in the Harlem River. A newspaper stated that the craft "seemed to embody an airship with a practical watercraft."[110] As reporters were kept away and no other information seems to have surfaced on this subject, we can only wonder if some of Keely's equipment may have survived to the extent that it was employed or tested by Astor on his strange craft.[111]

The whereabouts of only two of Keely's original devices are known today, which is but a mere fraction, considering the 20 large crates that went to Kinraide's cave at Jamaica Plain, and the devices that were in storage in Philadelphia. The Franklin Institute had one,[112] which was sold in the 1970s,[113] and currently there is a device on display in the American Precision Museum in Vermont. It is labeled

"Keely's Etheric Force Main Stator." This engine was donated around 1984 to the museum.[114] The device is quite useless for research purposes, as it lacks certain parts and is incomplete.

It is also a mystery that of all people in the world, Moore acted as the spokesperson for Kinraide. For it was he who had at one time battled over his father's will with his mother. He had always been antagonistic towards Keely, and had a substantial role in what has become known as Keely's exposure.

6

Anatomy of an Exposure

"That Keely used the ordinary forces of nature, electrical, magnetic, chemical, pneumatic, hydrostatic, in his experiments has long been charged, but never surely proved."

"The Keely Mystery," *The Press*, January 9, 1899

"There was danger that he would go down in history as a mystery. There is now no risk of that."

"The Keely Motor Exposure"
Public Ledger and Daily Transcript, January 30, 1899

"Keely's discovery would lead to a knowledge of one of the most occult secrets, a secret which can never be allowed to fall into the hands of the masses."

H.P. Blavatsky, *The Secret Doctrine*, 1888

During Keely's life, the accusations of an enormous fraud constantly loomed over his head as reflected by the endless quarreling and viewpoints in the contemporary press. But there was always the enigmatic inventor himself, ready with an explanation or another of his stupendous demonstrations. There was also the Keely Motor Company; its members had invested large sums in the whole affair, and it must be taken into account that they had every interest in refuting the accusations. But with Keely dead and the Keely Motor Company in disarray, for it seemed that he was influential in keeping the company in a harmonious state at least on the outside, the apparent discontent now fractured the company and rendered it impotent.

So when the machinations were about to develop that led to the tragedy, there was nothing that could oversee the coming investigations that led to his alleged exposure. And it must be said, in all fairness to Keely, that those who investigated the matter after his death were prejudiced from the start.

All began harmlessly enough, and it seems, by accident. On January 6, a large cast iron sphere was discovered hidden under the flooring of Keely's workshop. "Mechanics who examined the strange contrivance are at a loss to know for what it was used." Hopes sprang up, for it could be that this sphere was the ultimate key to Keely's secret. The sphere was located underneath the floor of the main workshop, buried in the ground, and special efforts seemed to have been made to conceal it. It weighed over two tons and rested on a solid stone foundation. The object was hollow with "protruding brass connections that were evidently joined at one time to brass pipes that ran beneath the floor and led to different parts of the building." One unusual detail was reported: "When struck by a hammer the sphere emits a series of strange sounds."

The sphere was discovered when the owner of the building, Daniel Dory of 1716 Spring Garden put his men at work cleaning out the building. A short time before this, the Keely Motor Company had already relocated the motor, tools and machinery to another part of the city, and nothing was left, or so they thought. It was while they were engaged in doing the work that one of them accidentally broke through the flooring and discovered the strange piece of iron. The owner of the building said he was willing to allow the people interested in the motor to make a thorough investigation.

But apparently the cleaning up delivered more accidental discoveries: "The floor of the workroom contains trap doors which are fitted to their places so tightly that they can be lifted only with a chisel. There is no cellar under the workroom and the purpose of the traps is a mystery. Another curious fact concerning the interior of the laboratory is that the floor of a small addition to the workroom, which was built in later years by the inventor, is raised three feet above the main door of the building, thus leaving a hidden space, the contents and uses of which are not known."

There was one man though, who knew about the sphere and the traps and the false floor. This man was Jefferson Thomas, of 1932 Mt. Vernon Street, who was for many years vice president of the Keely Motor Company. He laughed when he spoke about them in spite of the fact that he "had the grip." Thomas claimed that "The trap door was Mr. Keely's coal bin. ...Dorey, the landlord, came to me last night surrounded with an atmosphere of mystery and informed me of the finding of the sphere. I dispelled the mystery. That sphere, which, by the way, weighs about 6,000 pounds, was used by Mr. Keely something over ten years ago in his experiments in levitation."[1] Perhaps Thomas' statement may help to explain one of many claims that Keely was a fraud, which was published 18 years before. At that time, it was written that his secret was "nothing more nor less than the key of the door to the cellar underneath the room where Mr. Keely exhibits his motor."[2]

A day after Thomas' statement to the press, more people came forward to explain the existence of the enormous sphere. One of them was Schuellerman. He also knew of its existence and said grumbling that, "I have known seven men to be hung from a single pine limb in Montana for crimes smaller than that committed by the reporter who told the story of the iron globe."[3]

After having stated that the story of "that sphere of steel having been carefully buried under the floor of Mr. Keely's laboratory is a fabrication," Schuellerman also gave more details about the object: "...the sphere of cast steel...was made many years ago by the Chester Steel Casting Company, and was made to serve as a reservoir for an invention that Mr. Keely was then constructing called a multiplicator or generator. The ball whose diameter is 42 inches is hollow, having a space inside 30 inches in diameter, where the power which was formed by the generator was stored. The weight of this steel sphere is 6,625 pounds, and the entire weight of the multiplicator, including the reservoir was just 6,800 pounds."[4]

Schuellerman then told how Keely worked from time to time on the perfection of this device, until "at last an entirely new invention on the same plans as the old machine, but much smaller in size, was constructed." Keely sold the machine as scrap iron, but kept the globe "as a weight, to be used in tests with other inventions."[5]

It would be very convenient for opponents to say that Keely planned this all along and that he was merely masquerading the construction of a compressed air chamber by the so-called building and then discarding the device, but keeping the globe. But Schuellerman had more to say about the object. According to him, the globe was never really hidden, it just stood in the back room of the two apartments which he used as his laboratory, and about "sixteen or seventeen years ago, Mr. Keely used it in showing the wonderful lifting powers of a new invention which he called a vibratory lift. I saw him myself raise that ball, weighing so many hundreds of pounds, upon an iron tripod six feet high, without the slightest trouble, and many others viewed this remarkable feat at the same time."[6]

Schuellerman also claimed that after the experiments, the sphere remained in the backroom for several years, "covered with dust and surrounded with broken bits of iron, until at last, Mr. Keely gave another exhibition of the strength of this new discovery, the vibratory lift." According to Schuellerman, one afternoon before an audience of scientific men, instead of lifting the heavy sphere, "the direction of the strange force was reversed, and forced the heavy ball into the ground below the flooring of his laboratory, and since that time the sphere has been in that position, with about four inches of its surface protruding through the floor in plain sight." Schuellerman also claimed that Keely often told him that he was planning on lifting the object again. The trap doors in the floor were another matter: "I want to say that the only trap door in that floor leads to a hole under the floor where Mr. Keely stored coal, and can easily be lifted, for it is a crude affair at best." Schuellerman added, almost as an afterthought, that "When the multiplicator, of which I spoke, stood in the room it was found that owing to its immense weight, the ground under its base was sinking, and an excavation was made and the dirt banked up."[7]

Hill, Anna Keely's attorney, also made a statement concerning the globe:

"The piece of machinery buried in the ground in Philadelphia, and recently unearthed by workmen, is a portion of Mr. Keely's first engine. That was the first engine he ever made. It was made in 1874, and was of enormous size. The machine was built by Mr. Keely at an expense of $60,000. The original scheme to obtain power by the disintegration of water was abandoned. A new system was adopted, and with the innovation came the utter uselessness of the old machine. It was so big and unwieldy that he was at a loss where he could deposit it. For various reasons he did not want other machinists to get hold of it, so he determined to bury it. He dug a big hole in the ground and covered it all but the top with earth. There was no intention to hide it. It was simply put in the ground to get it out of the way. When, after his death, I went to Philadelphia with Mr. Kinraide, we readily found the buried multiplicator. We sounded it all around and came to the conclusion that it was too big to be transported to Boston. I can positively state that it had no wire communication with any other apparatus in the Keely works."[8]

A letter also professed doubts about the compressed air theory: "...what I wish to say in regard to the strong cylinder or globe, or whatever it may be, that is said to have been so carefully hidden. I remember distinctly that newspapers 10 or 15 years ago told of the making of such receptacles for the wonderful ether. There was then no secret about his using such appliances."[9]

Not surprisingly, Schuellerman's viewpoint was also shared by Collier, who also told a reporter that the original device, of which the globe had been a part, cost no less than $60,000. An anonymous person who "has been associated with the Keely interests until recently," alleged that he too knew of the iron sphere, and that Keely had used this ball "many years ago as a storage battery for his new dynamic force."

He also detailed Keely's levitation experiment, while giving the wrong estimate of its weight: "...he would attach a string of wire to the immense mass of iron. Then he would lift it from its trunnions and cause it to settle gently into an iron tripod six feet high." After the construction of smaller devices, Keely simply "twisted his wire and caused the six-ton weight to bury itself in the ground."[10]

Thomas, who the day before made his statement, now also "mentioned the names of men who, with him, had seen the act performed." He added a strange detail: "He said in explanation of the curious sounds heard when the sphere was struck sharply that the globe contained a vibratory mechanism inside that might produce the startling noises that frightened the workmen out of the building."[11]

A reporter who gained entrance to the workshop noted that the floor was torn up over the iron ball, which still rested in its bed of earth. "The mysteries of the various trap doors and underground arrangements of the building are still unsolved, as no one is allowed to search the deserted laboratory of the late inventor." That had to wait until Dorey, the owner of the building, received a reply from President Ackermann of the Keely Motor Company. The company had leased the building for that year, but the lease was "recently canceled by mutual agreement." Around that time, Keely's devices were already stored on North Broad Street and the workshop was more or less empty.[12]

The opponents had their say as well; a Carl Herring of 929 Chestnut Street —although admitting that he had never visited Keely's workshop, and stating that he based his judgment on "quite a number of descriptions" given to him by "able engineers,"—thought that Keely's machines "could have been run by compressed air." He thought one of his devices was easily recognized to be "nothing more than a Cartesian diver," and another of the machines, which was described before the Engineer's Club was, "of a nature that it could well have been operated by compressed air."

Naturally Herring thought that the hollow sphere, "evidently designed to withstand very great pressures," and considering that "a very little air at a very great pressure can develop considerable power" that could be led through "extremely fine tubes," was in effect a compressed air chamber.[13]

In speaking of the finding of what was now referred to as a "spherical tank," Professor H.W. Spangler of the University of Pennsylvania, was in his own words, "not at all surprised." He claimed that he had attended quite a number of Keely's experiments but, "I could never see anything strange in what he did on any of the occasions." Surely, Spangler was willing to give his opinion to the reporter of the Philadelphia Press. It not so much reflected his thoughts about the strange globe, but rather his vaporic gun: "I remember...when he fired a cannon by so-called hitherto unknown and mysterious force. ...Keely had the machine moved along the platform as a practical proof, as he called it, that it was in no way attached to anything in the floor. ...The machine was moved all right and Keely took some water, which he said was the force used, and emptied it into the machine; but at the same time I saw him deftly lift from the floor a little piece of what looked like wire, but which in reality was a tube, threaded at the ends to screw into a larger tube which was on the floor underneath where the machine rested, and before he fixed the shots, he stooped down and put this tube in position. This may have been put in a tube running from the very spherical tank. ...However that may be, I believed and still believe that the cannon was fired by means of compressed air and not by the mysterious force, a knowledge of which he said possessed."[14]

When additional details were released, contradictory elements cropped up. While at first the newspapers related how Dorey's workmen had found the globe, it was later written that "It was stumbled upon by Mr. Williams, who went to the property, at 1422 North Twentieth Street, with its owner, Daniel Dory."

It was further remarked that, "Why such a marvelous globe should have been ignomously buried as of no further use, is difficult to conceive. Doubters think it had a use just where it was found, no matter what it had been originally used for." And it was Mr. Williams who investigated the hollow globe more closely: "Mr. Williams says that it was tapped in several places with a small hole, reamed out at the surface to admit the flange of the plug, showing that it had been used to hold compressed air or gas..." and so the newspaper concluded that "it is more than probable that this globe has been in constant use ever since it formed a part of the laboratory equipment."[15]

Lingering doubts were rapidly growing out of hand, but possibly the best summary of the situation was given by a Philadelphia newspaper which noted ironically that, "The cast steel cylinder...may or may not have contained compressed air. The mere fact that Keely said it was a 'multiplicator' does not settle anything, nor is the mystery solved by the belief of various members of the Keely Motor Company that it was a 'multiplicator' or storehouse of energy, or mere dead weight of convenient size to demonstrate the value of his 'vibratory lift.' That a large number of Mr. Keely's associates knew the globe was there is altogether likely, but, as it seems such a handy thing to have around if you were going to work off a few 'apergy' experiments with a little compressed air, it is not to be wondered at that physicists and engineers may view its resurrection with considerable skeptic interest."[16]

This point of view was not always shared; a short letter sent to a newspaper professed its doubts about the compressed air explanation in connection with Keely's experiments; "I have frequently read in the newspapers of our scientists belief (some are positive) who have seen 'Keely's motor' that it was operated by compressed air being delivered to his motor through a 'thin' platina wire. The pressure exerted, I have seen stated, was anywhere from 20 to 100,000 pounds per square inch. I have not seen any statement, however, that Keely had any air compressing apparatus in his workshop, or conveniently located nearby."

How was it possible, the letter remarked shrewdly, to have manufactured such a hollow and thin wire of "considerable length" and "of such small diameter as to be flexible, as evidently Keely's wire was," and how could such a wire, be it made from platina, steel or any other material, have contained such a huge pressure "without bursting"? It also questioned how it had been possible for Keely "to compress air, without any apparatus, to such an enormous pressure, or cart it to his workshop for so many years without being discovered?" The answer by the editor was that it would be "foolish to speculate about things that might have happened, but did not, the maxim obviously applies to statements based upon pure guesswork. There is no evidence that Keely used compressed air, or that he manufactured it on the premises and concealed his apparatus." As for the wires, the editor gave the puzzling answer that "The imaginary 'wire' appears to have been made of 'compressed gas,' as an evidence of what fanciful theories people can make up when they have no facts to base them on."[17]

But the finding of the mysterious globe was only the beginning.

Around the same time when the discussions about the globe had become quite arcane, Clarence B. Moore, the son of Bloomfield-Moore and the spokesperson for the mysterious Kinraide, announced that he had taken a six months lease of Keely's laboratory on 1422 North Twentieth Street, for the purpose of making an investigation to determine, if possible, the truth of the Keely assertions. "The investigation is already under way, but the most important discovery, Mr. Moore says, is the great steel globe. ...Mr. Moore says that the chances of making a satisfactory investigation were greatly lessened by the dismantling of the machines previous to taking some of them to Boston."[18]

Moore had never been an impartial person, and in the midst of the preparations for his examination he confided to a reporter that he never had "any faith in the Keely pretensions, and this investigation has been started by me to see whether or not I was right in my opinion." The part that his mother, Bloomfield-Moore played in the Keely history, Moore declared, was his motive for undertakimg the investigation of the workshop.[19] When he realized that the memory of his mother was linked too closely to her interest in the Keely project, to which he had persistently been an antagonist, he resolved that "now or never was the hour for exposing the trickery which he was convinced lay at the bottom of the thing."[20]

And while lamenting the fact that the upcoming investigation would be greatly hampered since the place was dismantled, "if there was any connection with the machines it had been broken," we perceive another motive, quite like the one that had played a role in the withdrawal of Bloomfield-Moore's funding some years before: Moore was asked whether his investigations indicated that there was going to be a litigation over Bloomfield-Moore's will. To this he answered that Bloomfield-Moore had written several wills, and he did not know which was the last. He was sure though, that in one of her wills she left Keely her house at 1718 Walnut Street, valued at $60,000. "Of course she had time to make a new will after his death. I shall certainly not shun litigation, and rather than let the estate go into the hands of one incompetent to manage it I shall certainly combat the will," Moore said.[21]

There is more reason to suspect that Moore was not really that interested in Keely and his inventions, or in unmasking him as a fraud; of the coming investigation, he held the opinion that "in the near future disclosures will be made that will absolutely vindicate the course which he took toward the motor, and which led to serious disagreements between his mother and himself."[22] So perhaps Moore had considered the inventor as a mere pawn in the intricate power struggle that existed between him and his mother over his father's will and estate.[23]

What uncertain factor Keely represented to Moore, and how Keely in his turn aided and advised Bloomfield-Moore, we will probably never know. After all, Keely was well accustomed to a climate of slander, libel, and legal affairs, and it is quite possible that in the course of their acquaintance, more than matters of business directly relating to Keely's inventions were discussed. At this point, however, it is well to remember that Moore's investigation of Keely's workshop was the only one that was undertaken, and on which the verdict of history has come to depend. When Moore was asked who the people were with whom he would investigate Keely's workshop, he "declined to give the names of those who had been chosen to assist him in making the investigation."[24]

In the meantime, the controversy surrounding the mysterious sphere that was found in Keely's workshop, and the compressed air theory raged on. Even if the three-ton spherical shell was capable of resisting a bursting pressure of from

20,000 pounds to 100,000 pounds per square inch, an engineer questioned, "How did Keely get air under such enormous pressure in the sphere? Did he use a pump? If so, he must have used a boiler to generate the steam to run the pump. But he did not use a pump for the reason that a plant capable of delivering air under a pressure of 25,000 pounds to the square inch could not be well concealed, and even had it been concealed, there is nothing to indicate that either boiler or pump ever existed." The engineer also pointed out the difficulty of obtaining a pump capable of pumping against a pressure of 25,000 pounds to the square inch. The manufacture of such an extraordinary device weighing "at least a ton, and, perhaps, three" would have hardly stayed secret, if such a pump could be made at all.[25]

About the tubes through which Keely supposedly supplied the compressed air, the engineer noted that Hoadly, one of the country's leading compressed air experts, stated that steel bottles in his experiments with compressed air would not stand a pressure higher than 14,000 pounds to the square inch. At that pressure the steel bottles burst into complete disintegration, so that not a trace of them could be found. "Mr. Keely must have been way ahead of Mr. Hoadly, for apparently he not only used a much higher pressure and carried it in absolute safety, but he did it over 25 years ago, when there was no means known to science by which such a high pressure could be measured," the engineer, who nevertheless considered Keely a charlatan, ironically remarked.[26]

Scientist or charlatan, that was what Moore was going to find out, at least that was what he told the press. On January 16 and 17, he was "carefully inspecting the premises." The place was left guarded by private detectives on Monday night, and on Tuesday, the inspection continued.[27] The employment of the private detectives was probably necessary, since "an inquisitive crowd" hung around the place every hour of the day, "as if expecting to see the house crumbling or blow up."[28]

Moore now gave the names of those who assisted him: Professor Carl Herring, who was an electrical engineer; Professor Arthur W. Goodspeed, assistant professor of physics of the University of Pennsylvania; Professor Lightner Witmer, professor of experimental psychology at the same university; Dr. M.G. Miller, who was a specialist in exploration and mound digging; and Coleman Sellers, Jr.[29]

Moore also invited George W. Arnold, who was a carpenter and gas fitter, and engineer E.A. Scott, who had read a paper concerning Keely and his discoveries to the Engineer's Club of Philadelphia a year before.[30] Scott knew the location of "every piece of apparatus exhibited there by Keely, and the object was, if possible, to see what evidences of hidden appliances had been left by the Keely Motor people, who carried off the machines.[31]

The workshop was devoid of any machinery, since at that time these had been removed to Kinraide's laboratory, while the heavy appliances were stored in a building at Broad Street and Fairmont Avenue. The investigators found "nothing but blank walls, board partitions, bare floors,"[32] and of course that huge sinister looking sphere that still lay silently amidst the rubble. The team also allegedly found "many letters received by Keely," which for some reason had not been removed by the Keely Motor Company. The letters were written by "those who considered themselves his (Keely's) dupes and berated him for deceiving them."[33]

Every bit of flooring was ripped up and every nook and cranny explored in the floors, walls and ceilings, and it was found that the building was honeycombed with traps, holes or piping, etc.[34] The team discovered "many small tubes imbedded in the walls and concealed under floors," and were "little larger than wires" and were "said to be designed for standing high pressures." Goodspeed readily stated that the sphere could withstand a pressure of 20,000 to 100,000 pounds to the square inch, and the finding of the tubes once again revived the compressed air theory. "One tube that we found," said Goodspeed, "was nearly under and in immediate proximity to where the heavy lever testing device used to stand, by means of which I myself have seen Mr. Keely give very convincing evidence of the existence of a pressure of 20,000 or more per square inch."[35] These tubes led the team to the conviction that either compressed air, as was Herring's verdict, or compressed gas as Goodspeed thought, was used.[36]

Schuellerman ridiculed the idea of compressed air, but he made the surprising statement that Keely at one time "did dabble somewhat in compressed air, but it was not in connection with his motor, but in a gun which was tested near Shawmont some years ago." The air for that experiment was compressed by the S.S. White Company since Keely never had an apparatus "for doing it himself."[37]

The tubing found in the building was "small drawn copper stuff, none of it more than three-eighths of an inch in diameter and having less tensile strength than iron pipe of the same size would have." Schuellerman added that "It was not used for compressed air but for Keely's mysterious vapor." The discovery of the tubing came as no surprise to Schuellerman; in fact he wondered why they hadn't found more "of the stuff in the place, as there was lots of it there and Keely sometimes ran it through brick walls from one room to another."[38]

But that was not all that the search party had found. Besides the brass tubes in the brick work and under the flooring, heavy steel tubes "of larger caliber and thicker metal" were found "under the joists of the first floor, in the earth." The tubing was designed to withstand great pressures, and apparently such tubes were made in "the early 70s by the Philadelphia company Morris & Tasker." Many "double thick" tubes made by them for Keely were returned burst, and a special set of tubes was made for him. On one occasion a foreman of Morris & Tasker took a pressure gauge to Keely's workshop which would register up to 2,500 pounds to the square inch. "Keely attached it to his apparatus, and," said the foreman afterwards, "it immediately sent the index to the top of the scale, and I was so scared I lay flat on the floor to escape the shock, expecting the place to be blown up."[39]

The team also discovered what they thought was evidence that the huge sphere indeed had been used as a storage for compressed air, although rust on the heavy steel tube, and the fact that there was no connection with the sphere led the scientists to believe that "the tube had not recently been in use." But connections with the sphere had "evidently been made through one of the two trunnions or projections on opposite sides of the sphere." A hole through the trunnion had been closed with a plug screwed in and "planed off smooth." And another hole through the trunnion, "near where the end of the tube stopped," was still open. The team concluded that this hole was "about the proper size for a connection from the tube." Another smaller hole in the side of the sphere, placed midway between the trunnions, was just the right size for connecting one of the small brass tubes found running into the second story, and under the doorway of the first floor through the brickwork.[40] Around the same time a person came forward and explained to the press that the sphere had not always been located in Keely's workshop: the man who was formerly a pupil in the school just north of his workshop said that "this great ball" used to lie on the pavement in front of the laboratory, and it was the wonder of the small boys. It was concluded, however, that "It could have performed its office just as well in that position as in any other, as the fine tube connections could readily have been made to it and concealed."[41]

Gasfitter Arnold also had interesting things to say; according to him he ran certain gas and water pipes in the building years ago. There were three rooms on the ground floor, a front, middle and back room. The sphere was found in the middle room where "no stranger was ever allowed to enter." In this middle room Arnold continued, there was a large cylindrical tank in the southwest corner. In the back room, which was situated in a one story annex, there was an electrical battery, but "he could not tell what it was connected with." He did say that there were gas pipes with wires running through them. Also, two years earlier he had made a gas connection with a gas motor, which stood in the front room, and "showed where it had been disconnected and the end of the pipe capped." The gas motor had stood at the rear end of a large workbench on the north wall.[42]

More, even stranger things were discovered that would render a nice and quick compressed air solution to the Keely mystery more difficult, and that would strengthen doubts as to this theory. In the back room of the workshop a small trap was found in the wooden ceiling in the northeast corner that opened into a space about one foot high. A half brick had been cut out of the wall of the two-story building, through which a "silver polished steel rod, 7 feet 3 inches long and five eight inch in diameter, was trust." This strange rod was running east and west over the ceiling of the room, and projecting through a hole under the eaves. What the steel rod had been used for, the team was unable to say. It was remarked that "at one point on the rod, which was considerably corroded on top, there was a polished place where the rod had evidently rested on a bearing, and the bright surface ran half way around, showing that the rod had a rocking motion."[43]

In a line with this rod under the floor of the rear room in the second story, they found "a thick metal plug with a long screw" that was screwed through a joist. The plug was holed, and on one side a small tube was connected that was cut off. The plug and tube stood directly under a place where one of Keely's devices had stood, the "revolving ball," that "would roll over just as many times as the visitor would name, Keely sitting in the middle room and looking through a small window between the two rooms and sounding a mouth organ."[44]

All these mysterious findings led the team to conclude that not only compressed air, but also water and electricity had been used by the clever Keely: "In the northeast corner of the back room there was a water connection and sewer connection. Everything had been torn away." It was suggested that "There may have been a water motor there," since it was "hardly supposable that water closet appliances would have been removed by the people who dismantled the laboratory," and "several holes were cut through the brick wall at this point."

Moore and company also found a "disconnected insulated electric wire, which came in over a window on the second floor." This showed that "Keely had at one time been connected with an electric light station, the outside fixtures being still in place," and it was alluded to that a station of the Columbia Company was but "two doors above the laboratory."[45] But all these findings did not resolve the issue. A Philadelphia newspaper remarked that the disclosures added nothing to what had previously been said, namely that Keely never demonstrated by exclusion of the use of known forces his discovery of a new force, "nor do they furnish as strong demonstration of the use by him of compressed air." Instead, the findings were considered by the newspaper to be "cumulative evidence that the mysteries of the Keely laboratory were mysterious not because of any new discovery, but because he concealed beneath the floors and in the walls the tubes through which he conveyed power through his motor."[46]

The most controversial and puzzling statement about the finding of the tubes, the strange rod and the traps came from Kinraide. Around this time he was in his cave at Jamaica Plain sweating over Keely's devices and trying to learn his secret. Upon learning of the exploits of Moore and the scientists, he said that their discoveries "amounted to nothing." Kinraide had been in Philadelphia a week before Moore started to dismantle the workshop, and while there had made a visit to the workshop. "It did not look any different to me then than it has at any time since I first saw it," and he added that, "I told the owners of the building at that time that it would be unwise to leave the mass there, as it would create a lot of theories among those who might visit the building with a desire to make a sensation out of a small and unimportant matter."

Regarding the question about the tubing that presumably had been used in connection with the large steel sphere, Kinraide answered that he could not make "any reply to this direct question," as he had his position with the Keely Motor Company to consider. He did, however, remark that "the members of the stock company who have been putting in their money to carry on this work knew all about this force, and if they thought it was a swindle, as stated in this story from Philadelphia, they would not be long in saying so."[47]

At this stage of the Keely exposure, the discovery of the tubing in Keely's dismantled workshop did more harm than good to Moore's assertion that he was a fraud who operated his engines on compressed air.

A day after the investigation, a Philadelphia newspaper put it in the plainest words possible: "The discussion about the power used by the late John W. Keely to run his machines, started anew by the finding of brass and copper tubing in the dismantled laboratory...has been taken up by practical machinery men, who stamp as the sheerest nonsense that compressed air or gas could have been employed, if the alleged evidence collected by Clarence B. Moore and other investigators were to be taken as proof." Instead, the compressed air theory "must be put aside by the people who are hunting for ways by which they claim Keely fooled the world," as this was done "years ago." Furthermore, while stating that they did not believe that Keely had "the knowledge of a heretofore unknown power," they had been with experts who witnessed Keely's numerous experiments, always on the lookout for evidence of such deception, and that, if Keely was guilty, this would have been discovered "many years ago."[48]

A reporter also shrewdly noted that Consulting Engineer Herring retraced his steps: Herring, "who says he is not familiar with the uses of compressed air," made the statement in the press around the time of the finding of the globe, declaring that he was "satisfied" that Keely had used "highly compressed air." But at the time that Moore's team was dismantling Keely's workshop, he suddenly felt unsure and he told a reporter that "he did not mean to say it was absolute certain that compressed air was used, but that there was strong circumstantial evidence to that end." Herring also showed the reporter a piece of the tubing that came from the workshop. "It is very old, and Mr. Herring admitted that the Keely people could be right in their claim that the tubing was part of that used by him before he abandoned his vapor or ether force for the 'sensitizing' or 'vitalizing' power."[49]

"Perhaps Herring remembered a statement allegedly made by Keely some thirteen years before. At that time a reporter wrote that Keely had told him 'by laying little tubes underground connected with his engine, if he built a large one, he could run all the machinery in every factory in Philadelphia by simply drawing his fiddle bow once every morning and letting the sound in to the copper globe.'"[50]

One of the pieces of tubing had been found hanging over a hook in the laboratory, so the reporter asked Herring "if he did not think it strange that Keely should leave anything that would cause suspicion lying around." Herring answered lamely that "the spot where the tubing hung was not visited by others than Keely." He did admit though, that it was "evident" that the tubing found was not used on the machine Keely was operating when he was taken ill, as he did not believe that Kinraide, who took the machinery to Boston, would "sever tubes attached to the machines and neglect to follow up the tubes, on order to enlighten himself on the Keely mystery." Herring did believe that compressed air could have been forced through some of the tubing, and he had a solution to the question of how Keely could have driven compressed air in the sphere; according to him, no "heavy machinery was necessary, unless the 20,000 pounds pressure was to be stored in a minute or very short time, when powerful compressors and a powerful engine would be required." No, the answer was quiet simple: hand pumps. The sphere could be filled up to the 20,000 pounds pressure by the use of hand pumps if enough time, "say a week," was taken for the task. "A start could be made," Herring suggested, "with a large hand pump, then smaller pumps could be used down to a specially constructed pump with a 1-16th of an inch piston."

William F. Rudolph, the expert mechanic who had built Keely's latest engine and "frequently took apart and put together other machinery," did not think so. In response to Herring's suggestion, he said bluntly: "I never heard such nonsense. I would like to see the person who could pump up a pressure of more than 300 pounds in a 60-gallon reservoir, the size of the old sphere found, with hand pumps. It's plain that the force of the air inside would be too great to pump in more air by hand. If that's the case what is compressing machinery for. Why, to compress 20,000 pounds in a 60-gallon receptacle powerful compressors and an engine would have to be worked half a day. Then the tubes found would not stand the pressure of the stored air. I repeat that none of the Keely machines can be operated with compressed air." He also pointed to the problem of the exhaust of quantities of air used in operating the machines: Any exhaust would "make a noise, and a tremendous one if 2,000 pounds pressure were used. No such noise was heard at the Keely exhibitions in late years," Rudolph claimed. Rudolph gave the reporter a demonstration of what he meant by compressing 20 pounds of air, "which he allowed to escape through a tube of the same diameter as that found in the laboratory and about 10 feet long. The noise made by the escaping air could be heard 20 feet away," the startled reporter wrote.[51]

The steam and air machinery company of Borton & Tierney stated that to store air to 20,000 pounds pressure in a 60-gallon tank powerful compressors and a 40 horsepower engine would be needed. And to direct the compressed air to any device, tubes "much larger and stronger than those found would be necessary."[52]

Pieces of the tubing were shown by Herring at the annual meeting of the Engineer's Club, where they were examined by "quite a number of members," and were put on display in the office window of a Philadelphia newspaper where they were gazed at by hundreds of people. Meanwhile Moore's team claimed to have uncovered "other tubing and high pressure joints" between the floor and ceiling.[53]

Professor Hugh A. Clarke of the University of Pennsylvania, who made "a special study of vibratory physics," came to the conclusion that Keely played on harmonicas and sounded whistles, not to set up sympathetic vibration, "but to notify a confederate in another part of the building when to turn on or shut off the compressed air, that is believed to have been the motive force employed."[54]

And on the respectable Harvard University the discovery of the tubing was the "leading topic of university conversation here since the story appeared," but nobody wanted to say much about the whole affair, since "few of them had seen the motor, and as they had no definite knowledge of the matter, their statements were very conservative." The general opinion was that Keely was a "fake and as such unworthy of scientific notice." Professor John T. Trowbridge, the "electrical wizard whose whole life has been spent among motive forces," said that, "Keely did not invite the inspection of scientific men, but maintained the utmost secrecy in all his movements."[55] But on several occasions Keely had specifically invited scientifically inclined persons, and his workshop had as its guests people such as Professors Leidy and Lascelles-Scott. Also, another newspaper wrote that "Many investigators, scientific and otherwise, have seen his experiments. United States government experts have witnessed them."[56]

Trowbridge himself never saw the motor in operation or even a photograph of it, but in his mind the whole Keely venture was "on a par with the Jernegan scheme to extract gold from the sea." His opinion was that all great discoveries in the past had come slowly, and under the watchful eye of the world, so it simply would be "entirely out of precedent for a man to make such an overwhelming discovery almost without an effort." Professor Marks, a steam expert, intimated that he thought the whole Keely question "a humbug."[57]

The remarks of another professor, Professor Lightner Witmer, on the findings in Keely's workshop clearly show something of the deep-rooted prejudice that Keely suffered even in death, for "The external evidence of reservoirs and tubes was hardly necessary to demonstrate the delusional character of Keely's theories. Even if these objects had not been found, the writings of Mrs. Bloomfield-Moore, the pseudo scientific jargon of Keely and the official reports of the Keely Motor Company would have furnished, upon critical examination, indisputable testimony to the unsoundness of Keelyism. ...Little more is needed to bring Keelyism to its proper place in a museum of pathological mental products."[58]

President Ackermann of the Keely Motor Company issued a statement on behalf of the Board of Directors in which he claimed that the sphere and the tubes were "simply pieces of machinery connected with an altogether different device abandoned by Keely in 1887." The existence of the tubing was "fully known to the directors of the company and to scientists not financially interested in Keely." Besides, he added, such complicated machinery for compressed air did not even exist at any time in Keely's place, and the electric wires were "simply the remains of a burglar alarm."[59]

But what about the "jagged parts of burst iron pipes, plates of iron pierced through as with cannon balls, heavy plank perforated with shot, and other evidences that some powerful energy had been at work to produce these manifestly imposing results?"[60]

Those old tubes that were lying about in Keely's workshop, and which had led the stockholders and general public to believe that they were burst by his etheric vapor, were nothing more than remains of tests done by him and two employees of Morris Richmond Iron Works, they said. According to those two employees, Eugene Calwell and William Rickert, various tubes and parts of devices that were made by Morris Richmond Iron Works were tested on their strength with a hydraulic pump. Since the company did not have a pump of sufficient strength, theirs only making a pressure of 14,000 pounds, Keely ordered one specially made for him. This pump could deliver a strength of up to 30,000 pounds per square inch. With this pump, the tubes and parts were tested "up to the limit of the pump." The steel sphere was tested up to 28,000 pounds, and steel tubes were also tested, numbers of them splitting "in two." When "tubes of nine inches outside diameter, with a bore of only three inches, were split like reeds, Keely instructing the men to keep putting on the pressure."

The employees remembered one occasion, while testing a chamber of the generator with a number of New Yorkers present, a heavy plug that was screwed into the device "blew out with such a force that it broke the marble of a dwelling on the opposite side of Twentieth Street." The hydraulic pump remained at Keely's workshop all the years, a fact that was corroborated by others, in spite of Ackermann's statement that no such machinery "sufficiently powerful to compress air" was ever in Keely's workshop.[61]

Enters Nikola Tesla. Tesla the sphinx, was in many respects as enigmatic as Keely and as little understood, even though he gave the world alternating current, and his inventions led to more patents. In their strange and wonderful visions, Tesla and Keely matched each other perfectly. Tesla arrived in America when Keely was already 10 years involved in his wondrous experiments, but had grown over the years into what was to become one of the truly great and misunderstood geniuses of the 20th century. Tesla had corresponded with and visited Bloomfield-Moore. He signed his contract with Astor some 10 days before he chose to break the silence, amidst the furor that drew headlines in the press: "When the reservoir and the pipes were found I knew that the surmise I had long entertained was correct. I would like to think that Keely was not a dishonest fellow, and believe him simply to have been a man who erred so that he would have accomplished no great thing had he lived a dozen lives. Although he evidently used compressed air in his experiments, it does not follow that he did this deliberately to deceive." Then the genius concluded with a statement as mysterious as himself: "Acting on my conjecture, I have performed most of the experiments reported and still more wonderful ones to the lay mind."[62]

In spite of all the statements to the contrary, the Keely Motor Company kept its head cool and proclaimed at a meeting of the company, on January 26, that its directors reaffirmed their faith both in Keely and his invention,[63] and that the directors were going to meet Kinraide two days later. At that time, Kinraide was said to be "confident of his ability to complete Keely's inventions," and the directors of course expressed their "implicit faith in him."[64]

On January 29, the opposing forces focused, and their verdict was out. On that day, a huge multipage article was published in the *New York Journal* subtitled the "First Official Confession and the Only True and Authorized Explanation of the 'Miracles' of the Great Keely Motor."[65] The article contained statements of

Hill, Anna Keely's lawyer, Smith, who had been given the opportunity to learn "Keely's Secret," and elaborate cross-section drawings of the double floors and ceilings which contained shafts and belts that lead to a water motor in the basement. This motor was discovered when the disintegrator was taken apart. "In taking down the posts which held the stationary axis on which revolved the hub of the motor, with its arms, the first fraud was discovered. This framework had no apparent connection with the engine, beyond serving as a support for the stationary shaft or axis which passed through the hub of the motor. A false box, a hollow post and a hole extending down through the floor led to a careful investigation. Under the floor, between it and the ceiling of an unused store room that was always kept locked, an iron shaft with a small pulley on it was found running through the timbers supporting the floor. The pulley and the hole in the floor were directly under the hollow post of the engine."[66]

The team followed the iron shaft to the side of the wall, where it ended in another pulley. Directly beneath this pulley, but just above the ground floor of the room, another iron shaft came through the wall, also connected to a pulley. The team found a "small, well-worn belt which fitted over and exactly connected these two pulleys." In the small rear room that was filled with "old junk," and the floor of which was raised considerably above that of the middle room, a trap door was discovered beneath a box and an oilcloth. The trap opened over the shaft which came through the wall. "Here it was found that the shaft connected with a small water motor of peculiar construction, the water being supplied by a lead pipe coming in from the outside of the building. A small rubber tube extended from this water motor. When Moore's team attached a rubber bulb to this tube, the water motor could be started by pressing the bulb, and it would stop when the pressure was released."[67] The water motor was transferred to Kinraide's cave on Jamaica Plain.[68]

When it was taken apart by Kinraide, Keely's Globe Motor revealed a "strong spring," also described as a "heavy coiled spring" with gearing that could be fitted into the globe. This spring was connected by a diaphragm that pressed against the shell, acting as a brake. The spring, "having been previously wound up was inoperative to move the globe until the brake was released. This was done by screwing up the diaphragm in the transmitter and the globe would revolve. A very small tube was used to connect the transmitter and the brake in the motor."[69]

The globe also contained more diaphragms, "some flexible and in order, others hard and not easily moved." Although Kinraide failed to explain how, it was "evident that these had been in some way used to operate the Globe Motor."

Keely's mysterious compass, with a match instead of a needle, was found to contain a "false bottom which concealed a piece of iron like a needle. When revolving this the 'match needle' would revolve." Keely's transmitter was found to contain yet another "simple diaphragm," that, when pressed by an outside screw, produced an air pressure which could be transmitted by a tube running to the machine when operated.

Keely's famous floating weight that had so astounded Professor Leidy and many other people nine years before, was a "very light, hollow box, with an opening in it, so arranged that when air pressure was exerted on the top of the water in the jar the water would be forced into the box, and, being made heavier than the water, would sink." When the pressure was taken off, the box would rise or float. The cover of the jar concealed yet another diaphragm and was connected with a small tube with a diaphragm in the transmitter, and this "explains the mystery."

The Vitalized Disk was found to have been made partly of iron and partly of brass. When the brass side was held against a magnet, it did not attach itself, but naturally the iron side did. The iron parts of the disk were "gilded to match the brass, giving the whole the appearance of brass." Keely's Musical Sphere also contained a coiled spring and a diaphragm "similar to that described in the Globe Motor." The wire was connected with the mouth harmonica and the sphere was a tube that led to "a bag of India rubber," that would deliver air pressure when pressed, "sufficient to release the diaphragm brake and the sphere would revolve." The awesome disintegrator was also found to contain a "tubular iron reservoir containing compressed air."[70]

The article was to be the death blow for Keely and those who had believed in his inventions. Other newspapers of course quickly summarized the article, and the story was carried across the country.[71] Although Kinraide, as we have seen in the previous chapter, accused Bridge of breach of faith, and the examination of Keely's engines was still ongoing at Jamaica Plain, as far as the general public was concerned, that was that, and there the matter ended.

Not so for the Keely Motor Company! After all, there was still hope that Kinraide would prove the contrary. On this a newspaper remarked that, "The explanation of Keely's secret that is said to come from Mrs. Keely's counsel bears all the marks of authenticity and is entirely in accordance with what has been generally believed by those who have seen the motor in operation. Still, we do not understand that it is actually an official statement, and we suppose that it will be disputed by the stockholders and managers just as earnestly as previous explanations have been. The kind of faith that has been imposed in Keely is hardly to be disturbed even by ocular demonstration. ...there are some people who still believe in the cabinet trick...and there are those who have believed so long in Keely that even a detailed description of his air tubes and water motor, discovered beneath the flooring of his laboratory, will be rejected by them as a device of the enemy."[72]

And so it was; Collier and Thomas both were "loud in their denunciation of the statements and alleged exposé of the principles of the motor made by Charles J. Hill, counsel for Mrs. Keely and J. Ransom Bridge." Collier and Thomas held a conference with "several other stockholders of the company and decided to issue a statement." Collier said, "The statements attempting to prove Keely's inventions to be tricks and frauds have not shaken my faith in his works. The alleged exposing of Keely's methods, which have come forward so plentifully lately, are all answerable...," and he expressed his conviction that the joint statement would "restore faith in the Keely motor. The statement will be carefully prepared and will answer specifically the charges against the inventor."[73]

Whatever they were going to state, it would have to be good, since a reporter wrote that "those who are professing to carry on the Keely scheme must take themselves, in the public estimation, out of the class of dupes and into that of accomplices."[74]

Whatever Collier and Thomas first planned, it probably was something different from the statement that was issued at the meeting of the directors on February 29, perhaps because of the veiled allusion to illegal practices made by the reporter.

The directors intimated that "there was deception in some of Keely's methods." Collier said that it was the "consensus of the meeting today that we have to come down to a point or two and discard some of the methods whereby Keely made his invention so wonderful to his audiences. The musical globe machine, the appliances with which he claimed to have set a-going his machinery, or, rather parts of it, is an impossibility undoubtedly." And according to Collier, "among our directors a few tried to hold out for a while, not for the purpose of aiding a deception, but simply that we might weigh everything in the scale of possibility. Well, the musical feature, the theatrical appurtenance, has been weighed and found wanting."[75]

Collier did not know the reason for such theatrics, those were "best known to our late friend," but he also declared that "coming down to the disintegration of water, on this point, the Board of Directors stood a unit," for "We have all of us seen what he accomplished with his lifting machines. We believe the force is there."[76]

However, the following months would reveal that the force was sadly absent, at least in the cave at Jamaica Plain where Kinraide explored Keely's engines further. After his damning statements in July of that year, as seen in the previous chapter, the Keely Motor Company fell in a sorry state of disarray. In the months that led up to Kinraide's statement, doubt lingered as was predicted by *Scientific American,* which wrote, "the investigations were so thorough and the results obtained so satisfactory that it is to be hoped that, once and for all, the Keely motor may be considered to be exposed, though we have no doubt that, like the scotched snake, the tail may still continue to wiggle."[77] And indeed a magazine wondered: "Was it compressed air, hydraulic power or electricity? The recent exposures of trickery on the premises point strongly to the former conjecture, but the probability is that the exact modus operandi will never be absolutely established..."[78]

What then had been Keely's motive? Was it money? While this was the explanation preferred by his adversaries who painted him as a fraud who earned huge amounts of money with his inventions, even a cynical Keely-opponent admitted that, "Keely, as a matter of fact, lived fairly well, but neither lavishly nor ostentatiously, and he spent far more time during those 20 years in the dingy little shop, with its wires and cylinders and dismantled relics of previous experiments, than he did at his own hearthstone."[79] And another remarked, that Keely "spent the money which he obtained in experimental investigations cannot be denied even by the most strenuous of his opponents at that time or now."[80] What then happened with all the money that had poured into Keely's research? "That is a mighty difficult question to answer" a newspaper wrote, for while his "personal habits are not expensive," there was "no doubt" that "vast sums have been expended on useless machinery, devices and tools."[81]

It was this puzzlement that could not be taken away, and that prompted the author of the passage above to muse: "If an impostor, he certainly was an extraordinary one. Few men could have maintained so successful a game of trickery for twenty-five years, not only enlisting the sympathetic interest of such distinguished gentlemen as the late Professor Joseph Leidy of the University of Pennsylvania; George H. Boker, late minister of the United States to Turkey; and the late John Welch, minister to the Court of St. James.[82]

The puzzlement that lingered a quarter of a century later put Charles Fort, the great collector of the unexplained, in a pensive state when he summarized the aftermath as follows: "anybody who has ever tried to keep a secret for 24 hours, will marvel at this story of an impostor who, against all the forces of revelation, such as gas men, and coal men, and other persons who get into cellars—against inquisitive neighbors, and, if possible, even more inquisitive newspapermen—against disappointed stockholders and outraged conventionalists—kept secret, for twenty four years, his engine in the cellar."[83]

This puzzlement and this lingering feeling of doubt has remained ever since. Although elaborately drawn cross-sections of Keely's workshop were published,[84] and elsewhere photographs of the sphere, the trap doors and the tubes,[85] there unfortunately exists no such thing of the water motor "of peculiar construction" that was found and then quickly and conveniently hauled off to Kinraide. No photograph exists of the thing, and no triumphant explanation of its fraudulent nature.

Perhaps the water motor was a relic of the days when Keely wrested his force from the disintegration of water and was therefore quite incomprehensible. This would also explain the "peculiar construction" of the device, a construction that was not further explained or elaborated upon. Although Kinraide possessed the water motor, he made no statements about it. In fact, the water motor became so forgotten that the only orthodox explanation that has survived today is that Keely solely used compressed air.[86] In addition, there would not be any public demonstration or display of Keely's devices or the water motor.

As far as is known, there were no photographs made of Keely's dismantled engines and the parts that Kinraide allegedly found inside the machines. The only descriptions are the written statements, the cross-section drawings and sketches of some of his equipment.[87] These sketches included a cross section of his latest transmitter, showing the hidden rubber diaphragm, the piston and his harmonica with a rubber tube and a hollow wire. A problem was in the timing of the events; the exposé was published at a moment when Kinraide, considering his own statement, strangely disapproved. Why Kinraide did this, and then waited several months to make his statement that tallied with the exposé was never resolved.

And who was this mysterious but pivotal T. Burton Kinraide? While contemporary sources offer only tantalizing glimpses and details, any further documentation on this character is strangely absent from the pages of history. All that we know of him now is that he was very wealthy, and that on his estate he pursued research not unlike Keely in a special cave hewn out of solid rock. His laboratory, also described as "large and finely appointed," was called the Spring Park Laboratory, and it is alleged that in this laboratory Keely and Kinraide "delved and chatted." Kinraide, who was 35 years old at the time of Keely's death, was "quietly pursuing the life of a scientist, without attracting much attention outside of professional circles." Kinraide also brought out a new induction coil that was widely used in connection with x-ray equipment. The coil is called the Kinraide coil.[88]

Kinraide appeared several years before Keely died and subsequently became a constant visitor to the workshop, sometimes staying there for long periods of time. The two got on very well. Keely's last request had been to place all the material and data in Kinraide's hands. He had been in the house between the death and the funeral and he had accompanied Keely's widow in the carriage to the cemetery. Kinraide even paid for his grave, but after the exposé and Kinraide's statements several months later, he dissolved in the mist of time, never to be heard from again.

Considering all this, we could easily present another speculative scenario: Moore and his team uncovered what they interpreted as evidence of fraud, and Keely's workshop was carefully guarded by private detectives, so that nobody could verify their claim anyway. Kinraide—who also took precautionary measures by barring his cave to reporters[89]—shrewdly stated what Moore wanted to hear; everything that Keely had invented amounted to nothing. Moore was satisfied and Kinraide knew very well that through Moore, the press would be informed. Months later, Kinraide's own damning statements followed. The attention of the general public was diverted and what certain people saw as the real nature of Keely's discoveries was carefully hidden from public view. Thus Keely's work could be continued elsewhere in secret, a possibility that is considered true to this day in certain occult communities.

In the meantime, Hill had successfully refuted the claims of the Keely Motor Company, which was already weakened by internal quarreling and fractures. After the exposé, the status of the company was further weakened and rendered impotent. After all, the members of the company now also faced the possibility of accusations that they were accomplices in a scheme of fraud and deception. Interestingly, the exposé was first printed in a New York newspaper and not in a Philadelphia newspaper. The New York newspaper was owned by William Randolph Hearst, whose mysterious role will be examined more closely in chapter 12.

Or was Moore, who once faced accusations of the mismanagement of his father's estate, hoping to find something in Keely's workshop that would make him rich beyond his imagination? Was he simply diverting other people's interests by proclaiming loudly that all was a fraud? Or were Moore and his mother indeed the victims of an elaborate conspiracy, as the unnamed Englishman wrote to Bloomfield-Moore? All of this we will probably never know, but the feeling that the exposé could have been carefully orchestrated with a special purpose in mind prompted Fort to conclude: "It made no difference what else came out. Taboo had, or pretended it had, something to base itself on. Almost all people of all eras are hypnotics. Their beliefs are induced beliefs. The proper authorities saw to it that the proper belief should be induced, and people believed properly."[90]

The Keely Motor Company went the same way as Keely's engines and his alleged manuscript: into oblivion. But what about the sphere, that monstrous and silent device that had started it all, but which purpose still is as mysterious as the day it was found? A newspaper suggested that the sphere be brought to the laboratory of the University of Pennsylvania, "as a curiosity for exhibition."[91] This was never realized, and a year later a small notice in a Philadelphia newspaper appeared: "The big, mysterious iron sphere which John W. Keely, of motor fame used in some of his experiments, is now lying in front of the little shop, 1422 N. 20th Street, which Keely once used. It is covered with advertisements."[92]

7

To Understand the Art
Keely's Discoveries

"My system, in every part and detail, both in the developing of this power and in every branch of its utilization, is based and founded on sympathetic vibration."

Keely in, *Keely's Secrets*, T.P.S., 1888

"In the course of his long career Keely may have broken a number of laws, but he left the first and second laws of thermodynamics, which forbid perpetual motion machines, inviolate."

New Encyclopedia Brittanica, 1981

There were two periods in Keely's research and experiments. The first period that ran from 1874—or possibly earlier—involved the production of force by the disintegration of water, which was later called etheric vapor. This he accomplished with his Liberator, and we have several accounts that serve as an illustration of his experimental activities during that time. It was explained that "the etheric vapor was obtained by letting the water into the Disintegrator or Liberator at a certain pressure. There were certain chambers which it must enter first, and there were certain valves which were opened or closed next."[1]

A recent and more detailed suggestion of this procedure is that his hydro pneumatic pulsating vacuum engine, built at the very beginning of the first period, worked by using the infinite pressures developed from water hammer and cavitation, also referred to as implosion. The circuit of vibration energy in his Disintegrator engine could be compared to today's signal generators and synthesizers. Its aim was to disintegrate water into etheric vapor.[2] Glocker, who constructed both this engine and the multiplicator, described the hydro pneumatic pulsating vacuo machine as having cylinders of 3-inch bore and 3-inch stroke, with a fly wheel of 200 pounds weight, that was able to revolve at 300 revolutions per minute.[3]

The etheric vapor, also called vaporic substance, that Keely discovered by accident while experimenting with his hydro pneumatic pulsating vacuo machine in 1873,[4] was a most mysterious substance with almost legendary attributes. It was this etheric vapor that could drive a train of cars from Philadelphia to New York and propel a steamer across the Atlantic without fuel or sails, and could be obtained from a handful of water. "People have no idea of the power of water," Keely said, "a bucket of water has enough of this vapor to produce a power sufficient to move the world out of its course."[5]

The vapor had no smell or taste. Collier claimed that he frequently inhaled it without any perception of smell or taste. When he put a candle to it, the substance would not burn and did not smother its flame; "there was no heat about it—no galvanism—no electricity—no chemicals—no preparation of any kind excepting the mechanical appliances."[6]

Sergeant, who had also handled the vaporic substance, claimed that it was "pleasant to the touch," and furthermore he had "swallowed all I could get of Keely's vapor, and I find it is pleasant. You can live on it. ...It turns back to water, and water that I can drink."[7]

Expressing his amazement of the vaporic substance, Babcock mused that, "Comparing it with steam it is as different in action as it is opposite in origin. Steam is derived from heat or combustion, and so may be said to have a chemical origin; the vapor is a production of mechanical action, a spontaneous energy. Vibration, whether considered as an energy or a motion, is an inherent property or concomitant of matter, and therefore spontaneous."[8]

While this new and strange substance was heralded as a new source of untold power, and it could be won from water of which there was plenty, it was also sourly noted that statements about the etheric vapor were vague, consisting more of what it was not than of what it was. This, of course, would "hardly do for an adequate description of a new 'vaporic substance' of immense power, which is capable of rendering steam a thing of the past."[9]

The second period in Keely's research ran from 1882 until he died. He had for 10 years demonstrated the liberation of the energy on which he "stumbled over," while experimenting on vibrations in 1872. His efforts to construct a perfect engine failed. He explained that he made "the mistake of pursuing his researches on the line of invention instead of discovery; all his thoughts were concentrated in this direction up to the year 1882." Frequently, explosions occurred and he scrapped engine prototypes and sold their remains as old metal in his constant failures to construct an engine that would keep up the rotary motion of the ether.[10]

In this period of constant frustration, Keely discovered a force derived from the vibration of an unknown fluid or substance, locked between the atoms of the ether. This he called the vibratory force[11] or the etheric force. It was this force that he claimed was not like steam, electricity, compressed air or galvanism.[12] While during the first period he obtained the force through the disintegration of water, in the second period he could develop his force "in the air, in a vacuum, in the ether itself,"[13] or by the vibrating of hydrogen, which he first attempted around 1884 on the suggestion of Bloomfield-Moore.[14]

That year, he confided to a reporter that, "Stripping the process of all technical terms, it is simply this: I take water and air, two mediums of different specific gravity and produce from them by generation an effect under vibrations that liberates from the air and water inter atomic ether. The energy of this ether is boundless and can hardly be comprehended. The specific gravity of the ether is about four times lighter than that of hydrogen gas, the lightest gas so far discovered."[15] The vibratory force was produced by vibrations of the ether that "pervades the universe," and he claimed that these were so rapid, "like those attributed to light, that no cognizance can be taken of them by human sense." Objects which were "vitalized" or synchronized so as to vibrate in the ether in a certain definite relation to each other would together exert a force which, "if it can be applied the everyday business of life," would "supersede all other forms of energy, and at practically no cost..."[16]

Until 1888, Keely pursued "the wrong line of research," still trying to construct a "perfect engine" that could hold the ether in "a rotating ring of etheric force." Towards the end of 1888, he entirely abandoned his concept of the "perfect machine." Up to this time he practically built his research equipment himself. Then around 1888, he was to be provided with "the best instruments that opticians could make for him after the models or designs which he furnished."[17] Around 1890, he also made the remarkable statement that he succeeded in linking his machines on the polar current, "sympathetically," by sensitizing his devices "as to be able to operate the machinery from that force." This polar current was described as magnetic currents that envelop the earth as "an orange is in its rind." Keely claimed that his force was the result of an interference with this magnetic rind.[18] He later stated that "the polar forces included magnetism, electricity and gravital sympathy; each stream composed of three currents, or triune streams, which make up the governing conditions of the controlling medium of the universe. ...These sympathetic streams from celestial space, percussing on the dense atmospheric environment of our earth by their infinite velocities, wrest from their atomic confinement the latent energies which we call heat and light."[19]

The machines that Keely built during his first period were of "megalithic proportions—one weighing 22 tons." Most were implemented by the Atlantic Works and the Delaware Iron Works of Philadelphia, and were scrapped in turn as they were superseded by smaller, more sophisticated models.

His Generator of 1878, "undoubtedly the strongest mechanism in the world," as Babcock mused, weighed about three tons and would stand "freely in a space five feet long and high by two feet wide." It contained small spherical chambers, "mathematically differentiated in size," connected vertically by "tubular processes of iron, and irregularly by smaller ones of copper. One quart of water fills all the chambers and tubes intended to be filled." The Generator, upright in position, had five distinct parts or columns, called the central column, two side columns, and the front and back stand-tubes. The stand-tubes, although similar in appearance, were opposite in action. The two side columns were alike. The central or main column was larger than the other four combined, and was "more complex in structure." Air was water-locked in some of the chambers and tubes, where, "by its elasticity, introductory impulses" were given to the water when "the equilibrium" was disturbed. This disturbance was effected by the movement of an outside lever operating a four-way valve within. There were no other metallic movements inside, except the working of three independent valves. The apparatus was therefore considered "practically without wear and not liable to get out of order."[20]

The Generator used but one quart of water to produce 54,000 pounds per square inch pressure. "No heat, electricity, or chemicals were used. ...Output remained constant regardless of work effected."[21] Babcock stated that he "has had to use tons of metal where others required only pounds."[22] The Generator was one of two mechanisms that Keely built during this period, the other being the engine. The Generator produced the force that the engine used, and these two devices were what is commonly referred to as the Keely motor. The vaporic substance was the medium of the force that it carried.[23]

During the second period, the miniaturization of his engines continued for a considerable degree. The Generator was a structure six feet long and correspondingly wide and high, "...which resulted in the production of a machine in 1895 not so large as a lady's round work table which he named a Liberator. Continuing his labor of evolution, within one year, he made astonishing progress...combining the production of the power, and the operation of the cannon, his engine and his Disintegrator in a machine no larger than a dinner plate, and only three or four inches in thickness. This instrument was completed in 1886."[24] When he began experimenting with his discovery of another principle—which was one of the reasons the Keely Motor Company took him to Court—his engines became even smaller in size: "and the size of the instrument used now, in 1888, for the same purposes is no larger than an old-fashioned silver watch, such as we see in Museum collections."[25]

Another explanation for the miniaturization of his engines was his utilization of the new technology: "Before discarding the use of water in the production of his force, twelve 'generators' were constructed varying in weight from 175 pounds up to 34,000 pounds, with a 'receiver' weighing 7,000 additional and by which he was able to develop the enormous power of 30,000 pounds to the square inch. For this 'generator' seven or eight engines were constructed, with varying success in their operation. They all would 'run,' but not to the satisfaction of this indefatigable worker. At last, about four years ago, the discovery was made that air alone was better than the combined air and water employed before. This at once resulted in the important changes in the mechanism. The clumsy Generator of several tons weight gave place to the lighter Liberator. Of these, three have been constructed, each one more slight than its predecessor, until that at present in his laboratory weighs less than 150 pounds, while the inventor has in process of construction the fourth and last one, which 'is a perfect machine of its kind,' weighing less than seventy-five pounds, and with which he expects to produce a greater force than has ever before been shown."[26] Between 1872 and 1887, no less than 124 different machines or engines had been constructed in experimenting with one Liberator.[27]

In his productive career, Keely built between 129 and 2,000 experimental devices, but as far as is known, never built more than one of each.[28] This enormous number was reached because of the special requirements that he had; his engines needed a perfect construction, and if a device possessed but one little unevenness or imbalance, he considered the engine worthless. One of the reasons for this was that he worked with enormous pressure forces. In his search for perfection, vast sums of money were expended on machines which sometimes were never used: one device cost $40,000, but was rejected because he claimed there was "a flaw in it." He "thought nothing of spending $20,000 for a piece of machinery, and a few weeks afterwards throwing it aside as useless."[29] Other devices were made of special metal alloys; his Generator was made of Austrian gun metal "in one solid piece." It would hold about ten or twelve gallons of water and was four or five inches thick, made "to handle the very heavy pressure of 20,000 to 30,000 pounds of vapor to the square inch." Other parts were made of welded iron, "of great thickness and strength."[30]

Then there were the adjustments made on the machinery, which was a difficult affair; "When it is considered that Keely experimented with 129 machines during his career as an inventor, and that one after another of these was made to special measurements, to be thrown into the scrap-pile...Many of these machines called for the most delicate adjustment and the most consummate skill of the artisan."[31]

Although Keely constructed and built his engines himself at first, he was later supplied with "the best instruments and devices for research that opticians and other skilled makers could produce, after models or designs which he furnished."[32] Parts of his later devices were made by various companies like I.P. Morris's Richmond Iron Works,[33] the Atlantic Works, the Delaware Iron Works[34] at New Castle, Delaware, and at an unspecified company in Hartford, Connecticut, where "two immense wheels of copper tubing" had been made at "enormous cost."[35] Keely's proposed "new and large 250 horsepower engine was constructed by Newsham & Co. in Philadelphia,[36] and I.P. Morris's Richmond Iron Works made 'many pieces of apparatus for Keely.'" He spent much time at the Iron Works, putting the parts of his engines together himself.[37] During this early period he was trying to get a powerful receiver, and I.P. Morris made him a cast steel shell about 24 inches in diameter and 30 inches high which was capable of sustaining a pressure of 25,000 pounds. "It was cast in halves, flanged and bolted together, the bolts being placed quite close to one another. The joint was made with lead, through which the bolts run."[38] However, when this enormous pressure was "put upon it, the water which it contained fairly foamed as it came through the joint, looking like milk; it even came through the pores of steel, although it was three inches in thickness."[39]

It was also stated that the huge steel sphere, although not made by the Richmond Company, was "brought there to be finished after it was cast, and that it was put on the boring mill." Furthermore, "a great many valves" were made at I.P. Morris's Richmond Iron Works, "of a character to sustain heavy pressure, if of a large size; the great majority were small valves." The tool department of I.P. Morris's Richmond Iron Works made a "small engine" for him, with leather packing, "as if for water." The device was "vertical and would make 1400 turns a minute" and was connected with his Generator.[40]

Keely needed expert help to install his machinery at the laboratory, and for this he obtained the services of Albert Chance, an employee of I.P. Morris's Richmond Iron Works, who took another employee with him to work at Keely's workshop. The two also ran several tests on the Liberator, which was made of phosphor bronze. This remarkable device was "about seven feet in height and weighed at least 10,000 pounds."

Originally it had a large cylinder on top but this was replaced with a sphere. There were two globes on one side and three different chambers, and two heavy glass water gauges on the sides. The globes and the chambers were tested for strength, during which the lower chamber of phosphor bronze burst.[41] While it was not properly explained, it was stated that "towards the last there was some hesitancy in accepting Keely's orders." After he died, a steel plate, marked "Keely" was still standing at the Richmond Company's premises, "about two inches in thickness, and six feet square."[42]

Of all the engines that Keely built, photographic evidence exists of only a mere fraction. The devices on these photos have wonderful names such as the Compound Disintegrator, the Vibratory Globe Machine, the Resonator, the Rotating Globe which worked through human magnetism, a number of Vibratory Discs, the Spirophone, the Pneumatic Gun, the Provisional Engine, the Globe Motor, the Vibrating Planetary Globe, the Wave Plates, the Planetary System Engine, the Liberator, the Vibrodyne or Vibratory Accumulator, the Vibratory Switch and the Sympathetic Negative Transmitter. It is the same with written descriptions of his devices; for these we have to rely largely on eyewitness accounts, scattered throughout contemporary newspapers.

Around 1894, Keely operated his Provisional Engine, which was described in a report of an annual meeting of the Keely Motor Company as "very successful" and that apparently "developed great power." This device was "not operated by the force derived from disintegrated water," but by a Sympathetic Transmitter, a "wire from which to the engine caused the latter to be put in motion."[43] While the device itself has not survived, an equally puzzling description has: The Provisional Engine was composed of "the motor proper, and the transmitter, the machine rested on a heavy brass base. Here, too, was a hollow brass sphere or ball. Between the engine and the transmitter ran a series of wires, and along the base of the transmitter an array of steel rods. ...These steel rods were responsive to the touch, and compared to the ordinary musical scale, which is subject to the tuning-fork. The interior of the globe almost defied description, but out of the complex mass brass tubes and adhesive plates stood prominently. This was the shifting resonator, as Keely termed it. The tubes and plates took up the vibratory sound and carried it along with rapidity. Of these vibrations there were seven distinct kinds, said Keely, and each of these seven capable of infinitesimal division. The motor itself consisted of a heavy iron hoop, placed firmly on the plate. Within this hoop ran a drum with eight spokes. When it was once in operation, the movements of the drum were exceedingly rapid."[44]

The Sympathetic Negative Transmitter is also lost, and only a description remains that was made during the dismantling for an observation by Professor W. Lascelles Scott of London. On the request of Bloomfield-Moore he made a "careful investigation" of Keely's inventions. "Imagine a globe in which is a vibrating disk which Keely calls a 'cladna'; also, a series of tubes which, under certain circumstances act like small organ pipes..." was Scott's puzzled description. And while he failed to get from Keely a "connected account which satisfied his English sense as to what these were," he discovered that "the apparatus appeared to be regulated upon something like a definite order or plan." He discovered that the Sympathetic Transmitter was sensitive to chords as B flat, D natural, F, D, F sharp and A. Keely considered the first three notes and their combinations as "having a tendency in one direction, which he called the polar force," and the other three notes "a tendency in an opposite direction, which he called a depolar force."

Keely's Vibrodyne, which suffered an equal fate as so many of his other devices, was apparently so strange that the Professor "would not call it a motor, an engine or source of power, but certainly it was capable of revolving." The device possessed "spokes" which in turn contained "pipes" similar to those in the Sympathetic Negative Transmitter, "with which they were in union." When the Transmitter was connected to the Vibrodyne by a single wire, nothing happened "until the right note or series of notes" were struck on the Transmitter. Then the wheel began to revolve. A galvanometer showed no evidence of electricity, and Scott stated that he knew of "no electrical or mechanical current that could be transmitted over a single wire."[45]

Perhaps we should excuse poor Professor Scott, who did not know what to make of Keely's Vibrodyne, as Babcock, while musing over Keely's devices from his first period, stated that "Keely's inventions...are so entirely original, and so unlike any other devices that have been constructed, that there is nothing in the annals of research to afford a starting point for the understanding...Keely's instruments are no more like electrical apparatus than they are like the machinery used with steam, the product of crude molecular dissociation of water by heat."[46]

At the time of Keely's death in 1898, a new engine was in its completion stage, which he expected to have in running order the next year. The machine had the same shape as his Globe Motor, only larger. It was made out of copper with a globe of about two feet in diameter and weighing 75 pounds. The final mechanism was three feet in diameter and built of decarbonized steel, weighing

600 pounds. After the machine was taken to Kinraide, its whereabouts became unknown.[47]

Unfortunately, no photograph of his Vibratory Microscope exists, which, according to theosophist R. Harte, worked "by means of three wires placed across the lens of a microscope...its magnifying power equal to that of the great telescope in the Lick observatory—the largest in the world."[48] Likewise, there are very little details of the Vibraphone, which was "fashioned after the human ear," and which collected the waves of sound and made "each wave distinct from the other in tone when the 'wave-plate' is struck after the sound has died away."[49]

When studying the photographs of these machines, it is with a sense of awe and wonder. Keely's devices are so strange in concept, sometimes sturdy and almost barren in their industrial appearance, at other times possessing a poetic quality, a beauty that touches art and transcends that of mere functional equipment. Whatever the principles are that the machines operate on, these principles can by no means be those of conventional science. It is quite easy to see in these wonderful devices something of an artful nature. Years later, for instance, Hungarian constructionist artist Laszlo Moholy-Nagy would build his "Light-Space Modulator," a mechanized sculpture consisting of steel, plastics and wood. Moholy-Nagy's device that was meant as a work of art bears a striking resemblance in certain aspects to some of Keely's devices, and evokes the same feeling that an attempt was made to visualize unknown principles, forces and energies.[50]

Keely's science was in certain aspects a synthesis of various opposites; the part that involved his use of the force that he variously named ether, apergy or inter-atomic force, became more and more the domain of the occult underground of that time. The other part, which consisted of his use of notes, chords, tones and sound in connection with his inventions, had a more scientific bearing. Such a synthesis might at first seem unusual but in fact it was not; the association with art and music with science started centuries before in the school of the Greek philosopher Pythagoras. The Pythagoreans devoted themselves to mathematics and believed that everything could be expressed numerically. Since they observed that the pitch of musical notes depends on a numerical ratio in the length of the chords struck, they concluded that this ratio corresponded to the distance of heavenly bodies from the center of the earth,[51] in Keely's time poetically named the "wheelworks of nature."

Especially during the last century, scientific research on sound waves and the nature of sound rapidly progressed. This interest in sound was heralded by the work of Ernst Florenz Friedrich Chladni (1756-1827). For most of his life, Chladni was devoted to the study of sound, perhaps because he was an amateur musician,[52] as was said of Keely. Chladni was an interesting character with wide-ranging interests. In 1790 he invented the Euphonium, a strange musical instrument that was made out of glass rods and steel bars which produced sound by the rubbing of a moistened finger, or the transmission of vibrations by friction. His Euphonium, which is classified in the category of friction instruments, and the Aiuton, invented by Charles Claggett around the same time, were the first in a series of models of friction instruments. Some had piano keyboards and horizontal friction cylinders or cones that acted on upright bars, and others with bars stroked by the player's fingers or bowed by a continuous bow.[53]

Chladni also discovered the longitudinal waves and the so-called Chladni figures. Since nobody had studied the vibrations of plates, Chladni developed a technique in 1802 of supporting a plate of glass or brass at one central point, covering the surface with fine sand, and then stroking the edge with a violin bow. The sand then collected along the nodal lines of zero movement and thus formed "striking geometric patterns." He visited the Académie des Sciences in Paris in 1808 to demonstrate this technique and its results, and subsequently the visible display of vibrations in solids, and even in gases, was used in teaching.[54] One of his other interests was meteorites, which he collected. He was one of the first to theorize that they fell from the heavens. In 1794, he acknowledged the theory that meteorites were debris from space, a theory which he would expand upon in his in 1819 book *Über Feuer Meteore u. über die mit denselben herabgefallenen Massen.*[55]

Herman von Helmholtz (1821-1894), who formulated the resonance theory of hearing, devised an instrument that could test the presence or absence of a particular harmonic in a given musical tone. This was called the Helmholtz resonator or resonance globe, and was a hollow globe made out of thin brass, with an opening at each end across the diameter. The Helmholtz resonators enabled sounds to be analyzed into their constituent frequencies, with benefit to music and speech theory. Helmholtz also synthesized sounds, reproducing these by "combining the individual sounds composing it, as shown by his resonators."[56]

Like Chladni, Helmholtz was deeply interested in meteors, and in 1871 he wrote that meteors and comets might disseminate "germs of life wherever a new world has reached the stage in which it is a suitable dwelling place for organic beings."[57]

Karl Rudolph Koenig (1832-1901), a brilliant student of Helmholtz, moved to Paris in 1851 and took the unusual step for an academic of apprenticing himself to a violin-maker by the name of Vuillaume. Koenig made an unrivaled reputation for novelty and accuracy, and much fundamental work was done in laboratories in Europe and America, employing his equipment.[58] During the 1876 Philadelphia Centennial Exposition, when Keely had an operational Globe Motor, Karl Rudolph Koenig was awarded a gold medal for his work in acoustics. Koenig's greatest contribution to precision in the study of sound was the clock tuning fork, which "acted the role of a pendulum, and it could then be used as the standard for comparison with other forks."

Some years earlier, in 1862, on the International Exposition in London, Koenig displayed his newly invented manometric flame apparatus, "a bank of eight Helmholtz resonators, with their small ends each connected to a separate manometric capsule which acts on a gas jet. When the rectangle of mirrors is rotated, it is easy to see the flame that is vibrating, so that a sound can be analyzed."[59]

There are several similarities between the research of those early scientists and some of Keely's experiments; it is also interesting to note similarities between the Helmholtz resonators with their tuning forks, and parts of Keely's devices. Keely was also keenly aware of the works of these scientists, and at least one of them was aware of his work.

Koenig had been present during one of Keely's experiments. In a letter written around 1891, he critically "ventured a suggestion as a test of the nature of the force that Mr. Keely is dealing with." This involved Keely's claim that a needle of his compass revolved as a result of negative polar attraction. Koenig remarked that since Keely had written that gold, silver and platina were "excellent media for the transmission of these triple currents," but since everybody knew that these metals were "unaffected by magnetic influences," he should make a compass needle of one of these three metals.[60] To this request Keely replied that this was impossible, but he nevertheless took up the experiment and reportedly succeeded in making a needle of the three metals and let it rotate by "differential molecular action; induced by negative attractive outreach, which is as free of magnetic force as a cork."[61] According to Bloomfield-Moore, Koenig stated: "I not only think Mr. Keely's theories possible, but I consider them quite probable."[62]

Keely undoubtedly visited the great Philadelphia Centennial Exposition as so many other Philadelphians did, and may have seen Koenig's work there. His favorite author was Chladni, whose writings he read in his youth and whose philosophy made a "deep impression on Keely."[63] He even named a certain part of his Sympathetic Negative Transmitter a cladna: "imagine a globe in which is a vibrating disk, which Keely calls a "cladna."[64] Chladni plates were also used by Keely, one of these being a large steel plate of 20 inches in diameter, which he struck with a "tiny hammer."[65] He was also quoted as saying: "I am now having a compound siphon made by Helmholtz, near Paris. I like his work."[66]

There also was an esoteric strain in Keely's philosophies. For this we have only to look at his terminology. It has been established that most of his terminology was unique, meaning that it had no precedent in either scientific jargon or occult parlance or concepts. Every new invention or discovery brought with it a new nomenclature and vocabulary. But his terminology for the force that he claimed to have discovered is a different case.

In 1875, he called the medium that delivered his force "vapor"[67] or "vaporic substance,"[68] although it is suggested by Bloomfield-Moore that around that time this substance was also named "Keil."[69] From this the term "vaporic force" evolved. In the following years, we also find terms in his philosophy such as "inter-atomic force," or "dynaspheric force."

Around 1881, Babcock even suggested that Keely's force should be called "Keal," meaning "Keely vapor."[70]

Around 1887, Keely's force was being compared with Bulwer-Lytton's vril, and we once again encounter new phrasings. Now he called the force "vibratory force," "sympathetic etheric force" and "sympathetic vibratory force."[71] At that time, Keely mingled cabalistic and theosophical concepts while speaking of his force, that to him could be "best described as coming nearer to the primal force or will-power of Nature...it is that primal force itself...the breath of life which God breathed into man's nostrils at the creation of the world—the motion of the spirit on the face of the waters; the wave motion of the brain."[72]

Two years after this beautiful imagery it was alleged that he claimed to use an "ethereal force"[73] and the same year elsewhere it was stated that "Keely's force is vibratory sympathy."[74] In 1890 he called his force "vibratory force,"[75] a term that was still in use in 1895,[76] while in the same year the term "apergy" became grafted on Keely's force. In 1896, it was announced that his newest motor was "supposed to be energized by pure vibration, instead of "etheric vapor."[77]

This changing phraseology demonstrates Keely's personal intellectual evolution in thought and a maturing process of ideas. But his use of the term "ether" stemmed directly from occult tradition. There, before and after the idea became a topic for research in conventional 19th century science, the concept of unknown and invisible forces or energies that surround us daily were firmly rooted. Significantly Tesla kept using the term ether after its abandonment by the scientific world.

This notion of an all-pervading energy, a universal life force, has a long tradition and different names in various cultures: The Chinese called it Qi, the Vital Energy which played an important part in Sheng Fui. To the Japanese it was Ki. The Hindus called it Prana and the Polynesians and the Hawaiians, Mana. Hypocrites called it Vis Medicatrix Naturae, Galen called it Pneuma. In the writings of Hermes Trismegistos the force is Telesma.

Sixteenth century alchemists like Paracelsus and Van Helmont called it Munis and Magnale Magnum, respectively. To Franz Anton Mesmer the force was Animal Magnetism. Von Reichenbach called it Odic or Odylic Force, and the radiesthesists, Etheric Force. To Eliphas Levi and to Blavatsky, who saw all these connections, it was the Astral Light.[78]

In the twentieth century L.E. Ehman called it X-Force, Wilhelm Reich called it Orgone and the Nazi dowsers knew it as W-Force. To modern-day ley hunters it is Ley-energy or more poetically The Dragon Pulse, reminiscent of Levi's Great Serpent. Soviet parapsychologists called it Bioplasmic Energy or Psychotronic Energy.

"What is the primordial chaos but Ether? The modern Ether; not such as is recognized by our scientists, but such as it was known to the ancient philosophers...Ether, with all its mysterious and occult properties, containing in itself the germs of universal creation...," Blavatsky wrote.[79]

In a climate of new scientific discoveries and the rise of esoterism, ether became a widely researched concept. For instance, spiritualist Sir Oliver Lodge lectured on the ether and its functions in 1882 for the London Institution. He described ether as "an undivided substance that fills all space, that can vibrate as light, that can be divided in positive and negative electricity that, in vortex movement, creates matter and that through composition and not pressure, transmits all actions and reactions, that the substance may yield. This is the modern view on ether and its functions."[80]

But centuries before, the early alchemists gave the name ether to the Quintessence which to them was the fifth element, a power or essence that bound in a unity the otherwise separate four elements. Quintessence was synonymous with elixir, mercury of the Philosophers and etheric. The Quintessence was said to be semi-material and visible to certain persons. The four elements, the fifth named Quintessence and two other unnamed elements formed the Seven Cosmical Elements. These were held as conditional modifications and aspects of one element, the source of Akasha itself.

Akasha or Akasa was used in occultism and theosophy as an equivalent of the ancient term "aether," according to Hinduism the fifth and most subtle element. The word is the Sanskrit term for "all pervasive space." Akasha is also called Soniferous Ether. Theosophical doctrine links it to Quintessence. According to Blavatsky, the Akasha forms the anima mundi, the soul of the world. Through it, divine thought was allowed to manifest in matter. The anima mundi constitutes the soul and spirit of mankind. It produces mesmeric, magnetic operations of nature.[81] Blavatsky introduced the concept of Akasha in the early twentieth century and connected it to the other notions of the universal life force, such as the Sidereal Light of the Rosicrucians, Levi's Astral Light and the Odic Force of Von Reichenbach. Astral Light is held by the occultists as a manifestation of the Aether, which is not to be confused with the ether of the modern physicists. Aether however can be linked to the Etheric of the occultists. Etheric was to them the force or energy that gives life within our cosmos. The influence of the etheric forces on inert matter creates the diversities of natural phenomena.

The Akashic Records were the supreme records of everything that has ever occurred since the very beginning of the universe. Theosophy claims that these records exist as impressions in the astral plane. Edgar Cayce asserted that he often consulted these records, as did Mary Baker Eddy and Rudolph Steiner, who called the records the Akashic Chronicle. With it he produced his detailed descriptions of the mythical lost civilizations Atlantis and Lemuria, as theosophists Scott-Eliott and Leadbetter did before him.

While the term ether crops up in Keely's phraseology around the time that it was widely researched to do so, Bloomfield-Moore admitted that it was from a particular book that both she and Keely discovered that "it was the ether that he had imprisoned, and for years after he confined his attention to efforts to keep the ether in an engine, supposing the ether itself to be the energy induced."[82] This book was *A Sketch of a Philosophy,* published in 1868, and written by Angus MacVicar. Rev. John Andrew, who was a friend of the late MacVicar, brought the book to Bloomfield-Moore's attention. According to her, Andrew "maintained great interest in Keely's researches since he first heard of them."[83] Bloomfield-Moore introduced MacVicar's concepts to Keely in 1884. A compilation of MacVicar's *A Sketch of a Philosophy* entitled *Ether the True Protoplasm* was sent to Keely, after which his attention turned to researches of the ether.[84]

Shortly after, the book *Harmonics of Tones and Colours Developed by Evolution,* written by Charles Darwin's niece F.J. Hughes, was sent to Keely. This led him into the line of experiment that would enable him to show on a disc the various colors of sound, each note having its color, and to demonstrate in various ways Mrs. Hughes' own words "that the same laws which develop musical harmonies develop the universe."[85] Possibly this line of experiment was the origin of what would be known as Keely's remarkable musical charts.

Keely was not a seeker of perpetual motion, as was sometimes alleged.[86] The search for a device that without the external supply of energy would deliver everlasting labor has always been a dream of mankind, although it is considered impossible to attain. Interestingly, the idea for such a device originated in ancient India. Rotations, cycles, rotational images and cyclical processes play an essential part in Indian philosophy. A central doctrine of all Indian religions is the concept of Karma; the continuation of good and evil deeds of man that defines his fate in this life and in the cycle of future rebirths. This Karmic process is without beginning and end. The perpetuum mobile originated as the symbol for this everlasting cycle. Thus the Hindu astronomer and mathematician Bhascara described at least two of the earliest perpetuum mobile designs around 1150, one consisting of hollow stakes filled with mercury.

This idea of a perpetuum mobile was adopted by the Arabian scientists and philosophers of the 13th century, and in their writings of that time we again find numerous references to perpetuum mobiles with mercury-filled tubes. But where in ancient India the perpetuum mobile was considered more as a religious symbol, the Arabs saw in it a device that could be used for more practical ends, such as in waterlifts. The Arabian sciences would eventually influence Europe, and with it came the concept of the perpetuum mobile. Thus a disk-shaped perpetuum mobile appeared for the first time in 1235 with the French architect Villard de Honnecourt. In medieval Europe of the 13th century, with the increase of knowledge of magnetism and the magnetic needle, Guillaume d'Auvergne, bishop of Paris, proposed the idea that an analogy exists between magnetic induction and the transfer of movement capabilities of an outer heavenly body on a inner heavenly body.

In 1269, Pierre de Maricourt, also known as Petrus Peregrinus, wrote in a letter called "Epistula de Magnete" that solely treats magnets and their capabilities about experimental researches on the magnetic force, that he saw as Virtus Dei or the glory of God. De Maricourt saw the heavenly poles, not the poles of the earth, as the abode of the magnetic forces, that propel the magnetic needle. He proposed the idea that a globular-shaped magnetic stone would move with the rotation of the heavens. The axle should point towards the pole of the heavens. De Maricourt related in great detail of a magnetic perpetuum mobile with a magnetic needle fastened in its center.[87]

Later we find innumerable inventors, scientists, and even priests involved in the search for such an apparatus. One of these, the 17th century Jesuit priest Christoph Scheiner developed a magnificent idea. In 1616 he designed a device that would operate according to his idea on gravity. Scheiner thought that the center of the universe would also be the center of gravity. From this, he projected an axle to a given point, from which another axle in an angle of 90 degrees sprang. On the end of this axle he would fasten a weight. Since the weight would constantly try to move towards the center of gravity, an everlasting circular motion could be obtained. A sketch of his device which he named "Gnomon in centro Mundi" appeared in 1616 in his book, *Mechanica Hydraulico-pneumatica,* a title that by the way, bears a similarity with Keely's hydro pneumatic pulsating vacuo machine. Although Scheiner's idea would be discarded by later fellow-Jesuits, he was no deluded seeker; he built the first Keplerian telescope, and can be credited as the co-discoverer of the solar spots. He invented the pantograph and he researched the working of the human eye.[88]

Another Jesuit priest, Athanasius Kircher, proposed a perpetuum mobile in his book *Magnes sive de arte magnetica* that appeared in 1640-1641, that consisted of a wheel-shaped apparatus, built of a disk with iron rods that would turn itself around while under the magnetic influence of the static magnetic field of four magnets.[89]

Between 1812 and 1814, several centuries and innumerable perpetuum mobile designs later, Italian physicist Guiseppe Zamboni invented and designed a device which he called "Elettromotore perpetuo," consisting of two electrically charged pillars made of amongst others of 2,000 glass disks with two globes on their tops. Between these pillars, with opposing poles a pendulum was placed which then would swing between the two pillars.[90]

In 1887, Keely obtained continuity of motion in his engine,[91] but he never claimed to have invented nor sought the construction of a perpetual motion machine. This misunderstanding arose from other statements by him; one of these was that he claimed that he could "put in motion a power that can run a train without fuel of any kind or any other force for thousands of years continuously if desired."[92] Keely also expected that when the motion had been once set up in any of his machines, it would continue until the material was worn out. It was this claim which caused him to be classed with perpetual-motion seekers,[93] and which was perhaps the reason that Peretti[94] visited him. Peretti, the Vineland inventor, mason and machinist who claimed to have Keely's secret, confided to a reporter that he had tried to invent a perpetual motion machine "since his transaction with Keely...but it was a plan of his own, not Keely's."[95]

On this subject, Keely made several statements: "The nearest approach to a certainty is made through harmony with nature's laws. The surest media are those which nature has laid out in her wonderful workings. The man who deviates from these paths will suffer the penalty of a defeat, as is seen in the record of "perpetual motion" seekers. I have been classed with such dreamers; but I find consolation in the thought that it is only by those who are utterly ignorant of the great and marvelous truths which I have devoted my life to demonstrate and to bring within reach of all....[96] Perpetual motion is against nature, and it is only by following nature's laws that I can ever hope to reach the goal I am aiming to reach."[97] And in the concept of his vibratory engine, he "did not seek to attain perpetual motion."[98]

When studying Keely's research and trying to get a clear picture of his achievements, we are faced with a number of difficulties. The first problem, of course, is the almost total absence of his many devices and writings. Then there is his jargon, the language in which he described his discoveries. Although he had no formal school training after his 12th year, he somehow acquired an unusual command of the English language. This problem was already noted time and again in his day, and from someone who listened to him, we may obtain an impression of what that meant: "Mr. Keely talks with the rapid fluency begotten in thirty years study...while his thoughts are often clothed in words rarely joined together in framing a sentence...Now let the reader imagine the...statement poured into him at the rate of 250 words per minute, with no stop for refreshments, and he will experience solid relief to be assured by Mr. Keely that he is preparing for publication a complete explanation. ...It is simply impossible to reproduce more than a fraction of what is freely put at one's disposal in a chat of an hour with this remarkable man."[99] And a journalist, after having had the ordeal of listening to Keely's verbose explanations, shruggingly remarked that he "didn't understand it, but nobody ever understood differential calculus or even the Abracadabra in a day or two days."[100] And another person simply said that "The inscription on the Zuni tablets would have been quite as plain."[101]

Keely assured that he would explain all in a book or two, but these never saw print and we are left with only those parts of his theories that were published by Bloomfield-Moore and Colville, or are scattered in the various newspapers of the day. Bloomfield-Moore writes that, "Every branch of science, every doctrine of extensive application, has had its alphabet, its rudiment, its grammar...,"[102] but her book is mainly written as an apology. Though it presents enough theoretical matter, it sadly lacks a complete comprehensible glossary of Keely's terms or grammar. The problem with his grammar is that he used common words in an uncommon way, while he had a liking for uncommon words, which he used plentifully. A reason for this may have been that he was inventing a new terminology for his novel ideas, without taking the trouble to see if the terms were applicable or whether they had not already been used by others. In addition he had the ability of not only using uncommon words, but also of using many more words to express an idea than was in fact necessary.[103] To make matters worse, every new discovery or invention that he made brought with it a new nomenclature and a new vocabulary.[104] In all fairness to him it must be noted that he was not always misunderstood by his contemporaries. The accounts of those who understood his explanations differ from the accounts of those who didn't.

Faced with all this, the remaining difficulty is in understanding those parts of Keely's theories that were printed. When reading a sentence or a text, it may read grammatically correct, but the meanings are obscure and hermetic. Owing to the unique nature of his philosophies, we find his writings riddled with terms such as Sympathetic Vibration, Etheric Force, Dynaspheric Force, transympathetic, divine force differentiation, triune streams or Aqueous Disintegration.

Dale Pond, who has studied Keely's inventions for more than a decade, remarked that "While it is true that Keely used a lot of big and unusual words, he always used them in a masterful way....If one takes the trouble of looking these words up and studying them he will find they are used in ways outside of our normal awareness."[105]

Keely was asked to prepare a glossary of terms so that he might be better understood. Since such a glossary was never printed, it was erroneously held that he refused to do so,[106] which is not true since he—at least in theory—was planning on publishing his philosophy. At one time it was alleged that a Dr. Brinton, "acquainted with more strange languages, dead and living, than anybody else," made a "translation of Keely's philosophy into English, which is described as able, lucid and logical."[107] It is also asserted that such a glossary was published in the 1894 book *Dashed Against the Rock,* by Keely's friend, William Colville.[108] A short comprehensive glossary was recently published in the book *Universal Laws Never Before Revealed: Keely's Secrets,*[109] in which the reader may find a wealth of materials pertaining to a modern interpretation of the technical aspects of Keely's discoveries.

Keely's experiments spanned more than a quarter of a century and were wide-ranged and kaleidoscopic in their variety. What has been left to us—compared to what has been lost after almost a century—is but a fragmentary view of what Keely was actually aiming at and what he accomplished. He changed his methods many times, abandoning previously-held important lines of research, scrapping pieces of machinery, and renaming devices. It is even alleged that he sometimes destroyed his devices after a few runs, devices that were seen only by himself and his assistants.[110] At one time, during a spontaneous search by frustrated members of the Keely Motor Company in 1883, 12 discarded engines were discovered "wrecked" and "used in past experiments," carelessly thrown in a corner, upstairs in Keely's workshop.[111]

Each moment in Keely's long period of invention seemed to yield many more directions, and sometimes he would discover quite important applications by sheer accident. The view emerges that he was entangled in a sheer labyrinth of research, each important finding opening even more roads to take.

Indications for this labyrinthine maze, of which even his close allies like Collier knew little or nothing, are a slight remark that he was somehow able to weld metal without heat,[112] or of his "sensitizing process." About its origins we learn only that this new aspect in the complex history of his inventions began somewhere around 1894, and that Lancaster Thomas was present during the "sensitizing of several disks" and had assisted in its performance. Both Collier and Thomas stated that this sensitizing process was "absolutely new," although Collier assumed that this mysterious process could be patented.[113]

There are also casual mentions of a strange substance. Although this substance was pivotal in the working of Keely's force, we learn no more of it after the following slight references: "The only thing that caused any doubt seemed to be the exact composition of a peculiar metallic paste or amalgam without which the 'sympathetic attractive force' could not be produced." And although the ever-bolsterous Collier stated that he believed that his ally Thomas could produce this material, we hear no more of any attempt, or an explanation of what this substance was actually used for in Keely's inventions.[114]

The second reference is equally puzzling: "One of the secrets of the machine is the composition of the 'vitalized metal disks' and 'sensitized cartridges of metallic powder.'"[115] Elsewhere we learn more, although still not enough about this mysterious substance: "Aside from these disks, Mr. Keely has prepared a metallic powder, which, to look at, very much resembles iron or steel filings, but which lacks one essential feature of iron or steel—it will not respond to the attraction of a magnet."[116] No further explanation was offered upon the nature or composition of the cartridges of metallic powder, which were possibly the same as the peculiar metallic paste, or the unspecified "amalgam paste." In 1895, a puzzled reporter wrote that Keely said it took him three years of study before he could produce the substance.[117]

Two years later, G.W. Browne, "who has studied the subject" and concluded that it had to be "electricity" that propelled Keely's engines, tried his best shot at explaining what the strange metallic powder was. He wrote that parts of the Liberator consisted of "nine cylindrical boxes surmounted by vitalized discs of metal. In the center of these cylindrical boxes are placed small cylinders termed 'sensitized cartridges' loaded with a metallic powder only known to Keely, and which, he claims, furnished all the vitality which the discs possess." The reporter concluded that these sensitized cartridges must have been "in reality small batteries...They are almost filled with water when Keely puts into each a teaspoonful of a metallic powder, the composition which he claims as his secret, but which is known to the world as bisulfate of mercury (truly a metallic powder)."[118]

We've now established two major periods. The first period, prior to 1881, involved the creation of the force through the disintegration of water. The second involved the creation of the force out of almost anything. Keely also followed two major directions with his experiments. One was the constant search for a method of mastering his force and making it drive his various engines. After having mastered the force he said, it would be "adapted to engines of all sizes and capacities, as well to an engine capable of propelling the largest ship as to one that will operate a sewing machine. Equally well and certain is it that it will be adapted as a projectile force for guns and cannons of all sizes, from the ordinary shoulder-piece to the heaviest artillery."[119]

Over this direction, the shadow of the ever-vigilant Keely Motor Company looms heavily; and in this direction he constructed his more than 129 engines in search for the "perfect engine." The other direction, however, was to establish the laws of etheric force, which would lead to the earlier-mentioned glossary called "The 40 Laws of Harmony."[120] Over this direction shone the light of his supporter, Bloomfield-Moore.

He also claimed to have three systems, and he described them as follows: "My first system is the one which requires introductory mediums of differential gravities, air and water, to induce disturbances of equilibrium on the liberation of vapor, which only reached the inter-atomic position and was held there by the submersion of the molecular and atomic leads in the Generator I then used. It was impossible with these mediums to go beyond the atomic with this instrument; and I could not dispense with the water until my Liberator was invented, nor reach the maximum of the full line of vibration. My first system embraces liberator, engine and gun."

His second system he considered complete, "as far as the liberation of the ether is concerned, but not sufficiently complete as yet, in its devices for indicating and governing the vibratory etheric circuit, to make it a safe medium." His third system included aerial and submarine navigation.[121] Apart from a method of disintegration that he claimed to have discovered by accident while working on this third system, Keely also claimed to have discovered another application in which he was an unknowing forerunner of the radiesthesists, but was preceded by Franz Anton Mesmer—the healing of disease through the restoration of the inner balance. "One of Mr. Keely's discoveries shapes his theory that all nervous and brain disorders may be cured by equating the differentiation that exists in the disordered structure," Bloomfield-Moore writes, "When his system is completed, medical men will have a new domain opened to them for *experiment*" (her italics).[122]

Keely saw disease as "a disturbance of the equilibrium between positive and negative forces," and writing about mental illness, stated that, "In considering the mental forces as associated with the physical, I find, by my past researches, that the convolutions which exist in the cerebral field are entirely governed by the sympathetic conditions that surround them." In Keely's hard-to-understand thesis on this subject, owing to his hermetic prose, he nevertheless expounds on such original ideas as the differentiation of the molecular conditions of a mass of metal of any shape "so as to produce what you may express as a crazy piece of iron or a crazy piece of steel, or, *vice versâ* (his italics), and intelligent condition of the same," or the cause of insanity and sensory hallucinations, "such as flashes of light and color, or confused sounds and disagreeable odors, etc., etc." These, according to Keely, are the result of "a condition of differentiation in the mass chords of the cerebral convolutions, which creates an antagonistic molecular bombardment towards the neutral or attractive centers of such convolutions; which, in turn, produce a morbid irritation in the cortical sensory centers in the substance of ideation."

While the foregoing is again a typical example of original concepts packed in his difficult-to-understand prose, his ideas on the brain have a poetic quality. Since the normal brain is to Keely "like a harp of many strings strung to perfect harmony," cure is a matter of adjustment: "If we live in a sympathetic field we become sympathetic, and a tendency from the abnormal to the normal presents itself by an evolution of a purely sympathetic flow towards its attractive centers."

And while Bloomfield-Moore readily admitted like so many Keely researchers after her that "there are few who will fathom the full meaning of these views," his discoveries according to her "embrace the manner or way of obtaining the keynote, or 'chord of mass,' of mineral, vegetable, and animal substances," and she envisioned a universal appliance of Keely's inventions: "the construction of instruments, or machines, by which this law can be utilized in mechanics, in arts, and in restoration of equilibrium in disease."[123]

We are also left with several statements that, in line with this holistic view, he involved himself in researches that tried to link his devices to the waves of human brain, or to other bodily processes. We have the slight reference that his Rotating Globe "worked through human magnetism," and it is asserted that around 1882 he discovered something that was termed "the source of life and the connecting link between intelligent will and matter."[124]

This he explained three years later, "My researches teach me that electricity is but a certain condensed form of atomic vibration, a form showing only the introductory features which precede the etheric vibratory condition. It is a modulated force so conditioned, in its more modest flows, as to be susceptible of benefit to all organisms. Though destructive to a great degree in its explosive positions, it is the medium by which the whole system of organic nature is permeated beneficially; transfusing certain forms of inert matter with life-giving principles. It is to a certain degree an effluence of divinity; but only as the branch is to the tree. We have to go far beyond this condition to reach the pure etheric one, or the body of the tree. The Vibratory Etheric tree has many branches, and electricity is but one of them. Though it is a medium by which the operations of vital forces are performed, it cannot in my opinion be considered the soul of matter."[125]

Interestingly, there are reasons to suspect that certain parts of the concepts that were developed by those early searchers of perpetual motion were incorporated in Keely's highly intriguing researches on this aspect of his discoveries.

Bloomfield-Moore for instance, while digressing on contemporary researches into the nature of magnetism and how it could possibly affect the human nervous system, writes how Keely's experiments "show that the two (magnetism and electricity) are, in part, antagonistic, and that both are but modifications of the one force in nature ...Should it be that Mr. Keely's compound secret includes any explanation of this operation of will-force, showing that it may be cultivated, in common with the other powers which God has given us, we shall then recover some of the knowledge lost out of the world, or retained only in gypsy tribes and among Indian adepts."[126]

Elsewhere she assures her readers that while scientists "look upon the human organism as little more than a machine, taking small interest in researches which evince the dominion of mind over matter," it was Keely's research that had "shown him that it is neither the electric nor the magnetic flow, but the etheric, which sends its current along our nerves."[127]

According to her, the ultimate consequence of Keely's research results was that a "true coincidence" existed between "any mediums—cartilage to steel, steel to wood, wood to stone, and stone to cartilage—that the same influence (sympathetic association) which governs all the solids holds the same governing influence over all liquids; and again, from liquid to solid, embracing the three kingdoms—animal, vegetable and mineral—that the action of mind over matter thoroughly substantiates these incontrovertible laws of sympathetic etheric influence; and that the only true medium which exists in nature is the sympathetic flow emanating from the normal human brain, governing correctly the graduating and setting up of the true sympathetic vibratory positions in machinery necessary to success."[128]

Since we sadly lack coherent notes or any further documentation of any research results that Keely might have obtained, we now have only the assertions that along those lines he indeed constructed at least one of his devices according to that most innovating concept: "In the image of God made He man, and in the image of man Keely has constructed his Liberator. Not literally, but, as his vibraphone (for collecting the waves of sound and making each wave distinct from the other in tone when the wave-plate is struck after the sound has died away) is constructed after the human ear, so his Liberator corresponds in its parts to the human head," and in his Disintegrator, "the neutral center represents the human heart,"[129] Bloomfield-Moore explains.

However, since no documentation other than her writing and Keely's statements have survived, we can only speculate whether this line of research went beyond its planning stages. Did Keely build other devices according to those principles? A machine that one could merely think or wish into action? A technology that would be so avant-garde in nature that, like some of the more radical ideas in the fields of modern cybernetics, the differences between man and machine would gradually melt away into an indistinguishable whole? It is doubtful that the tools existed during his time to accomplish such a feat, and we are left to speculate on this particular point by absence of any other materials or documentation.

Keely developed the intriguing notion that his devices could be linked to the human brain, or to human magnetism by envisioning one force that permeated all things. Coincidentally the same year, his discoveries were compared to Bulwer-Lytton's vril-force, so aptly described in *The Coming Race*. Around that time Keely stated that his force resembled the "will-force of nature" and "the waves of the brain."

Details of this notion would eventually be transformed by others to the theory that Keely himself operated his devices through peculiar paranormal powers. The roots of this paranormal power theory emerged from theosophical quarters; there the aspects that we are surrounded by one force or power tallied more with Keely's ideas. From its beginning, theosophy had been looking for a reconciliation of modern Western science and ancient Eastern philosophy. To do so, theosophy borrowed deeply from the vast storehouse of age old Indian doctrine; with amongst others the idea of a perpetuum mobile symbolizing the everlasting cycles of rebirth. Meanwhile in the West, there was the astrological concept of the body parts of man that tallied with certain zodiacal signs and the influences of the planets, ultimately meaning that the blueprint of the cosmos was to be found in man himself. There were the endless Akasha-chronicles, and the sea of ether. There were also other Western concepts that preceded theosophy, such as those of the Odylic force, of Mesmerism, of Bulwer-Lytton's vril and the ether and of the fluidum and ectoplasm of the spiritists.

All this more or less evolved into the notion that man was a brilliant interface or controlling mechanism between this material aspect of existence and the invisible cosmic life force, the sea of light and of willpower, and the emanations of the one supreme being. It was a small step for an esoterist or a theosophist at the dawn of the 19th century industrial age with its bristling steam mills and in which science made a new discovery each day, to envision the mind as the link between the cosmic sea of life force and avant-garde technology, for the modification of man's surroundings and in the end perhaps the whole universe. One was either born with this capacity, or else one needed the development of one's willpower through initiation. And technology was not merely something that one would develop along rationalist, empirical lines; it was an aspect of man's inner evolution.

With this in mind, we must treat the following statements from Blavatsky, who considered Keely as one who obtained his results form the fifth and sixth planes of the etheric or astral force, the fifth and sixth principles of Akasha, accordingly.

She considered Keely to be gifted from birth. In her eyes, he was one of those mortals whose *"inner selves are primordially connected, by reason of their direct descent, with that group...*who are called *'the first born of Ether.'"*[130] Thus she reasoned that "...whenever such individuals as the discoverer of *Etheric Force*—John Worrell Keely—men with peculiar psychic and mental capacities are born,"[131] claiming that "It has been stated that the inventor of the 'Self-Motor' was what is called, in the jargon of the Kabbalists, a *'natural-born magician...'"*[132] it would follow that Keely's constant struggle to develop his system for commercial use was "beyond his power," for he could not "pass to others that which was a *capacity inherent in his special nature."*[133]

Blavatsky asserted that "the discoveries made by him will prove wonderful—yet *only in his hands and through himself.* ...The truth of this assertion has, perhaps, not quite dawned upon the discoverer himself.[134] ...no one, who should have repeated the thing done by himself, *could have produced the same results.* ...For Keely's difficulty has hitherto been to produce a machine which would develop and regulate the 'force' without the intervention of 'willpower' or personal influence, whether conscious or unconscious to the operator. In this he has failed, so far as others were concerned, for *no one but himself* (her italics) could operate on his 'machines.'" That Keely's organism is directly connected with the production of the marvelous results is proven by the following statement emanating from one who knows the great discoverer intimately. At one time the shareholders of the Keely Motor Company put a man in his workshop for the express purpose of discovering his secret. After six months of close watching, he said to J.W. Keely one day: "I know how it is done now." They were setting up a machine together, and Keely was manipulating the stop-cock which turned the force on and off. "Try it, then," was the answer. The man turned the cock, and nothing came. "Let me see you do it again," the man said to Keely. Keely complied, and the machinery operated at once. Again the other tried, but without success. Then Keely put his hand on his shoulder and told him to try once more. He did so, with the result of an instantaneous production of the current.[135]

The same year that Keely envisioned the holistic application of his discoveries, a woman told him that in her opinion he had "opened the door to the spirit-world," on which Keely politely answered: "Do you think so? I have sometimes thought I might be able to discover the origin of life."

But Keely himself maintained a certain degree of ambiguity surrounding this aspect of his discoveries; for according to Bloomfield-Moore, he at first gave "no attention whatsoever to the occult bearing of his discovery. It was only after he had pursued his research, under the advantages which his small Liberator afforded him for such experiments, that he realized the truth of this woman's assertion. It was then, in 1887, that 'a bridge of mist' formed itself before him, connecting the laws which govern physical science with the laws which govern spiritual science..." and a year later Keely allegedly told that he had made "such startling progress that he now admits...that if not in actual experiment, at least in theory he has passed into the world of spirit."[136]

Bloomfield-Moore's endorsement of Keely's views were in accordance with theosophy: "Paracelsus taught that man is nourished and sustained by magnetic power, which he called the universal motor of nature," she writes, and referring to Keely's ideas on his discoveries for the appliance of the treatment of illness, she refers to light therapy by Seth Pancoast, one of the founding members of the Theosophical Society. We again may glimpse a possible influence on Keely's Christian symbology, for while describing his discoveries Bloomfield-Moore states in reference to Pancoast that she has learned that "students, versed in Biblical lore, declare that the esoteric teachings of the Book of Job enunciate a system of light-cure."[137] But on the notion that Keely himself possessed paranormal capabilities, he thought quite otherwise. He once lamented, "I have been told that I am a powerful medium. But if there is in me any secret force such as resides in no other man...that force is in no way akin to spiritualism."[138]

Keely was never able to remove the suspicions of mediumship surrounding him and his inventions, nor was he able to refute the claims that he possessed a paranormally-gifted mind. Not only an unusual number of theosophists, mediums and those who were interested in spiritualism clustered around his person, in 1896 a reporter noted that Keely "made a very significant remark": "I am always a good deal disturbed when I begin one of these exhibitions," Keely allegedly said, "for sometimes, if an unsympathetic person is present, the machines will not work."

The reporter was quick to point out that "How many times have 'spiritualist' mediums said the same thing! How many times have they accounted for failure by explaining that the spirits were repelled by some unsympathetic person in the audience or 'circle'! ...It is reported, that the motive power he has discovered is 'the will of God,' and that the old man can move railroad trains and other things by word of mouth without the aid of any machinery whatsoever. ...Keely now admits that he is working on a new basis or hypothesis. Do not his latest allusions to the adverse influence of unsympathetic or unbelieving persons foreshadow a coming explanation that his occult force is really supplied from the spirit world?"[139]

And thus, in spite of Keely's denial, the press, but also Blavatsky, H.S. Olcott, her ally and co-founder of the Theosophical Society, anthroposophist Rudolf Steiner, Fort and others developed the theory that Keely might indeed possess paranormal capabilities. Blavatsky and Olcott named this "psychical abilities."

No wonder then, that as late as 1913, Dutch astrologer and theosophist A.E. Thierens wrote of Keely that: "man may have, through one's willpower, at his disposal endless forces in this physical world, of this J.W. Keely with his motor has given an at this time still not totally controllable but nevertheless highly curious example...There is...reason to believe that Keely found the key to that cosmic life force."[140]

Not always would these theosophical views excerpt their evocative power; all of the above led a researcher to grudgingly remark that "The greatest tragedy of the time was that Keely's ideas occurred at the time when theosophy was being evolved, and the ideas of the Hindu sages upon such varies matters as physics and chemistry tended to become grafted on those of such workers as Keely. It would be more correct to say that it was not so much the ideas which were erroneous as the interpretations, most of which time has shown to have been wrong."[141]

Steiner referred to Keely's endeavors as the result of "the new mechanical faculty," although his concept deviated from the theory that Keely was somehow a paranormally-gifted person. His idea was more in line with Keely's original concept; he developed a notion of a technology in which future machines might be driven by waves of the brain.

Fort however, with his witty flair, called Keely's supposed paranormal abilities a "wild talent."[142]

Musical Globe with Spiro-Vibrophonic attachment

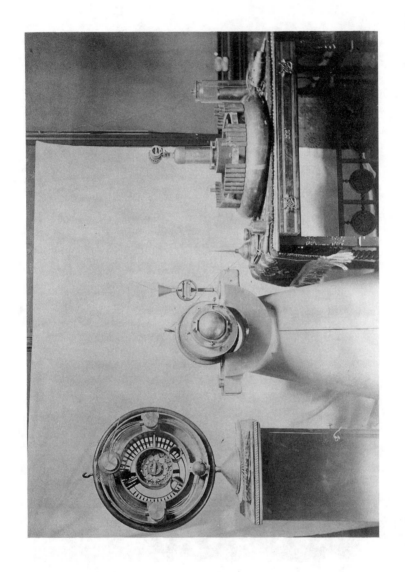

(Left to Right) Compound Disintegrator. Vibratory Globe and Resonator combined. Medium for testing vibration under different orders of evolution.

Liberator

Globe Motor and Provisional Engine

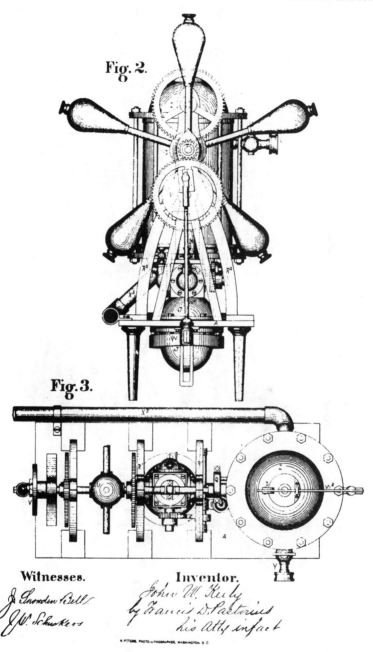

Fig. 2.

Fig. 3.

Witnesses.
J. Snowden Bell
J.W. Schuckers

Inventor.
John W. Keely
by Francis D. Pastorius
his Atty in fact

Hydro-Pneumatic-Pulsating-Vacuo Engine

The Transmitter

Compound Disintegrator (front view)

Compound Disintegrator (rear view)

Pneumatic Gun

"Test Medium"
Used for testing the sympathetic force of vitalized disks.

Vibratory Planetary Globe with wave-plate, fork and spirophone

Vibratory Globe and Accelerator

Revolving Ball

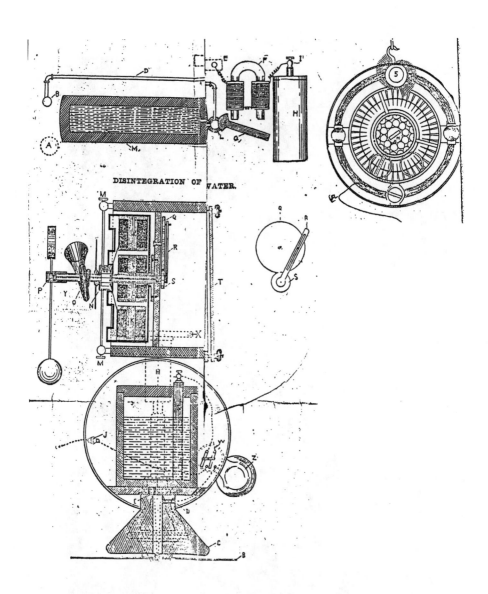

DISINTEGRATION OF WATER.

A rare cross-section cut of Keely's engines. It is not known if his devices were actually built this way since all of his drawings and blueprints mysteriously vanished.

Negative Attractor and Indicator

Glass containing weight which Keely claimed could be moved up or down by striking the zither strings.

Sympathetic Negative Attractor

Generator, built between 1881-1885

The mysterious sphere found under the floor of Keely's workshop. The object of much speculation, it's real purpose is not known.

One of the strange tubes (A) found in Keely's workshop. This and other unexplained findings prompted a search team to conclude that Keely was a fraud.

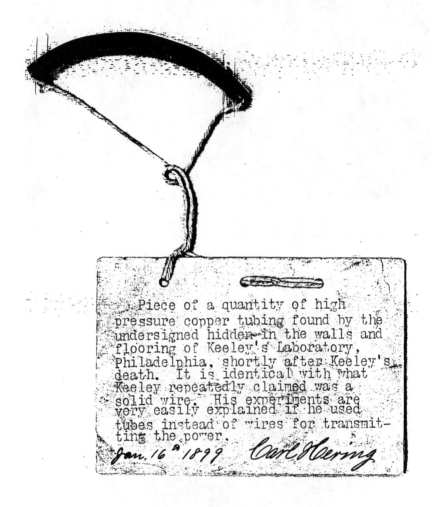

Piece of tubing with note by Carl Herring, one of the investigators of Keely's workshop after his death. Allegedly, this insignificant tube was able to direct compressed air under immense pressures.

8

Prisoners of the Neutral Point
Keely's Antigravity Experiments

"For obvious reasons, those who possessed the secret of apergy had never dreamed of applying it in the manner I proposed."
Percy Greg, *Across the Zodiac,* 1880

"...and after my patents are taken out, I will devote the remainder of my life to aerial navigation, for I have the only true system to make it an entire success."
John Ernest Worrell Keely, 1884

"...thanks to apergy, the force whose existence the ancients suspected, but of which they knew so little, all things were possible."
John Jacob Astor, *A Journey In Other Worlds,* 1894

During the quarter of a century that Keely experimented with his unique devices, it is alleged that he accidentally discovered a means to disintegrate matter, and built several devices for an astonishing series of antigravity experiments. Scattered throughout the articles are a few obscure references that he not only discovered a method to disintegrate matter, but at one time was involved in actual experiments with disintegration and had even constructed a device to do so at will.

One story by theosophist R. Harte has it that Keely, with the help of a small machine, was able to disintegrate a reef of quartz: "...twelve 'solid men'—millionaires—met by appointment in a laboratory in Philadelphia to witness an exhibition of the disintegration of quartz by a new method. They were mining magnates who had a tremendous interest in getting the gold out of quartz rock quickly and cheaply. The inventor obliged them by simply touching some blocks of quartz with a little machine he held in his hand, and as he touched each block it instantly crumbled into atomic dust. The specks of gold it contained stood out like boulders in a bed of sand. Then the twelve men solidly said: 'Mr. Keely, if you will in the same manner disintegrate some quartz for us in its natural place, we will each give you a check for — dollars.' So off they all went to the Katskill [sic] mountains, and there the twelve solid men pointed out a reef of quartz on the side of the mountain, as solid as themselves; and Mr. Keely took out his little machine and said: 'Gentlemen, please take the time.' In eighteen minutes there was a tunnel in that quartz mountain eighteen feet long and four and a half feet in diameter. ...All these men bound themselves to secrecy; and this is the first time that this incident has been made public. How was the quartz disintegrated? That is one of Keely's secrets."[1]

It was not true that this was the first time that this incident appeared in print; almost a year before, a newspaper had already published the tale of the alleged mysterious meeting at the Catskill Mountains, of which we get none the wiser: "Let me add that recently, in the presence of several gentlemen interested in mining operations, Mr. Keely bored with his engine eighteen feet in eighteen minutes sheer into the quartz rock of the Katskill [sic] Mountains."[2]

At one time, Keely allegedly also directed his disintegration research to experiments with dead tissue. Blavatsky assured her readers on two occasions that Keely somehow was able to disintegrate a "dead bullock" into a "heap of ashes" or even into "atoms,' in just "a few seconds."[3] While she states this as a fact, at the same time her statements are of the shortest and vaguest kind; for instance, she does not tell when this particular experiment took place or how Keely achieved it. It is therefore impossible to establish whether Blavatsky's statements were based on hearsay, were meant as an embellishment of an already incredible career, or were based on accounts of an actual experiment.

According to Bloomfield-Moore, in 1878 Keely discovered his means to disintegrate matter while he was working on his vibratory lift: "while experimenting on the improvised multiplication by this medium, he had occasion to put a piece of marble, weighing twenty-six pounds, on a steel bar to hold it in place, when then and there his first discovery of the disintegration of mineral substance took place."[4]

However, she dates the true genesis of this most unusual discovery as far back as 1875, when he was experimenting on his hydro-pneumatic-pulsating-vacuo engine, when, "accidentally, the first evolution of disintegration was made."[5] Some confirmation as to the accidental nature of his discovery comes from Harte, although he gives a different version: "The disintegration of rock is, however, a very small and accidental effect of that tremendous force that lies behind the 'secret.' Indeed, that particular application of the force was a chance discovery. One day the inventor was studying the action of currents of ether playing over a floor upon which he had scattered fine sand,—the ether was rolling the sand into ropes,—when a block of granite, which was used for fastening back a door, disintegrated under his eyes. He took the hint, and in a few days he had made a 'vibratory disintegrator.'"[6]

While Harte would want his readers to believe that Keely built his device in a few days, Keely himself stated that it took a much longer time: "I have been repeatedly urged to repeat my disintegration of quartz rock; but it has been utterly out of my power to do so. The mechanical device with which I conducted those experiments was destroyed at the time of the proceedings against me. Its graduation occupied over four years, after which it was operated successfully. It had been originally constructed as an instrument for overcoming gravity; a perfect, graduated scale of that device was accurately registered, a copy of which I kept; I have since built three successive disintegrators set up from that scale, but they did not operate. This peculiar feature remained a paradox to me until I solved the conditions governing the chords of multiple masses; when this problem ceased to be paradoxical in its character."[7]

After these intriguing but slight references and Keely's admission that he was not able to reproduce the disintegration process, we hear no more of this line of research, even though Keely admitted that he had solved the underlying problem. Although an elaborate theoretical exposé found its way into Bloomfield-Moore's book,[8] Keely very probably never pursued this line of research any further, and his disintegration of "rock into impalpable powder" became quite legendary in 1894, and was by then remembered as a "wonder" that he had done in the past.[9]

Likewise, his experiments with antigravity, then referred to as levitation, or "vibratory lift" or "vibratory push-process," as he called it, are shrouded in uncertainties. There is no indication of a specific date or year in which he started these experiments. A view of his early antigravity experiments (before 1890) can only be reconstructed with the greatest difficulty. If he took notes while experimenting, these are now lost. He built several devices with which to manipulate gravity, but although several of these sketchy descriptions have survived, it is not known how many antigravity devices he actually built. The only documentation that we have consists of inaccurate and superficial contemporary newspaper articles, several theosophical sources, a maze of entangled accounts—often gratuitously and distortingly copied by other writers—tantalizing hints and remembrances by people often trusted to paper as much as 10 years after the time period.

In addition, a suspicious cloak of vagueness and secrecy surrounds this area of his research. This vagueness and the dubious quality of the sources make a reconstruction of his antigravity experiments all the more difficult.

In 1878, Keely started experimenting with his system for vibratory lift, of which a method of disintegration branched. There is a slight but highly intriguing reference that three years later his system for vibratory lift may have led to a completed device of some sort. In October 1881, a reporter visited Keely's workshop, "at that time Keely had most of the lower apartment tightly boarded up, and the reporter, who thought an able mule or other source of mechanical power was concealed in the enclosure, tried in vain to induce Keely to open it. Keely said the enclosure contained a secret device that he invented for a California gentleman to lift heavy weights. Although this invention was said to be perfected at the time, there is reason to doubt the accuracy of Keely's statement, as no record exists unto this day (1885) of the application of the machine."[10]

Even if this is a correct account we may probably never know what Keely was hiding from view, who the unnamed gentleman from California was or why he felt the need to keep this particular device secret. We could speculate that this "secret device to lift heavy weights" was in fact a completed and fully operational antigravity device. We could further conjecture that, since the device was "secret," of course no record of its use existed. As we will see in chapter 12, there are several reasons to attach more value to this slight but puzzling reference.

Whatever the true nature of this device that he carefully kept secret, in 1884 he obviously foresaw the ultimate possibilities that his experiments might yield: "...after my patents are taken out, I will devote the remainder of my life to aerial navigation, for I have the only true system to make it an entire success in the vibratory lift and the vibratory push-process."[11] The term "push-process" evolved out of his philosophy concerning gravity: "As to the 'law of gravity,' it appears in the light of Keely's experiments, but one manifestation of a law of very much wider application—a law which provides for the reversion of the process of attraction in the shape of a process of repulsion."[12]

But there, Keely's visionary mind did not stop; the force that propelled his engines was "the sympathetic attractive,—the force which, according to his theories, draws the planets together. While in his system of aerial navigation, should he live to perfect it, he will use a negation of this force,—the same that regulates the motion of the planets in their recession from each other."[13]

In 1887, a newspaper mentioned that Keely "has been making a flying machine...," a description unfortunately so vague that nothing can be learned from it, although the article published "a part of his explanation of its principle": "This machine will be capable of making a sympathetic outreach of a distance great enough about itself to not only neutralize the effects of gravitation, but to permit the engine and its equipment, no matter how heavy or heavily burdened, to keep it."[14]

More reports of his antigravity experiments before 1890 can be found in a small number of theosophical publications. Blavatsky, for instance, wondered what it was that "acts as the formidable generator of invisible but tremendous force, of that power which is not only capable of driving an engine of 25 horsepower, but has even been employed to lift the machinery bodily?"[15] Although Blavatsky devotes several pages to Keely and his theories, on the point of his antigravity research, such as his disintegration of dead tissue, she unfortunately is not specific.

R. Harte, another theosophist, is specific though. We've already seen how Harte suitably forgot to mention that his account of Keely's disintegration was published before, and again he is vague to the point of total absence in citing sources or circumstances. In 1888, the same year that *The Secret Doctrine* was published, Harte referred to a demonstration: "One of Keely's scientific experiments is to put a small wire around an iron cylinder that weighs several hundred weight, and when the force runs through the wire, to lift the cylinder up on one finger and carry it as easily as if it were a piece of cork."[16]

Five years later, the continuation of Harte's description found its way into Bloomfield-Moore's book about Keely, but without the graphic description about iron cylinders attaining the weight of "cork" or the mention of Harte as the source, even though she asserted that the demonstration was witnessed by Ricarde-Seaver. Ricarde-Seaver, who might also have been Harte's source, saw how "...Keely, by means of a belt and certain appliances which he wore upon his person, moved a 500 horsepower vibratory engine single-handed from one part of his shop to another. There was not a scratch on the floor, and astounded engineers declared that they could not have moved it without a derrick, which to bring into operation would have required the removal of the roof of the shop. Of course it is but a step in advance of this to construct a machine which, when polarized with a 'negative attraction,' will rise from the earth and move under the influence of an etheric current at the rate of 500 miles an hour, in any given direction. This is, in fact, Keely's airship."[17]

According to Bloomfield-Moore, Keely also either built a scale model of his airship or used a model of an airship that was not of his design and conducted experiments with it: "In demonstrating what seems to be the overcoming of gravity for aerial navigation, Mr. Keely used a model of an airship, weighing about eight pounds, which when the differentiated wire of silver and platinum was attached to it, communicating with the sympathetic transmitter, rose, descended, or remained stationary midway, the motion as gentle as thistledown floating in the air."[18]

Unfortunately she too, cites no source and no exact date, nor does she elaborate on the exact design of the airship model. As far as is known, no photo was ever taken of his model, or of what must have been a spectacular demonstration. From her statement it is not clear if she herself witnessed the demonstration, or if it was just hearsay. We are thus left dangling in mid-air, and only have her word for it that this demonstration actually took place as described. No newspaper referred to this experiment. Since this wonderful description was bound to stimulate the imagination, writers later copied this passage without mentioning Bloomfield-Moore as their source.[19]

Around 1889, Keely was still lifting objects of immense weight. When the enormous metal sphere was found that signaled the coming exposure, Jefferson Thomas stated that Keely had used it for one of those experiments, "over ten years ago" before Thomas' statement was published by a Philadelphia newspaper in 1899. He gave a short description, analogous to Harte's account: "He attached a small wire to the sphere and connected the other end of the wire with his hat or the lapel of his coat, or some other article or apparel, on his person, where he always had a small vibratory apparatus. He would then, by this force, lift the 6,000 pound sphere."[20]

While no description of this demonstration was published in the 1889 newspapers, 1890 would mark Keely's often-publicized experiments with the lifting of weights-in-jars that contained water,[21] which, contrary to the demonstration with his model airship, were photographed. As far as is known, Keely would never repeat his demonstration with the small scale model of an airship, although Bloomfield-Moore considered his weights-in-jars demonstrations as a repetition of that particular experiment.[22]

Perhaps the start of his weights-in-jars experiments in the spring of 1890[23] prompted Bloomfield-Moore to print a pamphlet in London the same year, suggestively entitled "Aerial Navigation," meant for distribution among shareholders in the United States. She had always been fascinated with the prospects of achieving flight. She knew early aviation pioneer Hiram Maxim, and Keely readily acknowledged that she helped steer the application of the results of his antigravity research towards a system of flight.[24]

The pamphlet did not contain clarifications concerning Keely's antigravity research. It merely stated that "...he has now turned his attention to aerial navigation, for which the force he has discovered can be used without an engine." The remainder consisted of reprints of articles about Keely and the text of the contract between him and Bloomfield-Moore. Collier, however, refused to distribute the pamphlet amongst the stockholders.[25]

Also in 1890, Professor Joseph Leidy witnessed Keely's experiment in raising a weight by striking a "sympathetic tone": "Dr. Leidy was asked this question: 'doctor, is it true that this unknown force has actually before your eyes overcome the force of gravity with which we are familiar?' And the answer, slowly, deliberately, was: 'I see no escape from that conclusion.'" Keely stated in plain language that he expected "to solve the problem of aerial navigation...for I can already move a weight up and down in the atmosphere, or even in vacuum. ...What is the force with which I do this? The same sympathetic, attractive force that holds the planets together."[26]

At the beginning of the demonstration, Keely attached one end of his platinum-silver wire to his transmitter. The other end was fastened in the metal cap that covered a glass jar. "...This was a glass jar such as chemists use, set on a heavy wooden table about five feet away from the transmitter. Upon examination, the jar proved to have a solid glass bottom. It was about 40 inches high and 10 inches in diameter. It was filled with what looked, smelled and tasted like Schuyikilli water. On the bottom of the jar, clearly visible through the water and resting directly on the glass, lay three iron disks, a two-pound, a one-pound and a half-pound weight, such as one sees in the scales of a grocery store. The weights were examined and weighed by several of those present, and were found to be what they purported to be."

Keely took "a bit of string out of his pocket," and wound it around the little brass spindle in front of the cylinder on the top of his transmitter, then "jerked the loose end and set the spindle whirring." Then he "sat down by the transmitter and began striking the strings of what looked like a harp in the cupboard-like base of the transmitter. ...While one hand was playing on the strings of this 'harp,' the other was moving tentatively about on the resonant rods on tops of the base of the transmitter, just beneath the cylinder where the spindle was whirring. Keely claimed that when the same note on both rods and harp strings was struck, the 'force was at that instant...generated, or set in action...' All depended on the exact note, for perhaps there was a false note. The mass of weights at the bottom of the jar of water only quivered: one of them toppled off and fell to the bottom. Then all remained motionless.''

He again sounded the "harp" in the cupboard and the resonant bars on top of it, explaining to Leidy that he was trying "to get the mass-chord of those weights. Every aggregation of molecules or of matter has a sympathetic chord, through the medium of which I can operate my vibratory force." The chord was not found for some minutes. Again the spindle was spun by the help of the twine and "its whiz was distinct in the silence of the room. The search for the mass-chord continued on the 'harp' and the resonant rods. A deep, clear note resounded from both at the same time, and at the instant it broke on the ear the weight that lay on the bottom of the jar by itself quivered and then slowly but steadily moved up through the water, as if impelled by some irresistible force, until it impinged on the top of the jar.''[27]

A year later in 1891, Keely once again confidently stated that, "I see no possibility of failure as I have demonstrated my theories are correct in every particular, as far as I have gone and if my depolarizer is perfect, I will be prepared to demonstrate the truth of disintegration, cerebral diagnosis, aerial suspension and dissociation, and to prove the celestial gravital link of sympathy as existing between the polar terrestrial and equation of mental disturbance of equilibrium.''[28]

But the next years would not see Keely's depolarizer perfected, as the only experiment in this line that he conducted over the years was not disintegration, nor his far more interesting experiment with a scale model of his airship, but his experiment with the weights and the glass jars. In 1893, Jacob Bunn, vice president of the Illinois Watch Company, stated: "There can be no doubt...that Keely believes in himself and the wonderful examples he gave us of making heavy steel balls moving in the air at his pleasure by simply playing on a peculiar mouth organ.''[29]

A reporter witnessed such a demonstration in 1895, and remarked that, "The absence of acceleration in speed seemed to preclude the idea of magnetism, on the supposition that the disks at the top and the bottom attracted each other...Tesla's alternating current motor, run by one wire, did not seem to offer any explanation of the phenomenon.''[30]

But Keely explained that the disks were "sensitized," rendered "alternately attractive and repulsive by energy drawn from the ether of surrounding space by means under his control." He also used the terms "polar" and "depolar," with the attractive disks being polar and the repulsive ones being depolar. The platinum wire was connected to the device that would produce the force. He claimed that on two occasions, when an observer removed the disk from the top of the jar, the suspended weights "instantly fell, crashing through the bottom of the glass cylinder.''[31]

In 1896, Keely gave a demonstration of what he called his "polar-depolar force" or "the interchange of polar and depolar sympathy." The account relates how "copper globes were made to revolve in response to the notes of a tuning fork or a zither." The witnesses were given a look at his still-unfinished "new motor," only superficially described as "a great mass of steel." He claimed that the motor would "neutralize the force of gravity," so that "a child will be able to pick it up and hold it at arm's length." He also stated that with such a device, "no bigger than your hand," "a street car crowded to the roof" could be run.[32]

It is at this point, and with all these experiments, that fact and legend have blended into a fantastic whole. When unraveled, the components show that some of the legend was based on facts, while we may probably never know if these facts led to the fulfillment of the legend. The publication of the weights-in-jars experiments, coupled with those slight references in the newspapers and the statements by Keely, Harte and Bloomfield-Moore, led to big expectations: For if small weights could be lifted, Keely said, "the suspension and propelling of an atmospheric navigator of any number of tons weight, can be successfully accomplished by thus exciting the molecular mass of the metal it is constructed of; and the vibratory neutral negative attraction evolved, will bring it into perfect control, commercially, by keeping it in sympathy with the earth's triune polar stream."[33] He added confidently that "Gravity is nothing more than a concordant attractive sympathetic stream flowing towards the neutral center of the earth."[34] Bloomfield-Moore painted her vista in plainer language, but delivered an even more inspiring outlook: "After he has gained as perfect control of it as we now have of steam, airships weighing thousands of tons can be raised to any height in our atmosphere, and the seemingly untraversable highways of the air opened to commerce."[35]

While Bloomfield-Moore left something to the imagination, a year after the publication of her book Keely travelled even further with his visionary concepts. Thus he provided Colville with a fantastic outline of his ultimate plans. Colville's *Dashed Against the Rock,* published in 1894, a year after Bloomfield-Moore's book, provided the reader with every detail of what such an airship might look like, and where it ultimately might travel. Keely went far beyond Bloomfield-Moore's vista; in *Dashed Against the Rock* the skies were definitely not the limit. Colville's writings on this subject were a clear reproduction of Keely's visionary ideas and based on "authentic reports of interviews" with Keely.

"The vessel in contemplation, the aerial navigator," Keely confided to Colville, "will be over two hundred feet long, over sixty feet in diameter, tapering at both ends to a point, made of polished steel, and will be capable of being driven under the power of depolar repulsion at the rate of three hundred miles an hour. It can be far more easily controlled than any instrument now in use for any phase of transit. Another very remarkable feature connected with this system of aerial navigation is that the vessel is not buoyed up or floated in the air through the medium of the air, so that if there were no atmosphere it would float just as readily. Hence, under mechanical conditions most certainly capable of production involving massive strength of resistance to interstellar vacuity, this can be made capable of navigating even the remote depth of space, positioned between planets where polarity changes are being controlled by other adjuncts of concentration for that purpose. Safely enclosed within this structure, a man possessing the chemical knowledge that these new laws give, and with sufficient supply of material from which to make oxygen, could travel to other planets in this system of worlds as easily as the same ship could navigate the depths of the ocean."

According to Keely, all this would be accomplished by "A small instrument, having three gyroscopes as a principal part of its construction." This was used to "demonstrate the facts of aerial navigation."

These gyroscopes were "attached to a heavy, inert mass of metal, weighing about one ton. The other part of the apparatus consists of tubes, enclosed in as small a space as possible, being clustered in a circle." The tubes represented "certain chords, which were coincident to the streams of force acting upon the planet, focalizing and defocalizing upon its neutral center. The action upon the molecular structure of the mass lifted was based upon the fact that each molecule in the mass possessed a north and south pole,—more strictly speaking, a positive and negative pole,—situated through the center, formed by the three atoms which compose it. No matter which way the mass of metal is turned, the poles of the molecule point undeviatingly to the polar center of the earth, acting almost exactly as the dip needle when uninfluenced by extraneous conditions, electrical and otherwise. The rotation of the discs of the gyroscopes produces an action upon the molecules of the mass to be lifted, reversing their poles, causing repulsion from the earth in the same way as poles of a magnet repel each other. This repulsion can be diminished and increased accordingly as the mechanical conditions are operated. By operating the three discs, starting them at full speed, then touching two of them so as to bring them, according to the tone they represented by their rotation to a certain vibratory ratio, the weight then slowly sways from side to side, leaves the floor, rising several feet in the air, remaining in that position, and as the discs gradually decrease their speed of rotation the weight sinks to the floor, settling down as lightly as thistle-down."[36]

Keely's assurances and prospects incited the imaginations of Bloomfield-Moore, Colville and undoubtedly everybody else who read their books. But while his assurances and prospects still had to be realized, other things were giving rise to the legend. There exists for instance a photograph, taken in December 1889, of what Keely called a Planetary System Engine. No information on this device is available, and considering his constant flow of avant-garde ideas as well as the name of the device, it is inviting to speculate that it was part of his airship, perhaps even constructed for travel through space. If so, this would be a unique feature in itself, for Keely would have been the only person who at that time had actually built an engine for space travel. There is a delightful but in every aspect equally unsubstantiated rumor that the planetary system was some sort of atrolabium; the device would demonstrate the laws and forces holding the solar system together.[37]

Apart from allegations that Keely constructed a secret device to lift heavy weights for a man in California, he did construct a large and magnificent apparatus that he called the "aerial propeller."

The aerial propeller was meant to be an engine for an airship. There are no indications of the time period in which he was involved in the construction of his wonderful aerial propeller, but possibly around 1887 he started the construction of this device, since at that time a newspaper remarked that he "has been making a flying machine," which may have meant his aerial propeller. In 1894 Keely wrote to Bloomfield-Moore that "My strength has not reached the point whereby I can intensify myself to that degree requisite for the final sympathetic graduation on our aerial propeller, but I am rapidly reaching that most necessary condition, and am quite positive that by the first of June, I will be in perfect state to engage in this most fascinating work which I intend to push forward with my utmost vigor to completion."[38]

But in 1895 Keely was still engaged in the construction of the aerial propeller and stated that he considered his device to be "not yet in working order."[39] In 1897 Bloomfield-Moore wrote that "He still has to devise an instrument that will establish the connective link in the entire combination, whether in the terrestrial engine, or in the aerial propeller, before these machines can become obedient to the governing law of nature."[40]

More is known, although not much, about this aerial propeller that was as mysterious as its inventor: "The flying machine and the apparatus for lifting heavy weights were in the back of the second story room. ...The flying machine however, had no moving parts, but was used as an awe inspirer, Keely's bloodcurdling accounts of its capabilities being accompanied by an occasional unexpected stroke of an ear splitting gong, which formed a part of its equipment."[41]

The reporter who witnessed Keely's weights-in-jars demonstration, and who had looked with "considerable awe upon a machine filled with resonating tubes, which was said to be capable of lifting an immense weight by striking the proper chord," also studied Keely's aerial propeller. He looked with "some apprehension upon the engine built to navigate an airship, whose wheels were not made to revolve, but only to be set to certain combinations. It is not to be wondered at, therefore, when Mr. Keely struck the gong on one end of its shaft, which he said was the keynote of the machine, that the reporter shuddered at the thought that he might sail out into space on that machine without knowing its combination."[42]

Other accounts provide us with more details of this incredible device: "Keely is at present engaged on a large and complicated piece of machinery which he calls the propeller of an airship, and whose purpose is sufficiently indicated by its name. It is supplied with three 'resonators,' or broad sheets of thin metal, each set to a particular pitch, and each playing some part in the mysterious process of drawing power from space. In addition are sundry disk-shaped and other apparatus, the whole arrangement being highly complex."[43]

Perhaps the most elaborate description is the one that was given by Bloomfield-Moore: "Each disk of the polar and depolar groupings in the propeller of the airship contains seven pints of hydrogen. In preparing these disks, the hydrogen is submitted to a triple order of vibration. ...The space which the propeller of the airship occupies in Keely's laboratory comes within a radius of six feet square. A small space for so powerful a medium—distributing over one thousand horsepower, as tested by experiment."

According to her, "The aerial propeller consisted of more than two thousand parts, the most important of which were a positive graduating chladna that was guided by 'polar action toward the north and reversing by depolar action,' a sympathetic polar negative transmitter for operating and controlling 'the action of the machinery in producing polar and depolar power,' thus 'liberating the latent sympathetic power in twenty seven sensitized disks.'"

Other important parts were the polar and depolar intermittent accumulator that carried eight "focalizing disks" which received and distributed the sympathetic polar negative flow, a positive ring that was suspended on a small shaft with three points, meant to "preserve the integrity of the neutral center of the machine," and two resonating drums, one positive and one negative, multiplying the intensity of the sympathetic flow. Other parts of Keely's remarkable aerial propeller included 27 depolar triple groupings, nine in each grouping, consisting of three vitalized disks with resonators, replying "sympathetically to polar and depolar action," and a large polar ring that was "associated with the central resonators by nine resonating polar disks placed at equal distances." This, according to Bloomfield-Moore, was the medium for distributing the polar flow. The last part that she found important enough to describe, was a small negative ring, which was "the governor of the propeller, associated with a polar bar that oscillates from the polar field to the depolar field."[44]

While this may serve as an example of language struggling to capture concepts of a most avant-garde nature, Bloomfield-Moore also left us with a description of the steering-device of the proposed airship: "The instrument devised by Mr. Keely for bringing the airship under control in its ascend and descent, consists of a row of bars, like the keys of a piano, representing the enharmonic and the diatonic conditions. These bars range from 0 to 100. At 50 Mr. Keely thinks the progress of the vessel ought to be about 500 miles an hour. At 100 gravity resumes its control. If pushed to that speed it would descend like a rifle-ball to the earth. There is no force known so safe to use as the polar flow. Mr. Keely thinks that when the conditions are once set up, they remain for ever, with the perpetual molecular action as the result, until the machinery wears out. In the event of meeting a cyclone, the course of the vessel, he teaches, can be guided so as to ascend above the cyclone by simply dampening a certain proportion of these vibratory bars. The instrument of guiding the ship has nothing to do with the propelling of it, which is a distinct feature of itself, acting by molecular bombardment; moving the molecules in the same order as in the suspension process, but transversely. After the molecular mass of the vessel is sensitized, or made concordant with the celestial and terrestrial streams, the control of it in all particulars is easy and simple. In ascending the positive force is used, or the celestial, as Keely has named it, and in descending the negative or terrestrial. Passing through a cyclone the airship would not be affected by it. The breaking up of cyclones will open a field for future research, if any way can be discovered for obtaining the chord of mass of the cyclone. To differentiate the chord of its thirds would destroy it."[45]

Notwithstanding these descriptions, we will probably never know with any certainty what his whole system of aerial navigation consisted of, or what equipment it embraced. Keely also designed and built other devices to be used in his system of aerial navigation, such as his commercial engine of 1893-1896 that was never completed but was also intended for aerial navigation.[46]

In the visionary fever of all these remarkable descriptions and slight references, it is easy to forget that we cannot actually say whether Keely actually built or completed a full-sized airship that was capable of flight, and whether his entire system of vibratory flight and his aerial propeller were ever used beyond the experimental stage. Contemporary sources are vague and ambiguous on this point. Although Bloomfield-Moore's writing may sound like a suggestion that something of an actual, full-sized airship might exist, she also admitted that, "Not until the aerial ship is in operation will the world be able to comprehend the nature of Keely's discoveries."[47] In 1895, Keely declared that his system of aerial navigation still had to be "perfected and simplified before any patents can be taken out on the propeller of the airship."[48]

A typed, unsigned contract in the possession of Bloomfield-Moore, dated September 24, 1895,[49] but apparently signed by Keely "at the time John Jacob Astor and New York friends were talking of securing a large interest" in the Keely Motor Company,[50] also demonstrates that Keely was still trying to perfect his antigravity device towards the end of that year: "It is my desire to give to the world the benefit of my discoveries at the earliest possible moment, without waiting to protect myself by patents beyond what is due to the Keely Motor Company shareholders and to my interests in aerial navigation, the system of which has to be perfected and simplified before any patents can be taken out on the propeller of the airship. In accordance with the wishes of my co-worker, H.O. Ward—without whose assistance this system would never have had birth, and who has lighted up as by a pillar of fire all my dark and intricate paths—I have proposed to instruct Mr. Zak Samuels[51] in the taking up of dead lines and in the sensitizing of metal, so that it is acted upon by radiation on the same principle as in the human brain. As soon as I am ready to give this instruction I will notify Mr. Samuels (who is the only man I will have time to instruct) and this at the earliest possible moment."

Another confirmation comes from a theosophical source; in 1896, occultist W. Scott-Elliot while writing on Atlantis and its fabled vimanaas, digressed on Keely's proposed airship. The most important detail is that he too admitted that Keely still had not mastered his method of flight: "...the Atlantean methods of locomotion must be recognized as still more marvelous, for the airship or flying machine which Keely in America, and Maxim in this country are now attempting to produce, was then a realized fact."[52]

Based on such sources, and the only ones available today in which we can trace his antigravity ideas, we find that as of 1897, Keely was still trying to perfect his aerial propeller and did not have his system of aerial navigation in working order.

We are left to speculate as to whether he ever succeeded in getting his system of aerial navigation in working order, or if perhaps others completed his work. Whether Keely's ship was witnessed in his time, glittering in the sunlight as it traversed the very skies, is now open to conjecture. We find no trace of it in the surviving sources directly relating to him. It is possible that he carefully kept results of this line of research secret, as he had done with the alleged device for the man in California. Coincidentally, 1896 would be the beginning of unusual sightings of strange and never identified aeroforms in the skies over large parts of America that lasted well into 1897. Their origins have always been as enigmatic as the inventor from Philadelphia.

Some months after his death, a magazine revived a passing memory of what was perhaps the greatest of Keely's ideas: "It should not be forgotten that Keely predicted that the flying-machine, for which the century has been waiting, would come to pass the moment his motor stood completed."[53]

Nevertheless, it had been an original and innovative concept and such a legendary endeavor! And in time, legend it would become. Keely's antigravity experiments would be adorned with all the qualities and distortion that legends are made of. And so it became written, although suitably vague, that Keely "...successfully demonstrated a gravitational device to leading scientists of that time,"[54] probably meaning but without saying so, Keely's weights-in-jars demonstrations as witnessed by Prof. Leidy and others.

Sometimes the facts would be mixed together into an entirely different story: "In November 1874 John Worrell Keely exhibited in Philadelphia a fuel-less motor without any apparent source of power, now believed to have been a novel, antigravity device; he was said to have been mysteriously silenced, his invention destroyed."[55]

Recently it was written that he "demonstrated a free-energy flying platform to the bewildered representatives of the American War ministry in 1862...he could get out of his contraption free-energy propulsion..."[56]

Notwithstanding the enthusiasm of these claimants, it needs to be pointed out here that this time period is adequately reconstructed in chapter 1, and that Keely at that time was still a year away from his work on an engine that he called a "reacting vibratory motor." As far as is known, he never demonstrated a free-energy flying platform to United States representatives.

The allegation that in 1896 he demonstrated an aerial craft on an open field to representatives of the United States War Department and several members of the press is relative to time periods more on the mark, but based on the documentation that is at hand, unfortunately equally unfounded and at best speculative.[57] We have seen, based on contemporary sources, that Keely was still perfecting his system for aerial navigation in 1897, and 1896 would see no accounts in the press of such a demonstration, which, considering the state of affairs in the Keely history, would certainly have drawn headlines. Then again, the very same year would see the beginning of the great 19th century airship wave, during which thousands of Americans claimed to have seen those inexplicable aeroforms often equipped with bright lights. We will see in chapter 12 how this strange UFO wave is connected to Keely in a very subtle way.

What Keely did unknowingly accomplish was to become a forerunner and practical exponent of those who stood at the very base of antigravity research. While I have not discovered in the texts of these early theoreticists any proposition of applying or using their various principles for a system of flight—which sets Keely apart from this coterie—their search for a different approach of the phenomenon of gravity and the booklets that resulted from it form the theoretical foundations of what has become the quest for antigravity. This theoretical underground was a small, exclusive group that denied that gravity was a pulling force. It was in fact a pushing force, a quality that Keely had always maintained could be delivered from gravity.

Not much is known of this line of thought, except that its members were individuals. As far as I have been able to ascertain, they did not belong to any organized whole, although by naming them as a group might suggest otherwise. What they did was often refer to each other's writings, and by lack of huge amounts of like-minded studies to draw upon, sometimes traversed into the areas of fantastic literature where such concepts as antigravity or the manipulation of gravity was often used as a literary vehicle.

One of the first instances of the manifestation of this line of thought in printed form seems to be an obscure booklet published in 1905 in Holland called "The Push Force,"[58] by A.A.C. Belinfante. In it, Belinfante denies the theory of gravitation as we know it and substitutes this theory with the concept of "ether pressure from above."

"One may understand," writes Belinfante, "that the ether in the universe is no pulp, no fluid and no solid substance but the storage of all force."[59] In Belinfante's theory, the ether-atom is "forty times smaller than the hydrogen-atom,"[60] a viewpoint remarkably similar to Keely's, and ether is to be found in abundance at the poles, since Belinfante considers gravity on the earth as "the pressure of ether on the earth, reduced with the pressure of the centrifugal force against the ether-pressure."[61] And since the centrifugal force at the poles is "zero," it is there that the ether flows to: "Much ether penetrates the poles, which in the center of the globe is being intensified through the pressure of both the poles," Belinfante writes.[62]

Possibly Belinfante's endeavor was fueled by news of Keely's antigravity experiments, as he formulated his unusual theories concerning the nature of gravity. While Belinfante fails to mention Keely in his text, information about Keely and his antigravity experiments certainly was available in the Dutch language from 1893 through theosophical sources.[63] Belinfante writes that he began to declare his theory, verbally at first, in 1894 and subsequently in a small publication in 1897 which he sent to Dutch professors.[64] Also, a Dutch translation of Bulwer-Lytton's *The Coming Race* appeared 20 years before. That strange tale of vril, the exotic source of tremendous power which propelled flying machines and wings,[65] and which had a strange connection to Keely's work, we will treat in the subsequent chapter. In addition, occult initiate Rudolf Steiner, who was quite impressed by Bulwer-Lytton's novel, and who had much to say about Keely on several occasions, lectured in Holland in 1904.[66]

What makes it likely that Belinfante might have read or learned of Keely's theories about gravity, possibly through Blavatsky's or Bloomfield-Moore's writings, is that Belinfante was a close friend of J.K. Rensburg, who was deeply immersed in occult doctrine. Rensburg was a brilliant writer with opposing leanings who pondered over the meaning of the Grail, wrote about the existence of god-like beings on other planets and about radio-contacts with Mars. He referred to Belinfante's booklet in the foreword to a collection of poetry about Atlantis published in 1923.[67]

In connection with Belinfante's "Push Force" theory, Rensburg also mentioned Blavatsky's *The Secret Doctrine,* in which Rensburg, and through him possibly Belinfante, certainly learned of Keely. What makes this more plausible is that Rensburg, as he himself wrote, had "borrowed the idea of the existence of higher, superhuman beings" from Belinfante, an idea also present in both Blavatsky's and Bulwer-Lytton's writings. What Rensburg also wrote about was the booklet "There Is No Gravity" by Hans Goldzier, who penned it down under the nom-de-plume of "Th. Newest,"[68] clearly meant as a parody on Newton. Newest in turn referred to other booklets on the same subject, while supporting his theory.[69]

But apart from the theoreticists of antigravity, there is another stratum, another strange trail through which we may add to our knowledge of what Keely was ultimately aiming at. Aside from being a forerunner and practical exponent of a philosophical current that searched for alternative theories concerning gravity, he also influenced several writers of imaginative fiction of his time, a strain that in some instances is easy to discern. It is likely that one of those writers, Percy Greg, was influenced by news of Keely's researches, and in turn influenced William Colville and possibly Keely.

At the time that Greg's book was published, Keely's antigravity research was in its early stage, and news of his experiments or futuristic extrapolations would still be some eight years away. But in the light of the incredible 19th century social potpourri with its unique fin de siècle flavor that Keely and the writers were living in, and in which science, literature and occultism often met and sometimes mingled in a diffuse whole, the connection to Keely is certainly there in the instance of the book *Across the Zodiac*, written by Percy Greg (1836-1889). The book was published in 1880 in a two-volume edition by the same London firm that coincidentally published Bloomfield-Moore's book about Keely 13 years later.

The story of *Across the Zodiac* is considered to be an important early interplanetary novel that also features an ideal society,[70] and the book was destined to popularize the Mars theme in the early science fiction stories. Science fiction historian Sam Moskowits writes: "Never before had a space voyage ventured so far into the galaxy based on such sound scientific evidence. ...The book's major status derives from its detailed, meticulous exposition of some important scientific theories and speculations of the day, and to its explanations of the scientific rationale behind these ideas."[71]

Although it is stated that it is likely that Greg borrowed the concept of antigravity either from the 1863 book *A Voyage to the Moon* by Crystostum Trueman, or the 1827 book *Voyage to the Moon* by George Tucker,[72] the very use of the word "apergy," together with the graphic descriptions of the space vessel, the method of its propulsion and a very detailed description of the steering apparatus in Greg's book, point towards a close connection between Greg and Keely's avant-garde ideas. Interestingly, science fiction historian E.F. Bleiler notes the influence of Bulwer-Lytton's *The Coming Race* on Greg's book.[73]

Aside from uncovering the fact that a connection between Keely and Greg exists, it is impossible to establish the direction of the currents of influence with any exact degree. The trouble encountered while sorting out this genealogy of influences is with the usage of the term "apergy," a term that was applied for the first time in Greg's novel and which also became grafted on Keely's force.

Greg only vaguely writes that the word is derived from the Greek words "apo" and "ergos," that he translates as "work—as energy."[74] He further states that, although there were those who possessed the "secret of apergy," they had never "dreamt of applying it in the manner I proposed," meaning, as a source of antigravity to be used for space travel. "It had seemed to them little more than a curious secret of nature," writes Greg, "perhaps hardly so much, since the existence of a repulsive force in the atomic sphere had long been suspected and of late certainly ascertained, and its preponderance is held to be the characteristic of the gaseous as distinguished from the liquid or solid state of matter. Until lately, no means of generating or collecting this force in large quantity had been found. The progress of electrical science had solved this difficulty; and when the secret was communicated to me, it possessed a value which had never before belonged to it."[75]

This description brings Keely's procedures for the obtaining of vaporic substance to mind. Considering the time frame, Greg could certainly have read or heard about Keely's researches. *Across the Zodiac* contains 584 pages divided between two volumes, but the above passage is all that Greg chose to write about the force "apergy." Although Greg never refers to Keely, the passage could easily fit a description of his research.

From Greg's use of the very word "apergy" we also get none the wiser; while it is not readily discernible in Keely's jargon, and a letter to a newspaper about Keely's force stated that he "didn't call it 'apergy,'"[76] we read elsewhere that he sometimes did call his etheric force apergy,[77] although the only indications in contemporary sources lead us to assume that this happened no sooner than or around 1895. According to one newspaper account of a visit made to Keely on November 9 of that year, the "moving force" was referred to as "apergy."[78]

The term also appeared in a letter by Bloomfield-Moore published in a newspaper in 1895.[79] The term apergy became grafted on Keely's force and became associated with his discovery in more recent times.[80] But Bloomfield-Moore's letter was written during the time that she and Keely were negotiating with her acquaintance Astor, and it was from Astor that she obtained the peculiar term. She wrote: "John Jacob Astor stands alone as having penetrated nature's secret, when he named the counterpart of gravity, or its dual force as the force to be used in aerial navigation. Mr. Keely once gave me permission to name it and I propose to take Astor's name 'apergy,' and immortalize him if Keely contends."[81]

An article about Keely, also published in 1895 and titled "Apergy: Power Without Cost," again explaining the term as "the reverse counterpart of gravitation," also referred to Astor's *A Journey In Other Worlds*[82] - not surprisingly, since its writer had given Bloomfield-Moore the proof of his article that was originally titled "Keely's Discoveries." But, as she writes, "I have requested him to change (the title) to 'Apergy: Power Without Cost.'"[83]

It might very well be that Astor in turn obtained the term by reading *Across the Zodiac*, since science fiction historian Moskowits states that Greg's book indeed was an influence on Astor as witnessed in his use of the term "apergy."[84]

Greg's fictional *Across the Zodiac* conveniently begins in September 1874 with a visit to a spiritist seance in New York, just a month before Keely would give his historical demonstration in Philadelphia that led to the foundation of the Keely Motor Company. In Greg's tale, an Englishman received a coded manuscript found on an island somewhere in the Pacific. The manuscript was in a canister among the wreckage of an extraterrestrial object that struck the island with enormous force. The writer of the manuscript discovered the secret of apergy, here used as a repelling force that amounts to antigravity. He builds a space vessel called the "Astronaut" that is much like a ship. The space vessel has special features for maintaining life in space and chemicals for air supply: "On the further side of the machinery was a chamber for the decomposition of the carbonic acid, through which the air was driven by a fan. This fan itself was worked by a horizontal wheel with two projecting squares of antapergic metal, against each of which, as it reached a certain point, a very small stream of repulsive force was directed from the apergion, keeping the wheel in constant and rapid motion,"[85] plant life for purifying the air, a thick hull to protect the passengers against the extreme cold of space, etc.

The inventor then travels through the solar system with the help of the antigravity device that operates on apergy. On Mars, he meets its inhabitants, the "Martials,"as Greg calls them. These Martials have electrically powered, wheeled carriages that move along paved streets. There are black and white, and color photocopy machines, microfilm, telephones, plastics, conversion of waste to fertilizer, electrically driven balloons, boats propelled by a jet expulsion of water that are capable of submerging, telegraphs for communication, gas guns and a wealth of other technological inventions.[86] At the heart of the novel lies a secret society; the inventor learns of the existence of this society variously called "The Children of the Star," "The Children of Light" and "The Order of the Silver Star," into which he is initiated during a stunning ceremony. Many of centuries of study have provided its members with telepathy and the means of killing by thought force if their secrets are divulged to the uninitiated. The book is riddled with esoteric and mystical philosophies.[87]

Some passages from *Across the Zodiac* that describe the propulsion, the machinery and the steering method of the space-vessel are worth quoting, especially when compared to Bloomfield-Moore's and Colville's writings on the airship that Keely was contemplating: "I needed a repulsion which would act like gravitation through an indefinite distance and in a void - act upon a remote fulcrum, such as might be the Earth in a voyage to the Moon, or the Sun in a more distant journey. As soon, then, as the character of the apergic force was made known to me, its application to this purpose seized on my mind. Experiment had proved it possible...to generate and collect it in amounts practically unlimited. ...But the vessel had to be steered as well as propelled; and in order to accomplish this it would be necessary to command the direction of apergy at pleasure. My means of doing this depended on two of the best established peculiarities of this strange force: its rectilinear direction and its conductibility. We found that it acts through air or in a vacuum in a single straight line, without deflection, and seemingly without diminution. Most solids, and especially metals, according to their electric condition, are more or less impervious to it - antapergic. Its power of penetration diminishes under a very obscure law, but so rapidly that no conceivable strength of current would affect an object protected by an intervening sheet half an inch in thickness. On the other hand, it prefers to all other lines the axis of a conductive bar, such as may be formed of (indecipherable) in an antapergic sheet. However such bar may be curved, bent, or divided, the current will fill and follow it, and pursue indefinitely, without divergence, diffusion, or loss, the direction in which it emerges."[88]

"Therefore, by collecting the current from the generator in a vessel cased with antapergic material, and leaving no other aperture, its entire volume might be sent into a conductor. By cutting across this conductor, and causing the further part to rotate upon the nearer, I could divert the current through any required angle. Thus I could turn the repulsion upon the resistant body (sun or planet), and so propel the vessel in any direction I pleased. ...[89] In the center of the vessel was the machinery, occupying altogether a space of about thirty feet by twenty. The larger portion of this area was, of course, taken up by the generator, above which was the receptacle of the apergy. From this descended right through the floor a conducting bar in an antapergic sheath, so divided that without separating it from the upper portion the lower might revolve in any direction through an angle of twenty minutes. This of course, was intended to direct the stream of the repulsive force against the Sun."[90]

Greg also gives a very detailed description of the steering device of his ship, which, when compared with Colville's description of his "small instrument with the three gyroscopes,' again yields remarkable similarities in thought and concept: "My steering apparatus consisted of a table in which were three large circles. The mid-most and left hand of these were occupied by accurately polished plane mirrors. The central circle, or metacompass, was divided by three hundred and sixty five lines, radiating from the center to the circumference, marking as many different directions, each deviating by one degree of arc from the next. This mirror was to receive through the lens in the roof the image of the star towards which I was steering. While this remained stationary in the center all was well. When it moved along any one of the lines, the vessel was obviously deviating from her course in the opposite direction; and, to recover the right course, the repellent force must be caused to drive her in the direction in which the image had moved. To accomplish this, a helm was attached to the lower division of the main conductor, by which the latter could be made to move at will in any direction within the limit of its rotation. Controlling this helm was, in the open or steering circle on the right hand, a small knob to be moved exactly parallel to the deviation of the star in the mirror of the metacompass."

"The left-hand circle, or discometer, was divided by nineteen hundred and twenty concentric circles, equidistant from each other. The outermost, about twice as far from the center as from the external edge of the mirror, was exactly equal to the sun's circumference when presenting the largest disc he ever shows to an observer on Earth. Each inner circle corresponded to a diameter reduced by one second. By means of a vernier or eye-piece, the diameter of the Sun could be read off the discometer, and from his diameter my distance could be accurately calculated."[91]

A year later, Greg's book was followed by another, equally strange novel, written by William Delisle Hay (?), titled *Three Hundred Years Hence or a Voice From Posterity,* that like Greg's novel, was published in London. Several of its themes also show a similarity to Greg's *Across the Zodiac,* and to Keely's concepts. But little is known of Hay other than that he once lived in New Zealand, and wrote two scholarly treatises on British fungi,[92] and again, the personal details yield no clues as to any direct connection. Of Hay's book, opinions differ widely. The story may be looked upon as a "...solid prediction of the future, firmly based on the advance of technology..." making "more changes to the state of the human race than practically any other work of the 19th century..."[93] and as "...seriously intended prediction,"[94] but also as a "ruthless Spencerian survival of the fittest." [95] This is not so strange, since the elements that are of interest here are imbedded in a story of terrible wars, overpopulation and its gruesome solution: a One World Order and the extermination of whole races.[96]

A minor element in the story that serves as a roadmark towards the influence of Keely is Hay's description of a small, compact, easy and cheap-to-build device that requires no fuel. Called the "mechanical friction machine," it replaces coal for all domestic and industrial purposes.[97]

A major element of influence is Hay's description of the discovery of a new force called ballisticity, or the Basilic Force, that links vitality and matter, permitting direct mental and organic control of light, heat, motion, electricity, magnetism, chemical affinity and attraction. While Hay's novel features huge submarines and diving suits, earth tubes and subterranean earth-borers, the tapping of the geothermal heat and the warming up of the poles and the founding of enormous subterranean cities,[98] his descriptions of numerous types of airships and flying machines are its most interesting detail.

These airships are propelled by four distinctly different systems of motive power. One of these systems employs lucegen, an astonishingly lighter-than-air gas that alters its volume, and thus its buoyancy, by a factor of 10 when an electric current is passed through it. Canopies of balloons contain this extremely light gas; the canopy is positioned below instead of above the carriage: "The car was thus immersed, as it were, in a bladder covering it externally but leaving it open above; it sat in its balloon just as it might in water."

Hay foresaw that such a design might be unstable, so in order to prevent such a balloon from toppling over, powerful magnets are used that are attracted to the Earth's magnetic field. Stabilized, the "Lucegenostats" are able to carry large weights of freight or passengers. Yet these Lucegenostats are surpassed by even more powerful aerial machines, working on the three other systems. Their lift is being provided by basilicomagnetism for greater power and safety. The propellant comes from "generated heat and electricity," which causes a pair of fans, modeled on wings of birds, "extended along the sides of the craft from stern to stern" to flap. This type is known as the "alamotor," and is used for small, utilitarian crafts such as those used for agricultural purposes. The "spiralometer" is driven by one or more propellers of the "pusher" variety, normally placed at the stern. "Heat and electricity give the motive power, and this form is the most generally employed on aircraft," Hay writes.

The other airships, able to lift any weight, use a new and "utterly unsuspected force," which Hay names zodiacal force, clearly meant as a hint to Greg's *Across the Zodiac*. Hay describes the zodiacal force as "that which holds together the elements of the air." Driven by Zodiamotors, being powered by Zodiacal electricity, the airships travel a thousand miles an hour or faster, "though seldom employing more than half that rate."[99]

In 1882, a year after Hay's futuristic visions and just a short distance from Philadelphia, a most remarkable book was privately printed. The title of the book was *Oahspe*, and it would in many respects prove as puzzling as the Philadelphian inventor. The history of its conception was as outré as its contents. *Oahspe* was "automatically" written in the course of one year by John Ballou Newbrough (1828-1891). He was not a lonely literator who was starving in a garret, instead he was no stranger to the occult communities of the time. Newbrough had traveled widely in Europe and the Orient, lecturing to spiritualist groups while "wearing brightly colored Oriental robes."[100] His fame as a medium and automatic writer spread and, consequently, legends sprang up around him. He was said to be able to paint pictures in total darkness, using both hands in the process. It was said that he could close his eyes and read any book in any library. As with Keely, many claimed that Newbrough possessed some means to overcome gravity; he could easily lift weights of more than a ton. It was also claimed that his astral body was able to visit any place on earth.[101]

Newbrough was active as a medium for two years in The Domain, a small spiritualist colony in Jamestown, New York. He also acquired the ability to write automatically in long-hand. All this occult activity would have its repercussions on him. One night in 1870, he was visited by what he described as wingless angels, bathing in a soft light. A voice told him that he was destined for a special mission. Ten years later he was again awakened by the mysterious light, and the voice commanded him to buy a typewriter. Angels, the voice assured him, would guide his fingers while he typed. These automatic typewriting sessions began in the early morning of January 1, 1881, and would last until December 15 of the same year.[102]

After fifty weeks, in which Newbroughs' angels controlled his fingers for half an hour before sunrise, the controls finally stopped. He was told by the angels to read for the first time what he had written, and then to publish it as a book, entitled *Oahspe*. Newbrough printed the book himself in Boston in 1882 on a press bought with money from seven anonymous associates. No author's name was on the title page. A second, revised edition appeared in 1891, which was reprinted in London in 1910.[103]

Oahspe features an elaborate and complex cosmogony, with thousands of millions of gods, huge spiritual realms and higher dimensions. In this vast hierarchy, there are also countless numbers of heavens, one of these called Etherea. All heavenly bodies, suns, planets, and moons arise from and are sustained by rotating vortices of space-time, so reminiscent of what Keely termed "the celestial forces."

Mutual influence or similarity between the two we will probably never know. Perhaps *Oahspe* was handed over by that other medium and Keely's friend, Colville. Newbrough's manuscript of *Oahspe* was destroyed in a flood, and his social life during the years that he visited spiritualist circles and traveled widely before he wrote *Oahspe* was never the subject of any in-depth study. What is interesting in connection to the intellectual framework in which Keely lived, conducted his antigravity experiments and formulated his theories concerning gravity, is that a similar train of thought can be discerned in *Oahspe*: "Things fall not to the earth because of the magnetism therein...but they are driven toward the center of the vortex, by the power of the vortex."[104] And what Keely called the "sympathetic attractive, or the force that draws the planets together" and in another application would regulate the motions of the planets "in their recession from each other" is named by Newbrough "the force of the vortex" or "vortexya."

But Keely also admits that "All such experiments that I have made...resulted in vortex motion invariably, both sympathetically and otherwise. Vortex motion follows nature in all corpuscular action."[105]

Newbrough also mentions ether, but calls it "ethe": "There are two known things in the universe: ethe and corpor."[106] *Oahspe* further adds that "Attraction...existeth not in any corporeal substance as a separate thing. There is no substance of attraction. Nor is there any substance of gravitation. These powers are the manifestation of vortexya."[107] And Newbrough, or his angelic guides, warns sternly that to withdraw the vortexian power would be catastrophic; "the earth would instantly go into dissolution."[108] Keely gravely states that, "It is the sympathetic attractive force which keeps the planets subservient to a certain range of motion, between their oscillations. If this condition were broken up, the rotation of the planets would cease; if destroyed at a given point of recession, all planets would become wanderers, like the comets; if destroyed at another given point, assimilation would take place...meeting, would fuse into one mass."[109]

Newbrough's universe in *Oahspe* is not a cold and lonely place, forever stretching itself blindly and meaninglessly out into the void; angels in nonmaterial Etherean ships inhabit the universe. One ship is almost as large as the earth. Some are propelled by vibrations of colors, others by musical vibrations,[110] a concept that is also familiar in Keely's thoughts on aerial navigation.

But in this framework, this stratum of spiritualists wanting to escape the horror vacui of a void universe, of occultists, theosophists, early theoreticists of antigravity and writers of imaginative fiction, we find another minor connection between Newbrough and Keely. Beside the similarities in thought—although each expressed these in a unique, personal jargon—the geographic surroundings that Keely and Newbrough shared, and the obvious rubbing of shoulders that at one time or another could have occurred, the link is found in a spiritualistic magazine. This magazine was *Banner of Light*, one of the leading spiritualist publications of that time. Newbrough himself published the account of how he came to write *Oahspe* in a long letter in *Banner of Light* dated January 21, 1883.[111]

Keely's friend William Colville who was a prolific writer and like Newbrough a spiritualist, also submitted his writings to this magazine. And like Colville, Newbrough was a lifelong freemason. When Newbrough died in 1891 of pneumonia, his associate was in Boston, overseeing the second printing of *Oahspe*. Newbrough died before the book left the press. His body was interred in a Masonic cemetery in Las Cruces.[112]

In 1890, a year before Newbrough died, Robert Cromie's fictional *A Plunge into Space* was published. The second edition featured a foreword by Jules Verne, the only foreword that Verne ever wrote. In the book, Henry Barnett discovered how to control the ethereal force after 20 years of experimenting, "which permeates all material things, all immaterial space"[113] and that combines electricity and gravity: "...I have discovered the origin and essence of that law which, before me, never man did ought but name, or, at best, did but chronicle its known effects—the law which makes that universe of worlds a grand well-ordered army instead of a helpless mob of mutually destroying forces; when I tell you that within this ragged room, there stands a man who—grant him but ten years of human life—could sway a star in its course, could hurl a planet from its path? Man, I have discovered the mightiest secret of creation. ...I have discovered the origin-of-force!"[114]

The influence of Greg's *Across the Zodiac* on Cromie's novel has been noted.[115] But Cromie (1856-1907), a Northern Irish writer and journalist, possibly through his profession, also read about Keely's weights-in-jars experiments, which received widespread press coverage the same year that the book was published, and passages in *The Secret Doctrine* that was published two years before Cromie's novel, in which she mentions Keely's disintegration experiments. These accounts impressed Cromie to a considerable extent, and as a consequence we find his borrowing of these actual experiments in his fictional novel. Take, for instance, the scene in which an unfortunate native meets a horrible end: "The Indian took the metal bar. ...His hand closed on the bar and instantaneously his frame drew up rigid. For a second or two he stood stiff and deathlike, and no man spoke. Then before the eyes of the horror-stricken crowd, the man's body sank down into a shapeless mass of pulp." Apparently one of Barnett's devices "almost wholly destroyed the attraction of cohesion in the man's body."[116]

Cromie's description of an antigravity experiment also bears an uncanny resemblance to Keely's weights-in-jars demonstrations: "The ball, a tiny sphere, was lying in a tube of glass. This tube stood in an upright position on a plate of that strange gray material. ...Barnett laid the end of a thin wire on the gray substance. The ball within the tube flew to its upper end and remained there, suspended, as it were, by a magnet. ...The experiments with the glass tube show that the law of gravitation may be diverted, directed, or destroyed."[117]

In this, Barnett succeeds and a large, black and globular spaceship called the "Steel Globe" is secretly built in an inaccessible region in Alaska. "A spiral staircase wound round the interior circumference of the globe. This staircase, or rather sloping path, had one very curious feature. The handrail was duplicated, so that if by any superhuman means the enormous bulk could be turned upside down one could walk on the underside of the spiral...Across the center of the Steel Globe a commodious platform swung like a ship's lamp. On this a very large telescope was fixed...the platform was literally packed with astronomical instruments. Strange registers, the graduated lines on which were so fine as to be almost invisible without the aid of a magnifying glass, were set into the woodwork of a solid table in the middle of the swinging deck. Strongly made iron tanks filled a considerable portion of the interior space. ...These tanks contained compressed air. ...Innumerable windows pierced the whole circumference of the globe."[118]

With the Steel Globe, Barnett and his friends travel to Mars where they find utopia. The Martians travel around in luxurious airships, but also have other means of negating gravity at their disposal: The Martians "were instructed...in that strange exercise of what may be called—in default of a better name—animal electricism. This discovery enabled the Martians to regulate at will the attraction of gravity upon them so that they could move at any distance they wished from the ground."[119] Their discovery of a superior civilization prompts one of his friends to propose the erection of an "Inter-Planetary Communication Company Limited."[120]

Of the direct influence of Keely on Bloomfield-Moore's acquaintance, writer, inventor, and wealthy stealthy John Jacob Astor it is not necessary to make educated guesses, to circumvent any social terrain, or to fathom any possible connections through obscure philosophies or esoteric pamphlets, for there is ample documentation that the two had met.

In 1894, in the year that Colville choose to publish Keely's intriguing accounts and rudimentary blueprints for a spaceship, Astor published his remarkable book *A Journey In Other Worlds,* after spending over two years writing it. Of this even the biographer of the Astor clan, while missing the connection to Keely, seemed impressed: "In 1894 he published a Jules Verne-type science fiction novel...in which he predicted future developments such as aeroplanes, television and space travel."[121] Astor's novel with descriptions of an antigravity device working on apergy, a term that he had borrowed from Greg's book, was widely read and became a bestseller.[122] Notwithstanding the success of his novel, Astor would never return to that specific literary genre.

In Greg's novel, Martian astronomers think of Jupiter as "not by any means so much less dense than the minor planets as his proportionally lesser weight would imply. They hold that his visible surface is that of an enormously deep atmosphere, within which lies, they suppose, a central ball, not merely hot but more than white hot, and probably, from its temperature, not yet possessing a solid crust." Perhaps a Martian writer thinks Jupiter's satellites hold life "since the satellites of Jupiter more resemble worlds than the planet itself, which may be regarded as a kind of secondary sun."[123]

And while one of Cromie's protagonists wonders what the planet Jupiter is like,[124] the reader finds the answer to that question in Astor's very peculiar novel, for there, not Mars - already subject of so many visits - but Jupiter is the ultimate aim. *A Journey In Other Worlds* has a striking opening: "Jupiter—the magnificent planet with a diameter of 86,500 miles, having 119 times the surface and 1,300 times the volume of the earth—lay beneath them." While the traveling to the other planets in our solar system is an unusual and unique feature of Astor's novel, the positions of the axes of the planets of our solar system clearly obsessed him, and the straightening of the axis of the earth is its most important theme. A company called "The Terrestrial Axis Straightening Company" is concerned with the straightening of the axis of the Earth. This is done in order "to combine the extreme heat of the summer with the intense cold of the winter and produce a uniform temperature for each degree of latitude the year round."[125]

The newly discovered force of apergy is interestingly but only vaguely described as "a fluid," which brings Keely's vaporic substance into mind. Apergy is obtained "by simply blending negative and positive electricity with electricity of the third element or state." Antigravity is obtained by "charging a body sufficiently with this fluid, gravitation is mollified or partly reversed, and the earth repels the body with the same or greater power than that which it still attracts or attracted it, so that it may be suspended or caused to move away into space."[126]

Astor states that the secret of apergy was already known by certain persons; while Greg simply states that these persons were "those who possessed the secret of apergy." Astor equally vague calls them "the ancients." Perhaps Astor referred to the same "ancients" about whom John Keely once said that, "The ancients were far better schooled in spiritual philosophy than are we of the present age. Their mythological records, in their symbolic meaning, prove this fact. They recognized this latent element as the very breath of the Almighty; the sympathetic outflow of the trinity of force, the triple spiritual essence of God himself. Their concepts of Deity were greater and truer than our own. From them we learn that when God said 'let there be light,' He liberated the latent celestial element that illuminates the world: that when He breathed into man the breath of life, He impregnated him with that latent soul-element that made him a living and moving being."[127]

While Keely claimed that an airship "of any number of tons weight" could be suspended if it would be kept "in sympathy with the earth's triune polar stream" and Belinfante wrote that "ether is found in abundance at the poles," Astor hints that the qualities of apergy have something to do with the terrestrial axis. During work on the straightening of the axis, apergy became to be better understood: "I only regret...that when we began this work the most marvelous force yet discovered—apergy—was not sufficiently understood to be utilized, for it would have eased our labors to the point of almost eliminating them. But we have this consolation: it was in connection with our work that its applicability was discovered, so that had we and all others postponed our great undertaking on the pretext of waiting for a new force, apergy might have continued to lay dormant for centuries. ...With this force and everlasting spring before us, what may we not achieve?"[128]

In Astor's novel, water is pumped out of the Arctic Ocean and shifted to Antarctica. Thus they overweigh the southern hemisphere and change the Earth's center of gravity. "Blessed are they that shall inherit the earth. ...We are the instruments destined to bring about the accomplishment of that prophecy, for never in the history of the world has man reared so splendid a monument to his own genius as he will in straightening the axis of the planet. No one need henceforth be troubled by sudden change, and every man can have perpetually the climate he desires," Astor writes.[129]

And although with apergy it is possible to "someday visit the planets," why do so, "since the axes of most of those (planets) we have considered are more inclined than ours, they would rather stay here."[130] Nevertheless, a large cylindrical spaceship called "Callisto" is built. Traveling a million miles an hour, Jupiter and Saturn are visited. Many life-forms, such as the beautiful musical flowers of Jupiter, are encountered. Saturn is the abode of the spirits of the dead. The book is filled with remarkable predictions: solar and wind energy, advanced medicine with inoculations for all sorts of diseases, linear motor trains, wireless transmission of power, super-metals, electric automobiles, battery operated airplanes with screw propellers and weather control. Traffic policemen check automobile speeds with instant cameras and the rabbits of Australia are exterminated by an artificial disease.[131]

The novel also clearly showed that its writer had not forgotten his idea of creating artificial rain, although they had been rejected by the U.S. Patent Office. Rejected as they might be, Astor now wrote that, "Rainmaking is another subject removed from the uncertainties, and has become an absolute science."[132] Curiously, one finds a similar intellectual occupation in Cromie's novel: By means of a "powerful machine" the Martians are able to "electrically" disturb the atmosphere, thus creating rain conveniently only at night. "I hope, Mr. Barnett, you will bring one of these machines back with you when you are returning," a protagonist of Cromie's novel utters, "It is wanted badly in some places I know."[133]

During his college years Astor took courses with astronomy professor Pickering, who mentioned that the seasons were due to the inclination of the earth's axis off of the ecliptic. If the earth was not tilted away from the sun, Pickering suggested, the planet would have a uniform, moderate climate. Thus, a part of Astor's obsession with the axes of the planets is explained in his novel. In the same year that he started to write his novel, Pickering proclaimed to have discovered "lakes in great numbers on Mars. The Canals."[134]

Pickering also was a corresponding member of the Society for Psychical Research,[135] and perhaps this is why that in a novel of futuristic technological extrapolations, one encounters such an out-of-place concept as Saturn being the place of the spirits of the dead with whom Astor's protagonists communicate. A few months after the publication of his novel, Astor presented a copy to Tesla. Tesla did not seem to care much for Astor's book, although he promised "to keep it, as an interesting and pleasant memento of our acquaintance."[136]

Astor may have presented Keely with a copy, although this is not known, nor what Keely thought of Astor's novel, or any of the other tales that so aptly depicted his visionary ideas. Bloomfield-Moore read *A Journey In Other Worlds* as we have seen, although she made the strange remark that, "At present I do not know of any one who believes that Astor wrote the book, but I have convinced myself that he did write it and wish to do him the justice that he deserves."[137]

There were also other reactions; a Keely antagonist wrongly suggested that Keely shrewdly drew a connection between Astor's *A Journey In Other Worlds* and his discovery: Astor wrote "a book of scientific fantasy, in which he does wonders with a newly discovered force operating directly the other way from gravity, and which he calls 'apergy.' Well, shortly after Astors visit to Philadelphia Mr. Keely made public mention—meant to be taken as serious mention, mind you,—of harnessing Mr. Astor's 'force.' Keely didn't call it 'apergy,' but it smelled just as sweet. And its use showed which way the wind blew—Astorway for the time being..."[138]

Astor's book also drew the attention of the occult communities, but certain circles surrounding the Golden Dawn were not amused by his novel, a strange thing considering the rich polar tradition in the occult.[139] A review of *A Journey In Other Worlds* was published in *The Unknown World,* a turn-of-the-century occult magazine that barely lasted two years and was full of articles about the cabbalah, Rosicrucianism, spiritism and other subjects that interested the coterie surrounding Arthur Edward Waite, its editor and Golden Dawn member. In it, the anonymous writer cynically remarked that "Of this story it may be said that *le roi s'amuse* (their italics), and when a prince of finance unbends that would be a ponderous criticism which was needlessly serious. All this, notwithstanding, a scientific romance should have at least the complexion of possibility, and here it is distinctly wanting. The straightening of the terrestrial axis is a very large piece of absurdity, and the imaginative element throughout is somewhat forced and stilted. Moreover, it is not written in a readable style. A special faculty is requisite for the scientific romancer; it is possessed by Jules Verne; to a certain extent Mr. Maitland exhibited it once in *By and By;* there is a gleam of it in the first chapters of *The Goddess of Atvatabar;* it was plenary in Lord Lytton; it is quite wanting in Mr. Astor."[140]

Another who allied his imagination with Keely's concepts was Louis Senarens (1865-1939). Senarens is characterized as "a Brooklynite of Cuban descent and an extremely prolific writer with an enormous amount of wordage." Senarens corresponded irregularly with the great French author Jules Verne, and an exchange of ideas for their stories took place.[141] During his lifetime, Senarens was referred to as "the American Jules Verne," and a comparison of the work of both authors indicates the similarity. Senarens wrote stories of airships suspended in the air by helicopter blades, "helices," three years before Verne did so in his *Clipper of the Clouds* that was published in 1886. Even the illustrations were identical, with three of those in Senarens' story also used in Verne's *Master of the World.*[142] Not only that, one of Senarens' 1897 stories was named *Frank Reade Jr. and his Queen Clipper of the Clouds,* written by "Noname,"[143] a fanciful pseudonym clearly derived from Verne's character Captain Nemo. In 1897, the year that a puzzling airship-wave struck large parts of America, Senarens also wrote his *Across the Milky Way; or Frank Reade, Jr.'s Great Astronomical Trip With His Airship 'The Shooting Star.'* In it, an airtight spaceship is featured which is propelled by a rotascope and an attractomotor, a device that utilizes magnetic affinities among the heavenly bodies and establishes magnetic fixes on selected worlds. The return is possible by reversing the currents. By playing off such forces against the earth's gravity, the attractomotor enables space flight. Interestingly, an occultist is also aboard the spaceship, hoping to reach theosophical heavens. This detail in itself could well mean that the author was inspired by some theosophical writings or pamphlets about Keely. Perhaps he read Bloomfield-Moore's book with its reference to aerial navigation, which must have been most tempting. After all, a writer of such unusual tales would have feverishly sought for everything that would deliver him even more outré plots and ideas.

An unusual feature of Senarens' tale was that while in space, the explorers sometimes left the ship in diving suits and helmets. In his 1898 *The Sinking Star; or Frank Reade, Jr.'s Trip Into Space With His New Airship "Saturn,"* once again the rotascope and the attractomotor are featured.[144] According to Moskowitz, Senares was "the first to propose that an air vessel be driven by electric engines powered by storage batteries."[145]

Another story that was definitely influenced by Keely's inventions was also fitting for its irony. The story was titled "Edison's Conquest of Mars," written by Garett Putnam Serviss (1851-1929). Serviss was employed as a staff writer at the *New York Sun,* and as a journalist at the *New York Journal,* the newspaper that published the devastating Keely expose on January 29. Serviss' story, his first novel, was serialized in the *New York Evening Journal* where news about Keely was published as well, from January 12 until February 10, 1899, during the height of the Keely exposure. The story concerns the invasion of the earth by Martians, and was a rewrite of H.G. Wells *War of the Worlds.* In Serviss' version, the scientists team up under the leadership of Edison to build powerful weapons against the Martians. Ironically, Edison and not Keely is credited in the story for producing two remarkable inventions: the vibratory disintegrator and an electric repulsion antigravity device.[146]

We can even trace a certain influence of ideas in the writings of Jules Verne (1828-1905); *The Hunt for the Golden Meteor,* which was published posthumously in 1908, features a scientist who invents a "neutral helicoidal ray" with gravitational powers,[147] the use of the term "neutral" in connection to gravitational powers pointing towards Keely who used the same term in connection with the same force. Some years later, in 1914, a Frederick Robinson (?) published *The War of the Worlds, A Tale of the Year 2000 A.D.* in Chicago at his own expense. In it he describes how science has progressed enormously by the year 2000. Airplanes travel at three hundred miles an hour and moving sidewalks serve pedestrians in New York. The moon has been provided with oxygen and is colonized, and it is possible to travel to Mars using apergy,[148] a theme that by now sounds all too familiar.

Three years later, the same city would deliver another strange novel with the obvious Keelyesque influence. The book was called *The Wizard of the Island or the Vindication of Prof. Waldinger,* and was written by a Frank Stover Winger (1865-?), a writer of whom absolutely nothing is known. His novel employs huge flying devices, and a vibrometer that enables the operator to control almost everything vibratory, from sound to light and gravity. The vibrometer works not on the concept of atomic energy which its inventor refuses to accept, but instead on regulating the flow of the ether. With this device a small plane called "the electron" is propelled that can fly a thousand miles an hour.[149] Obviously Winger hinted that he knew what he was writing about; in a monograph on the physics in Blavatsky's *The Secret Doctrine* published a few years before, it is stated that "Some men of science speak of electrons as 'centers of power' in the substance of the ether."[150]

Then there is the bizarre novel written by Everard Jack Appleton, equally unknown because behind this nom-de-plume not one writer, but a syndicate and two writers hid.[151]

The first wrote *The Sound Machine* that was published in 1906, in which a free-floating balloon with a device hooked up to it is described, "some sort of apparatus consisting of little boxes, levers and wires." With this device one is able to turn sound into energy[152] - a strange concept that the writer undoubtedly borrowed from Blavatsky's *The Secret Doctrine*, where she treats Keely's vibratory force as something that is similar to the occult power of sound, and the force that in the ancient times was fixed "on a flying vessel, a balloon" and "reduced to ashes 100,000 men and elephants."[153] The second wrote a tale of which the title alone already evokes memories of Keely: the tale was named *Tom Swift and his Polar Ray Dinasphere,* and was published as comparatively recently as 1965.[154]

Long after most of these fantastic tales, which were what science fiction consisted of in the early days, Keely would even enter ufology in the famous contactee book *Flying Saucers Have Landed* by Desmond Leslie and George Adamski[155] through the same medium that had informed innumerable others of his existence: Blavatsky's *The Secret Doctrine*. For Leslie was an Irish occultist[156] and Adamski too had risen from the occult undercurrents. At one time and well before his career as a contactee, he had been a member of The Brotherhood of Tibet, a quasi-theosophical order. And as we will see, it was in the occult underground that Keely's discoveries became fully appreciated.

PART II

Secrets
of
Occult Technology

9

The Sorcerer's Apprentice
The Occult Connection

"It is the fact that Keely is working with some of the mysterious forces included under the name Akasa that makes his discoveries interesting to theosophists"

R. Harte (sec. T.P.S.)
introduction to Clara Bloomfield-Moore's *Keely's Secrets,* July 1888

"I had never heard the name NYARLATHOTEP before, but seemed to understand the allusion. Nyarlathotep was a kind of itinerant showman or lecturer who held forth in publick halls and aroused widespread fear and discussion with his exhibitions. These exhibitions consisted of two parts—first, a horrible—possibly prophetic cinema reel; and later some extraordinary experiments with scientific and electrical apparatus."

Howard Philips Lovecraft
in a letter to Reinhardt Kleiner, December 14, 1921

Philadelphia, the city where Keely lived, experimented, and demonstrated his devices almost all of his life, was a hotbed of esoteric activity, a focal point of secret societies. Philadelphia was an occult vortex where various strange and Fortean events happened before, during and after Keely's lifetime, such as automatic writing, poltergeist activity, spontaneous human combustion, falls of ice, rocks, sulfur and gelatinous substances from the skies, ball lightning, cases of teleportation, out-of-the-body experiences, encounters with the ever-elusive Men In Black, and UFO sightings, some events going back to the 18th century.[1]

From its founding, Philadelphia attracted mystics and believers of all kinds: groups of Quietists, Dunkers, Moravians, hermits, astrologers and magicians. An unusual number of ghost stories are to be found in the Philadelphia tradition.[2]

In 1693, a selected group of Pietists with their families led by Johannes Kelpius (1673-1708), gathered from all parts of Europe at one port, and set sail for America in their own chartered boat, the Sarah Maria. They arrived in Philadelphia on June 23, 1694, where they established their headquarters farther west on the banks of the Wissahickon River. Staunchly millenialist and communal and in possession of certain manuscripts of purported Rosicrucian origin, the group also practiced occult, healing arts and possibly cabalist studies, since Kelpius had met with Knorr von Rosenroth, the writer of the famous *Kabbala Denudata* that was translated by Golden Dawn co-founder MacGregor Mathers and published in 1887. After Kelpius' death, the group disbanded, but they are remembered as the originators of the Pennsylvania hex tradition. It is alleged that this group brought Rosicrucianism to America.[3]

Philadelphia was also the town where the first American Masonic lodge sprang into existence before 1730 and where Benjamin Franklin was initiated into freemasonry; where in its Masonic temple the Royal-Arch masons, the Knights of the Temple and the Knights of Malta held their regular meetings.[4]

In 1795, the first utopian work of fiction published in America was printed in Philadelphia.[5] In the 1840s the millenarian Millerite sect, one of many that flocked the Philadelphia streets, distributed the apocalyptic "Philadelphia Alarm."[6] Philadelphia's 19th century esoteric community consisted of a dizzying mixture of spiritists, cabalists, theosophists, and various occult lodges and orders, possibly a temple of the Golden Dawn called Thoth-Hermes,[7] a branch of the Societas Rosicruciana in Scotia established since 1879,[8] and a temple of the Rosy Cross, re-founded in 1895 by Freeman B. Dowd, a member of the Hermetic Brotherhood of Luxor.[9] Before that, in 1882, Dowd issued a booklet called "The Temple of the Rosicross," which became serialized between 1885 and 1888 in the occult magazine *The Gnostic* that was co-edited by John Keely's friend Colville.[10]

Isaac Meyer, the cabalist of whom it is rumored was once in possession of documents on Keely's devices, privately published his learned treatise on the cabbala in Philadelphia.

Charles Godfrey Leland (1824-1903), the American folklorist and journalist whose "immersion in gypsy lore and witchcraft influenced the revival of the latter in the 20th century,"[11] was born in Philadelphia. Family lore had it that one ancestor was a German sorceress and Leland always believed that he resembled her in an atavistic way.[12] Another story he told had it that his baby nurse was reputed to be a Dutch sorceress who had performed rites over him to ensure his development into a scholar and a wizard. It is further alleged that his mother's relatives took pride in one of their forebears, a doctor who had acquired a reputation for sorcery.[13]

Leland was attracted to the occult and folklore at an early age. During his youth he was often given to solitary walks through the woods and through the Swedish and Spanish neighborhoods of Philadelphia.[14] He studied in Germany and led a life of exotic travel. He penetrated the mysterious worlds of the gypsies, witches and voodoo and spent several summers with the Native Americans to learn their spiritual lore. He also discovered Shelta, the secret language of the tinkers, learned to speak Romany, the language of the gypsies, and was adopted in their society. He made his literary debut in 1856 with a curious book published in Philadelphia, in which he coupled his fascination with dreams and their explanation with poetry by others as examples of interpretation.[15]

In 1886, while in Italy, he allegedly met with a Florentine witch to whom he referred only as "Maddalena." He was introduced to other witches who divulged the secrets of the craft to him. The information obtained formed the basis of his book *Aradia, or the Gospel of Witches,* published in London in 1889.

After his death Leland was cremated and his ashes were returned to Philadelphia. He studied and wrote for half a century about things occult, and was also a journalist for fifteen years in Philadelphia, New York and Boston.[16] Thus Leland, with his deep occult interests, and a journalist living at the same time as Keely in Philadelphia, would almost certainly have heard or read about him, even though no documentary evidence of their mutual acquaintance exists. There is a small and puzzling sentence in Bloomfield-Moore's writings where she states that "Should it be that Mr. Keely's compound secret includes any explanation of this operation of will-force...we shall then recover some of the knowledge lost out of the world, or retained only in gypsy tribes and among Indian adepts."[17] It must also be taken into account that there are uncharted areas in both the private lives of Leland and Keely that have not been trusted to paper. Leland spent most of his life in Europe, and could easily have met with Newbrough, Colville, Hartmann or Paschal Beverly Randolph, as they too frequented European spiritist and esoterist circles.

Randolph (1825-1875) founded the Brotherhood of Eulis and stood at the base of the Hermetic Brotherhood of Luxor, which had a Rosicrucian temple in existence in Philadelphia in 1895 through its member Freeman Dowd.

Before spending the remainder of his life in Boston, Randolph traveled to Europe in 1855, 1857 and 1861. It is said that he met with Levi and Bulwer-Lytton.[18] In France, he mixed in the Mesmerist circles around Baron Jules Du Potet de Sennevoy and Louis Alphonse Cahagnet. There Randolph discovered that, unlike most American spiritualists, the French Mesmerists were well versed in the Western magical and occult traditions. What also made a profound impression on Randolph was their use of magic mirrors, crystals and drugs, especially hashish, during spirit evocations . In 1861, he traveled to the Near East, where he learned a different kind of magic from the wandering dervishes. On his return to America, Randolph publicly denounced spiritualism and his role as a passive medium. Interestingly in 1861 Keely was, according to his own admission, "exposing spiritualistic mediums" in St. Paul, Minnesota, and was "nearly run out of town for doing so."

Instead of spiritualism, Randolph began to teach a complete system of practice and theory in which he saw man as having the task to first become individual, then a divine individual and in the end to become a god and to travel endlessly through infinite universes. Randolph also beheld the universe as being filled with vast hierarchies of elementals,[19] as Newbrough would declare some twenty years later in his strange *Oahspe*.

It is asserted that Randolph, who was a close friend of President Abraham Lincoln, was also a member of the Societas Rosicruciana In Anglia, a branch of which was established in Philadelphia in the 1879 through the Societas Rosicruciana in Scotia. In 1880 the Philadelphia branch was renamed Societas Rosicruciana In United States, and afterwards Societas Rosicruciana Republicae Americae. The same year a college in Boston received a charter.[20] Randolph also founded the Fraternitas Rosae Crucis in Quatertown, Pennsylvania.[21] The Societas Rosicruciana Republicae Americae reportedly still exists today under the name Societas Rosicruciana Civitatibus Foederatis. The Masonic Rosicrucian order is, as was before, only open to master masons.[22]

Helena Blavatsky (1831-1891) resided in Philadelphia for a while. She would become one of the most influential figures in the occult world, and through her writings Keely would become known in esoteric communities everywhere.

Henry Steel Olcott, co-founder of the Theosophical Society, also lived in Philadelphia for a while, and along with Blavatsky investigated the claims of spiritists Jennie and Nelson Holmes. The Holmeses, through Robert Dale Owen, a former U.S. Congressman and foreign ambassador and leading spiritualist, appealed to Olcott to allow them to test their powers. The first seance was held on January 11, 1875, the last on the January 25. Olcott agreed to investigate and was accompanied by Blavatsky.[23]

Blavatsky resided at 1111 Girard Street. After the Holmes investigation she would stay in Philadelphia "for many months."[24] She might easily have met with Keely in that time period, although she never wrote of a meeting with him in her account of the investigation in Philadelphia.[25] While she devotes some space to new scientific discoveries in her 1877 *Isis Unveiled*, and while the quotation in chapter 1 clearly shows that a kindred spirit already existed, she does not mention or refer to Keely.[26]

This would change, for when *The Secret Doctrine* was published in 1888, she stated in its pages that, "In the humble opinion of the Occultists, as of his immediate friends, Mr. Keely, of Philadelphia, was, and still is, at the threshold of some of the greatest secrets of the Universe; of that chiefly on which is built the whole mystery of psychical forces, and the esoteric significance of the 'Mundane Egg' symbolism."[27] She also held it that "Mr. Keely's discoveries corroborate wonderfully the teachings of Occult Astronomy and other Sciences."[28]

Thus she not only admitted that "the Occultists" were interested in Keely's inventions, but she also linked Keely to the esoteric doctrines, and this assured that other occultists everywhere who read her book turned their attention to the Philadelphian inventor. In fact, Blavatsky referred to Keely more than once in her seminal 1888 opus. The chapter titled "The Coming Force," refers to Babcock's 1881 pamphlet "Exposition of the Keely Motor," the back of which read: "The Doom of Steam, or, The Coming Force." Both titles in turn referred to *The Coming Race*, by that other great and mysterious initiate, Bulwer-Lytton.

Blavatsky's interest in Keely's inventions was fueled by Bulwer-Lytton, with whom an age-old and secret tradition breached the surface. This high initiate, alleged Rosicrucian, politician and friend of the French magus Eliphas Levi did not only influence the course that Blavatsky would take—reading one of his novels started her career as author while still residing in St. Petersburg[29]— she would later catch a glimpse of her favorite author during the Great Exhibition at Crystal Palace in 1851.[30]

Not only Blavatsky's career as a writer began by reading Bulwer-Lytton, early theosophy was principally inspired by his writings. Blavatsky's fascination with Egypt as the fount of all wisdom arose from her enthusiastic reading of Bulwer-Lytton's books. Only in her second book *The Secret Doctrine*, the location of the "source of ancient wisdom" would shift from Egypt to the Far East.[31] Also, his most famous and most curious book *The Coming Race* would influence generations of occultists, and ultimately helped a secretive Berlin group formulate their ideas on the generation of free-energy in the 1930s.

Edward George First Baronet of Bulwer-Lytton of Knebworth (1803-1873) was a curious man with deep-ranging esoteric interests, and it is with him that we may clearly see the entanglement of occult lore and avant-garde science emerge for a brief but influential moment. Bulwer-Lytton studied at Cambridge, was elected to parliament in the 1830s, knighted in 1838 and made a baron in 1866. He was one of the most popular authors of his day with his collected works totaling over one hundred titles. Today his fame as an author is largely forgotten except for the book *The Last Days of Pompeii,* published in 1834.

In the fields of horror and science fiction literature, Bulwer-Lytton would earn fame with three influential novels by which he is still remembered to the present day. These were his 1842 Rosicrucian initiatory tale *Zanoni*, *A Strange Story* published in 1861 and his enigmatic *The Coming Race*.

These novels would form the foundation of a whole current of occult fiction. Thus, parts of *Zanoni* found their way in Randolph's *Mysteries of Eulis*,[32] who borrowed from it without acknowledging it. Later the current of occult fiction was popularized by Theosophist Marie Corelli (1855-1924). Bulwer-Lytton's novels also became essential reading for occult adepts, such as the initiated of the Argenteum Astrum, Aleister Crowley's renegade order founded after his fraction with the Golden Dawn, although *The Coming Race* is strangely absent from its required reading list.[33]

In these novels Bulwer-Lytton connected scientific reasoning with occult elements, based on contemporary theories of animal magnetism, hypnotism, space and time,[34] a curious fusion that was noted by Thomas H. Burgoyne, co-founder of the Hermetic Brotherhood of Luxor. Under the nom de plume *Zanoni*, after Bulwer-Lytton's 1842 novel of the same title, Burgoyne remarked that Bulwer-Lytton was "thoroughly convinced of the great value and importance of uniting ancient alchemy with modern medicine."[35]

Recently it has been noted that the descriptions of magical rituals in Bulwer-Lytton's *A Strange Story* are similar to means of contacting extraterrestrial inhabitants and other-dimensional entities, and that *A Strange Story* in fact offers a classification system of these beings.[36]

Bulwer-Lytton appeared to the audiences of his day as a successful and perhaps a somewhat eccentric writer, but in his private life he was deeply involved in occult pursuits. He became the member of at least two, possibly three, occult societies, not only in England but also abroad. According to Wynn Westcott, one of the founders of the Golden Dawn, in 1850 Bulwer-Lytton was appointed "member in absence" of the exclusive German high-grade Rosicrucian lodge "Karl zum aufgehenden Licht" based in Frankfurt. This ancient lodge was founded in the 18th century and was one of the last representatives of a German Masonic-alchemical Rosicrucian system.[37] One of its members was the well-known Masonic historian Georg Burkhard Kloss (1787-1854). Until the early 19th century, this lodge was involved in alchemical activities and members had their own alchemical laboratory at their disposal.[38] Bulwer-Lytton allegedly corresponded with this lodge and became intimate with their alchemical teachings and doctrines.[39]

It is also alleged that Bulwer-Lytton became the Grand Master of the Collège Métropolitain in 1871, whose members were closely allied to the Rosicrucian lodge "Karl zum aufgehenden Licht." The Collège Métropolitain consisted of the highest and secret class of members of the order of the "Chevaliers bienfaisants de la Cité-Sainte" that was founded around 1770 on Knights Templar tradition. The highest grade of the Chevaliers was named after the patron of the Merovingian empire, the Cité Sainte, or the Holy City of course being Jerusalem. Members of this order were prepared for theurgic magic, the intercourse with "spiritual, material, invisible and visible beings" and the "transcendent experience of the Rose-Cross," inside a special room, called the "Chambre d'Operation." The Collège Métropolitain also had an "inner order" that zealously guarded the "true secrets." The Collège Métropolitain had connections to the Order of the Strict Observance, also modeled after Knights Templar tradition.[40]

In 1871, Bulwer-Lytton also became Grand Patron of the Societas Rosicruciana In Anglia, the mother lodge of the Philadelphia branch[41] the year that *The Coming Race* was published. The Societas Rosicruciana In Anglia was founded in 1865 by Robert Wenthworth Little and was only open to master masons, the same being the case with its American branches. Members of the Societas Rosicruciana In Anglia included not only Randolph, but also Rudolph Steiner, Eliphas Levi, MacGregor Mathers, William Wynn Westcott and William Robert Woodman. In 1888, Mathers, Westcott and Woodman would found the Golden Dawn.

From *Zanoni* alone it is already evident that Bulwer-Lytton was well versed in Rosicrucian literature. He read that strange novel *Le Comte de Gabalis,* written by the equally mysterious abbe Montfaucon de Villars, which he quotes in *Zanoni.*[42] Since the occultists took Bulwer-Lytton's *Zanoni* very seriously and considered its writer a high initiate, other offers for membership would follow. Interestingly, there is little historical evidence that Bulwer-Lytton was further involved in either Masonic or Rosicrucian activities—although it is hinted that he was a freemason[43]—or that he belonged to any occult society except for his honorary membership in the Societas Rosicruciana In Anglia. John Yarker, the occultist and notorious promoter of bogus Masonic grades, and who had sold Blavatsky a Masonic diploma, also tried to interest Bulwer-Lytton in one of his Masonic systems. As with other letters by esoteric and occult orders, Bulwer-Lytton never bothered to reply.[44] Most historians of the occult consider this proof that he was only theoretically interested in the occult sciences.[45]

But considering his membership of one, possibly two very exclusive continental orders, the reason for his lack of interest in other occult societies of his time may very well be that he was far deeper in the occult strata than most of his esoterically inclined contemporaries, who held him in high esteem. There were also stories by those who claimed to have witnessed his demonstrations of his telekinetic abilities, strange rumors of how he became increasingly eccentric towards the end of his life, morbidly afraid of being left on his own, terrified of being buried alive, and tales of his curious nickname "the old sorcerer."[46]

Bulwer-Lytton also mixed in circles that included the great English mystic and visionary seer William Blake. As a result, he became well versed in the art of geomancy, for in this, he was initiated by William Blake's friend John Varley, and it is asserted that he and Varley "worked at astrology together." Later the two, including the English statesman and Prime Minister Benjamin Disraeli, were often given to debates on the pros and cons of witchcraft, spiritualism and "they plunged into discussion and experiment...they even tried their hands at crystal gazing."[47]

Disraeli (1804-1881) was a close friend of Bulwer-Lytton, and at one time Bulwer-Lytton drew up his astrological chart. Disraeli not only shared Bulwer-Lytton's deep-rooted interest in matters of the occult, but he also perceived the existence of secret societies as something very real, and he held this belief his entire life. Disraeli wrote of his belief in several novels involving secret societies and political conspiracies.[48]

In 1856 in the House of Commons, Disraeli not only warned against the threat posed by secret societies in Europe, but also of the danger of supporting Italian revolutionary movements because of the influence of the secret societies in these movements. Disraeli also said that "a great part of Europe—the whole of Italy and France and a great portion of Germany, to say nothing of other countries, are covered with a network of these secret societies."[49]

There is also Bulwer-Lytton's cryptic and somewhat similar reply in a letter to Hargrave Jennings, who sent him *The Rosicrucians, their Rites and Mysteries,* that was published a year before Bulwer-Lytton's *The Coming Race* shook the occult undercurrents to their very foundations. In this letter, Bulwer-Lytton wrote: "There are reasons why I cannot enter into the subject of the 'Rosicrucian Brotherhood,' a society still existing, but not under any name by which it can be recognized by those without its pale." At the foot of the letter, he added a postscript which read: "Some time ago a sect pretending to style itself 'Rosicrucians' and arrogating full knowledge of the mysteries of the craft, communicated with me, and in reply I sent them the cipher of the 'Initiate,' not one of them could construe it."[50]

Not only from his letter to Hargrave Jennings, but also from *Zanoni* it is evident that Bulwer-Lytton's occult knowledge was profound, highly detailed, and went beyond the limits of a mere theoretical nature. This could only stem from sources which carefully guarded that knowledge: the esoteric societies themselves. *Zanoni* was published years before his memberships of the Societas Rosicruciana In Anglia, the German Karl zum Aufgehenden Light and possibly the Collège Métropolitain. It is possible that long before he joined these orders, Bulwer-Lytton may have been involved in other alchemical and Rosicrucian circles, of which his contemporaries did not even know their names—real secret societies whose existence both Bulwer-Lytton and Disraeli alluded to.

There are speculations as to the identity of these secret societies; for instance a mysterious Rosicrucian order is referred to, an order variously known as the Fratres Luces, the Order of the Brotherhood of the Light, the Order of the Brotherhood of the Cross of Light, or the Order of the Swastika, that was located in Paris and that may have formed the model for the Golden Dawn. The English adepts came to learn of its existence through unconventional means, such as crystal gazing. In 1873, the contact with Cagliostro resulted from it.[51] Therefore allegations that Franz Anton Mesmer and Cagliostro were members of the Fratres Luces must be treated accordingly. Nevertheless, Mesmer was one of the sources of inspiration for Bulwer-Lytton who himself had gazed in crystals, and created his concept of the vril-force.

Another clue was found in Bulwer-Lytton's writings; the mysterious D., the proprietor of an occult bookshop in *Zanoni* very much existed in actual life. He was identified as John Denley, and his occult bookshop was, as Bulwer-Lytton admitted, "one of my favorite haunts." Denley's bookshop was visited by a shadowy group of people from the occult undercurrents. One of them who worked for Denley copying old occult manuscripts was the Rosicrucian seer Frederick Hockley (1808-1885).

Hockley developed an interest in spiritism and experimented with crystal gazing as early as 1824. In 1864 he became initiated into freemasonry and would join various lodges. In 1872, a year after Bulwer-Lytton's honorary membership, Hockley was appointed a member in absence of the Rosicruciana In Anglia, and in this he was advanced into the highest grade. In 1875 he became a member of the Metropolitan College. Two years later he was appointed a member of the Theosophical Society as an "honor" that was bestowed on him by Olcott. He also went to Paris, where he met with an "invisible power," and in one of his manuscripts he refers to an unnamed society in France, "followers of the Rosy Cross."

Perhaps as a result of this, Hockley was considered to be "a true Rosicrucian adept" and he himself believed to be working in the Rosicrucian tradition. But he is also claimed to have been a member of the Fratres Luces and it is asserted that this order indeed existed. The order was of a very exclusive nature; Westcott who was a member of the Rosicruciana In Anglia and would later be one of the founders of the Golden Dawn, tried to join but was refused.[52]

Another person who knew both Hockley and Denley well and visited the latter at his bookshop to acquire his source materials there, was the equally mysterious Francis Barrett, a self-described Rosicrucian whose home was the center of an occult group. Barrett, who lived and died in poverty, published the tremendously influential *The Magus, or the Celestial Inteligencer* in 1801. Strangely he had another passion—that of aeronautics.[53]

There is also Lytton's well-documented connection with the great French magus Eliphas Levi. Levi was known in English occult circles, but it was admitted that very little was known of his writings which were not translated into English at that time, even among members of the Societas Rosicruciana In Anglia[54] of which he was appointed honorary member in 1861.[55] Levi even commented upon the fact that, while visiting London in the spring of 1854, he encountered "amidst much that was courteous, a depth of indifference or trifling. They asked me forthwith to work wonders, as if I were a charlatan, and I was somewhat discouraged, for, to speak frankly, far from being inclined to initiate others into the mysteries of Ceremonial Magic, I had shrunk all along from its illusions and weariness."[56]

Yet Eliphas Levi (1810-1875), who was the most important occultist at that time and of whom it was claimed had direct contact with surviving members of the original Rosicrucians, found reasons enough to visit Bulwer-Lytton in London in 1853 or 1854. During his visit Bulwer-Lytton came to appreciate Levi's vast occult knowledge and became his friend.[57] Levi visited him again in 1861.[58] During his second visit, Levi presented him with three of his books, and inscribed these with long and friendly dedications.[59]

Bulwer-Lytton's son admitted to Golden Dawn member Arthur Edward Waite that indeed his father had known Levi, but that he thought that this acquaintance was first established in either Paris or Nice.[60] It is suggested that Blavatsky also met Levi while studying Mesmerism in France. Not only Bulwer-Lytton's, but also Levi's influence looms heavily in her later doctrines.[61]

When a chapter of the German Theosophical Society was founded in 1884 in the German town of Elberfeld, the ceremony took place in a "Chambre d'Operation" that was consecrated by Levi in 1870-1871, around the time that *The Coming Race* was published. This occult space was a sacred room like the Collège Métropolitain and the French magicians were known to use, and before them the 18th century Illuminates. Descriptions of such rooms are sketchy at best. It is alleged that such an occult space was "only scarcely lit by candles and was used for meditation and magical practices." Olcott was present during the founding of the German Theosophical chapter. Wilhelm Hübbe-Schleiden, who visited Blavatsky on several occasions, was appointed its president. On the outside the chapter, called Loge Germania, was modeled like a Masonic institution.[62]

Wherever Bulwer-Lytton and Levi first met, Levi's visits to London were of the utmost importance to him as well, for he explains that during his first visit, the physical manifestations of the invisible presented themselves to him for the first time during elaborate magical rituals that he conducted through the mediation but not the presence of Bulwer-Lytton.[63] During his second visit, Levi allegedly conducted a magical ritual with Bulwer-Lytton at the Pantheon in London.[64] It is held that Bulwer-Lytton's *A Strange Story* that was published a year later was heavily influenced by this occult experience.[65]

At that time, Levi developed his concept of "Astral Light" which he also called "the Great Serpent" or the "Great Dragon," that he envisioned to be the carrier of the cosmic life-force[66] and was regarded by some as the cosmic memory of the aether, the equivalent of Akasha. This astral light was called "Sidereal Light" by earlier occultists. The invisible and diaphanous region around the earth corresponded to the astral body of man, which Paracelsus called "ens astrale" or "Sidereal Body" and which he linked with the stars.[67] Blavatsky called it the "sidereal force," which in turn helped Dutch literator and grailseeker Rensburg, Belinfante's friend develop his philosophy of "Inter astral Siderism."

It is believed that Levi—any correspondence between Levi and Bulwer-Lytton after their second meeting is strangely absent[68]—provided the root of the idea of vril, the mysterious force that Bulwer-Lytton described in *The Coming Race*.[69] This could very well be so, since it is alleged by Bulwer-Lytton's son that in a letter by Levi, there was talk of "the existence of a universal force and its use."[70]

"What is vril?" the unnamed hero of Bulwer-Lytton's novel asks in astonishment. "There is no word in any language I know which is an exact synonym for vril. I should call it electricity, except that it comprehends in its manifold branches other forces of nature, to which, in our scientific nomenclature, differing names are assigned, such as magnetism, galvanism, &c...in vril they have arrived at the unity in natural energic agencies, which has been conjectured by many philosophers above ground...by one operation of vril, which Farraday would perhaps call 'atmospheric magnetism,' they can influence the variation of temperature—in plain words, the weather; that by other operations, akin to those ascribed to mesmerism, electro-biology, odic force, etc., but applied scientifically through vril conductors, they can influence over minds, and bodies animal and vegetable, to an extent not surpassed in the romance of our mystics. To all such agencies they give the common name vril."[71]

Vril, the "all-permeating fluid," can "destroy like the flash of lightning" or could "replenish or invigorate life," "heal or preserve" or "cure disease." Vril "enables the physical organization to re-establish the due equilibrium of its natural powers, and thereby to cure itself." With vril it is also possible to "rend way through the most solid substances." From it, light is extracted that burns "steadier, softer and healthier than other inflammable materials"[72] and thus flowers and foliage are more brilliant in color and larger in growth.[73]

Vril is evoked by hollow staffs, having in "the handle several stops, keys, or springs by which its force can be altered, modified, or directed—so that by one process it destroys, by another it heals—by one it can rend the rock, by another disperse the vapor—by one it affects bodies, by another it can exercise a certain influence over minds. It is usually carried in the convenient size of a walking-staff, but it has slides by which it can be lengthened or shortened at will. When used for special purposes, the upper part rests in the hollow of the palm with the fore and middle fingers protruded." Its power, however, is not the same in the hands of any carrier; instead it is "proportioned to the amount of certain vril properties in the wearer, in affinity, or rapport, with the purposes to be effected."[74]

Along with vril have come certain biological adaptations; the bodies of the underground race, the Vril-ya, are filled with vril, and one must have this hereditary adaptation to use it properly, which is a special nerve in the hand: "The thumb...was much larger, at once longer and more massive. ...Secondly, the palm is proportionately thicker than ours—the texture of the skin infinitely finer and softer—its average warmth is greater. More remarkable than all this, is a visible nerve, perceptible under the skin, which starts from the wrist skirting the ball of the thumb, and branching, fork-like, at the roots of the fore and middle fingers."[75]

Women are superior to men at handling these staffs, since the "female professors are eminently keen," owing to their "finer nervous organization" to the perception of vril,[76] and therefore have "a readier and more concentrated power over that mysterious fluid or agency which contains the element of destruction."[77]

In *The Coming Race*, numerous devices are described as airships, elevator-like lifts, automata that perform tasks that are animated and controlled by vril, and personal flying wings: "These wings...are very large, reaching to the knee, and in repose throw back so as to form a very graceful mantle. They are composed from the feathers of a gigantic bird. ...They are fastened round the shoulders with light but strong springs of steel; and, when expanded, the arms slide through loops for that purpose, forming, as it were, a stout central membrane. As the arms are raised, a tubular lining beneath the vest or tunic becomes, by mechanical contrivance, inflated with air, increased or diminished at will by the movement of the arms, and serving to buoy the whole form as on bladders. The wings and the balloon-like apparatus are highly charged with vril; and when the body is thus wafted upwards, it seems to become singularly lightened of its weight."[78]

The influence of *The Coming Race* was enormous. Greg's *Across the Zodiac* was partly inspired by it. The first novel of English writer C.J. Cutliffe Hyne, who fantasized about the legendary Atlantis, was modeled after Bulwer-Lytton's book.[79] Tesla himself admitted that he read *The Coming Race*.[80] The whole occult world noted it. In *Isis Unveiled* Blavatsky exclaimed that Bulwer-Lytton "allowed his readers to take it as a fiction," and she saw the similarity of vril with Von Reichenbachs Od, Levi's Astral Light and Akasha amongst others.[81] "Absurd and unscientific as may appear our comparison of a fictitious vril invented by the great novelist, and the primal force of the equally great experimentalist, with the kabalistic astral light, it is nevertheless the true definition of this force," she wrote.[82]

Golden Dawn member Waite also saw the similarity between vril and Levi's Astral Light, and we may only guess at the extent of its influence upon such orders as the Golden Dawn, which after all, sprang forth from The Rosicruciana In Anglia that had included Bulwer-Lytton amongst its honorary members. We have already seen how Bulwer-Lytton was mentioned in a scathing review of Astor's *A Journey In Other Worlds*. This not only is again another proof that Bulwer-Lytton indeed was read by those circles surrounding the Golden Dawn, but also that Astor's novel that was influenced by Keely had been looked upon by the same coterie. Interestingly, several writings of Rosicruciana In Anglia member and Golden Dawn founder Westcott were published by the Theosophical Publishing Society. It is therefore quite possible that the initiates of the Golden Dawn not only meditated on the deeper meaning of vril but also read Bloomfield-Moore's pamphlets about Keely which appeared under the same imprint, and later would have studied Blavatsky's statements about Keely in *The Secret Doctrine*. Colville expressed his admiration for Bulwer-Lytton's concept of vril when lecturing on the human aura: "...the word Vril, which was certainly derived from Vir, the superior man. The simple force of energy of life, the pure spirit of humanity, is Vril, and this it is which builds and heals and can exhibit power to command all combinations of varying elements to appear and disappear."

Burgoyne of the Hermetic Brotherhood of Luxor was highly impressed by Bulwer-Lytton's novel. He compared the manipulations of vril by the Vril-ya with the handling of the magnetic currents of the Earth and the Akasha by the first human race, "ethereal but sufficiently material to be objective and tangible."[83]

Later, anthroposophist Steiner would express his admiration. After World War I he confided to Guenther Wachsmuth, the German translator of *The Coming Race*, that Bulwer-Lytton "had seen what was possible in evolution, especially by the discovery of heretofore unknown forces of nature. The imagery of Bulwer-Lytton's novel is partly a memory of the now lost abilities of mankind in the earliest prehistory of the Atlantean era, but especially a vision of the coming phases of evolution, a very essential contribution."[84]

The many resemblances and similarities between *The Coming Race* and Keely's inventions were also noted by a newspaper a year before *The Secret Doctrine* was published: "It is no doubt a marvelous story, that told by John Worrell Keely, a fairy tale of a force which puts to shame the omnipotent 'Vril' of Bulwer-Lytton's fancy."[85]

When *The Secret Doctrine* was printed the following year, Blavatsky excitedly concluded that Keely had stumbled upon great things and, beside more wonderful analogies, she beheld the analogy with Bulwer-Lytton's vril: "Sound, for one thing, is a tremendous occult power. ...Sound may be produced of such a nature that the pyramid of Cheops would be raised in the air...that which he (Keely) has unconsciously discovered, is the terrible sidereal Force, known to and named by the Atlanteans MASH MAK...It is the Vril of Bulwer-Lytton's *The Coming Race*...It is the vibratory force, which, when aimed at an army from an Agni Rath fixed on a flying vessel, a balloon, according to the instructions found in the Ashtar Vidya, reduced to ashes 100,000 men and elephants, as easily as it would a dead rat. It is allegorized in the Vishnu Purana in the Ramayana and other works, in the fable about the sage Kapila whose glance made a mountain of ashes of King Sagara's 60,000 sons, and which is explained in the esoteric works, and referred to as the Kapilaksha—Kapila's eye."[86]

Theosophist Scott-Elliot too perceived this analogy with the vril-force that by now had reached mythical proportions: "In the earlier times it seems to have been personal vril that supplied the motive power—whether used in conjunction with any mechanical contrivance matters not much—but in the later days this was replaced by a force which, though generated in what is to us an unknown manner, operates nevertheless through definite mechanical arrangements. This force, though not yet discovered by science, more nearly approached that which Keely in America is learning to handle than the electric power used by Maxim. It was in fact of an etheric nature."[87]

Since *The Coming Race* was such an important influence on the occult undercurrents, various searches have been conducted to establish Bulwer-Lytton's sources of inspiration. On the surface one finds parallels between Baron von Reichenbach and Franz Anton Mesmer. Mesmer in turn was highly influenced by the renaissance alchemists and cabalists Agrippa and Paracelsus, thus forming another link between the Rosicrucian tradition and Bulwer-Lytton, beside Eliphas Levi and through him the French occult undercurrents. Mesmer's name crops up time and again in Bulwer-Lytton's esoteric novels and one author has even suggested that "without him the vril-concept of the English writer would hardly have been developed in this form."[88] Others have pointed towards the little-known writings of the French author Louis Jacolliot as the source for Bulwer-Lyton's idea of vril[89] and on the influence of Montfaucon's strange book by explaining that the term "Gabalis" is a Paracelsian term, meaning the vital energy that animates the world.[90] The case obtains another intriguing dimension when we read that it is suggested that Bulwer-Lytton obtained his idea of the mysterious vril-force from Keely.[91]

This theory was proposed because of the similarity in ideas between Bulwer-Lytton's aerial craft and Keely's inventions. However, at the time of publication of *The Coming Race*, Keely was still seven years away from his first rudimentary antigravity experiments, which around 1887 would lead to the construction of "a flying machine."

Bulwer-Lytton obtained the idea of aerial craft by reading *Comte de Gabalis,* for in Montfaucon Villars' most curious novel we find not only numerous references to "the subterranean people" and "the people from the air," but also the tale of the "aerial wanderers" who were said to have "fallen from aerial ships," a story that is often repeated in UFO-literature.[92]

Levi and Bulwer-Lytton dabbled somewhat with strange technology. It is therefore doubtful that Keely's concepts directly influenced *The Coming Race*. There is a tangible link though, however small, between Bulwer-Lytton and Keely, and this is to be found in the person of Bloomfield-Moore.

Clara Bloomfield-Moore (1824-1899) was an extraordinary and highly intelligent woman who developed a deep interest in matters scientific and occult. She corresponded with hundreds of scientists in all parts of the world,[93] including Tesla, whom she met at least once, as well as early aviation pioneer Hiram Maxim. At one time, she offered financial support to Maxim if he would go to America to consult with Keely and "become the custodian of the latter's secret."[94]

Around Bloomfield-Moore, we see an unusual clustering of some of the most famous, wealthy and influential people of her day, such as her wealthy acquaintance Astor. After the death of her husband in 1878, she would remain in London. A year later she would meet and develop a warm friendship with the famous literary poets Elisabeth Barrett Browning and Robert Browning. After she became Keely's supporter, she tried to convince Browning that his poem "Childe Roland to the Dark Tower Came" symbolized the whole Keely affair. But on this the poet politely answered that "Childe Roland was only a fantaisie, that he wrote it because it pleased his fancy." And as she interpreted to Browning the meaning in the light of Keely's discoveries, "he listened with interest and a smile of doubtful meaning played over his features, for Mr. Browning never expressed any faith in this modern Prometheus as to his commercial success."[95] But according to her, "One Christmas evening we were amusing ourselves by 'Keely's Discovery' as the subject. Much more expeditiously than I had written down the rhymes to which he was to confine himself in its composition, he wrote the sonnet."[96]

Bloomfield-Moore was also presented to the court of Queen Victoria. Both her daughters married nobility; one daughter married Swedish Baron Carl von Bildt, who at one time had been Secretary of the Swedish Legation in Washington. Her other daughter married Count Carl von Rosen, who would eventually become First Chamberlain and Master of Ceremonies at the Court of Stockholm, Sweden. It was the son from this marriage, Count Eugène von Rosen, who allegedly sent "Keely's secrets" to Sweden. Her home in London was often the resort of writers and artists.

There also was another side to her complex character. She gained a reputation with the public for eccentricity, possibly because her London home witnessed visits of those of the occult circles, such as Cheiro, Colville, Leland, Golden Dawn member Westcott, and theosophists Besant and Blavatsky. Bloomfield-Moore's writings are remarkable for their deep esoteric knowledge. In her 1893 book about Keely one encounters the cabala, theosophy, occult templarism, the Rosicrucians, Jacob Böhme and a host of other occult and hermetic doctrines. She was not a theosophist, according to Count Eugène von Rosen when he was asked this obvious question: "No. She was interested in the study of theosophy as a broad-minded woman. She was interested speculatively, but did not believe in it."[97]

If not a theosophist, and we have only von Rosen's word for this and he says nothing of the countless other esoteric societies to which she may have been affiliated, she certainly had occult leanings as her book and pamphlets clearly demonstrate. With Blavatsky, Bloomfield-Moore developed a relationship that is described as "long" and "intimate," and Blavatsky used selections from her writings about Keely in *The Secret Doctrine*. As a rather interesting detail, a son of the marriage of one of her daughters, Count Erich Carl Gustav von Rosen, founded the Finnish Air Force in 1918 by presenting it its first airplane. Von Rosen, a noted expediter and archaeologist, had his arms painted on the upper and lower surfaces of both wings of the aircraft. His arms formed the swastika, derived from his expeditions in the Orient. The swastika of course also formed an integral part of theosophical symbology before it became used for far more sinister purposes.[98]

Another of Bloomfield-Moore's acquaintances also cooperated in her quest to promote Keely. I inspected a copy of her 1893 book in a private collection, that is inscribed by her as follows: "Howard Hinton Esq, from the compiler in grateful acknowledgment of aid in making known Keely's discoveries. October 9. 1893," and this leads us to the chance discovery of a writer with remarkable ideas: Charles Howard Hinton (1853-1907).

Hinton was the author of many essays about the fourth and other dimensions in space and time. These were collected in his *Scientific Romances* published in 1886 and *Scientific Romances: Second Series,* published in 1902. His interest was partly inspired by Edwinn Abbott's curious 1885 novel *Flatland.* Hinton wrote a novel about a circular, two-dimensional world, *An Episode of Flatland* in 1907. Hinton's many essays were an attempt to find a scientific rationale for the existence of ghosts. He also tried to imagine a four-dimensional God from whom nothing in the human world can be hidden. We may trace influences of Hinton's ideas to H.G. Wells' novel *The Time Machine* that was published in 1895.[99] Interestingly, Hinton's 1904 book *The Fourth Dimension* is listed in an authoritative UFO bibliography in the group, "The search for extraterrestrial intelligence."[100] As with Bulwer-Lytton's novels, *The Fourth Dimension* was obliged study material in Crowley's Argenteum Astrum.[101]

As was self-described Rosicrucian Francis Barrett and her acquaintance Astor, Bloomfield-Moore had always been intensely interested in aerial navigation. But apart from the 19th century esoteric milieu in which she was involved, the direct link between Bulwer-Lytton and Keely is to be found in her acquaintance with Disraeli, with whom she was "on very friendly terms."[102] Disraeli had been involved in occult experiments and crystal gazing decades before with his friend Bulwer-Lytton.

Disraeli died the year that Bloomfield-Moore came to learn of Keely's existence by reading Babcock's pamphlet with its reference to *The Coming Race.*

Since their association was a very friendly one and given the nature of the intellectual pursuits of the two, Disraeli and Bloomfield-Moore may have discussed Bulwer-Lytton's strange legacy and undoubtedly in the course of their conversations Disraeli may have confided to her at least some of his beliefs in the existence of secret societies.

Would it be entirely coincidental then, that Bloomfield-Moore referred to the connection between Keely and Bulwer-Lytton when she wrote in 1893 that "When Bulwer wrote of 'a power that can replenish or invigorate life, heal and preserve, cure disease: enabling the physical organism to re-establish the due equilibrium of its natural powers, thereby curing itself,' he foreshadowed one of Keely's discoveries."[103] It is therefore entirely possible that the assumption that Keely was influenced by the works of Reichenbach, which were published in 1862, and Bulwer-Lytton's 1871 *The Coming Race* is a valid one. If this is the case Keely probably obtained this knowledge from Bloomfield-Moore.

Thus, the initiates Bulwer-Lytton and Levi passed on the Rosicrucian tradition and provided the philosophical foundations upon which occultists and notably Blavatsky would build their own repertoire. Blavatsky was an admirer of and influenced by both Levi and Bulwer-Lytton; Bloomfield-Moore had been closely associated with Disraeli, Bulwer-Lytton's friend and companion in occult endeavors and experiments. Bloomfield-Moore became Keely's supporter and friend. Bloomfield-Moore and Blavatsky would develop a long and intimate friendship. Keely found his way in the pages of Blavatsky's *The Secret Doctrine*.

Keely was rejected by the scientific establishment, and history was quite ready to forget him, but through Bloomfield-Moore's efforts he found a warmer welcome in that other sphere of reality, the occult underground, especially with the theosophists. The Theosophical Society was founded in New York on November 17, 1875, by Blavatsky, Henry Steel Olcott, William Quinn Judge, J.S. Cobb, Seth Pancoast, H.J. Newton and Emma Hardinge Britten, who allegedly also was a member of the Hermetic Brotherhood of Luxor and from her days as a teenage medium belonged to more than one secret society that included important figures of occultism.[104] Olcott was elected first president,[105] Pancoast was elected vice president.[106]

Like Levi's occult heritage and Bulwer-Lytton's strange tales, Blavatsky's ideas would reverberate through the occult world, and this would eventually even transform the very language: the word "occultism" was unknown in the English language before its appearance in 1877 in *Isis Unveiled*.[107] Through Blavatsky, who was one of the most influential people in the shaping of the 19th century occult framework, countless other occultists, theosophists and esoterists read about Keely and his discoveries.

The first edition of *The Secret Doctrine* sold out quickly, necessitating a new printing. William Thomas Stead, who received his copy from Blavatsky herself, wrote in a letter dated December 8, 1888: "I have read only your preface and the chapter on Keely, in whose discoveries I am much interested."[108] Like Hardinge Britten, Randolph, Blavatsky, Olcott, Colville and Newbrough, Stead was an advocate of spiritism. He also edited the spiritualistic publication *Borderland* between 1893 and 1897.[109] At one time, Stead invited Cheiro, who visited Keely in 1890, to inspect a haunted house.[110]

Much like Theodor Reuss, the co-founder of the Ordo Templi Orientis who was a police spy, Stead is said to have been an informant. He was also a journalist and as editor of the then-radical and sensational *London Pall Mall Gazette*. He gave heavy coverage to the gruesome Whitechapel murders committed by Jack the Ripper. In this capacity, he accepted articles on the Whitechapel murders by Robert Donston Stephenson.[111] In this strange, dark, and haunting figure we again see a spider-thin thread leading to Bulwer-Lytton.

Stephenson, who also called himself Dr. Roslyn D'Onston and had as his occult pen-name 'Tau Tria Delta,' led a mysterious life riddled with strange coincidences. He studied chemistry in Munich—where he could have met Hartmann who studied medicine there. Stephenson also studied in Paris where he could have met Kellner who studied natural sciences with a special emphasis on chemistry in the same city. What is certain is that Stephenson did have some doings with certain Italian secret societies—the kind of which Disraeli sternly warned against—as he fought with the revolutionary Garibaldi. Stephenson also pursued occult studies under Bulwer-Lytton, and in later years he allegedly lectured on the occult. In 1890, he was living with Mabel Collins, novelist and editor of *Lucifer,* the journal of the Theosophical Society. There are some vague allusions to his practicing ritual magic. Whatever the truth, Stephenson published a curious booklet in 1904 that consisted of comparative Biblical studies. After this, he just vanished from the pages of history.[112] Like Astor, Stead would die during the tragic Titanic disaster in 1912.[113]

Tesla and Edison also learned of Keely through Blavatsky's book.[114] Tesla had other sources of information as well, but both refused to visit Keely. Apparently this led to no hard feelings, since Bloomfield-Moore quotes both Tesla and Edison in her book. A quotation of the latter she borrowed from one of Blavatsky's writings: "I don't believe that matter is inert, acted upon by an outside force. To me it seems that every atom is possessed by a certain amount of primitive intelligence."[115]

The fact that Edison was quoted by Blavatsky is not that surprising, considering his membership in the Theosophical Society since April 4, 1878, a few months after Keely extended an invitation to him.[116] A little-known facet of Edison's character was that he was a believer in reincarnation[117] and was interested in other psychic matters as well.[118]

Blavatsky's statements about Keely would continue to appear, even after the publication of *The Secret Doctrine*. In an article, published four months after her death, her views on Keely were once again put forth, together with a hint of a possible disintegration experiment: "Add to this the forthcoming long-promised Keely's vibratory force, capable of reducing in a few seconds a dead bullock to a heap of ashes."[119] It demonstrates that like Bloomfield-Moore, she kept her faith in the reality of Keely's discoveries since the day that she first came to know of him.

Thus it was no mere coincidence that in 1896 during a theosophical conference in New York, "a high regard for Keely" was being exhibited. During the convention, acting president of the Theosophical Society J.D. Buck read a paper in which he stated that, "No one holding firmly to the mechanical theory of the universe has advanced a single step in any real discovery or apprehension of the essential truths of cosmic or human evolution. The single exception is J.E.W. Keely of Philadelphia. J.E.W. Keely seems to combine the intuitions of the seer with the practical knowledge of mechanics, and is at once a scientist and a philosopher. Though he has nowhere completely formulated the old philosophy to which I have referred, his conception of the constitution of matter and the correlation of force is in complete harmony with it. In his apprehension of the working powers of nature he has no equal in his generation."[120]

And a month later, during a marriage ceremony, English Theosophical President Hargrove alluded to Keely in connection with ancient Egyptian knowledge, which was now theosophical doctrine: "In those days they understood the meaning of vibration. ...Remember too, that the sounds you will hear...are vibrations, and they, too, belong to the magic of antiquity, which it will before long become our duty to revive."[121]

Eventually through Blavatsky's writings the whole occult world learned about Keely. Even today those wishing to learn more can read Blavatsky's comments side by side with Bloomfield-Moore's statements about him in *The Secret Doctrine*. The book remains in print and the passages about Keely are still to be found in its pages. The Theosophical Society was to become Keely's domain after his death, but his alleged exposure would cause ripples on the serene lakes of theosophical content even there.

Apart from contemporary newspaper accounts, it is in the slowly yellowing pages of theosophical pamphlets and magazines like *Lucifer* and *The Theosophist* that an interesting change in tone may be gathered from these writings about Keely. After his alleged exposure by Clarence Moore, Keely was to become temporarily degraded to a marginal and humiliating footnote in the writings of the theosophical superstars.

When news of his alleged exposure reached the theosophical camp, Olcott was quick to write an apology for Blavatsky's favorable writings about him, for these had become a big problem. The exposure led several French esoterists—and from this we have a confirmation that information about Keely was not only available but also studied by some of the French occult scene—to question the veracity of Blavatsky's writings; certain theosophists of the French section demanded to know how it was possible that the discovery of Keely's discredited inter-etheric force was treated in the *The Secret Doctrine* as "a great fact," whereas it was a complete swindle; and how far this contradicts the declaration that the book was "inspired, directed and corrected by the Masters of Wisdom."

Olcott wormed his way out of this embarrassing position, and thus providing us some insight into Blavatsky's sources of information by stating that: "Of her own knowledge she knew nothing about Keely and the validity of his pretensions, she got her facts at second and third hand, from Mrs. Bloomsfield-Moore, Mr. Evans, and other old patrons of Keely."

Olcott admitted that there might be something extraordinary about Keely after all "at least in the beginning, Keely possessed some extraordinary psychical powers, however much he may have cheated later, when possibly those forces in him were exhausted, than that he was a scamp throughout. ...It is sheer nonsense to say that such superior scientists as Prof. Leidy, Mr. Wilcox and others, and the master mechanics of railways and other skilled mechanics who examined and reported favorably on the Keely motor of Philadelphia, when his first syndicate was formed to utilize the invention for railways, were suddenly stricken blind and mentally paralytic."

Olcott then revealed the fact that Blavatsky and Bloomfield-Moore were on friendly terms for Blavatsky met Bloomsfield-Moore a decade later in London, and developed a relation that he describes as a "long subsequent intimate association," during which Blavatsky "deepened the first conviction" by learning more about Keely, after which she simply "sailed away...."[122]

Since Olcott considered Blavatsky's belief in Keely another easy target for her critics, he again vented his dissatisfaction with the whole affair as a bitter footnote, while praising Blavatsky's *Isis Unveiled* and *The Secret Doctrine*. This time he left out any information in favor of Keely: "I think she would have felt deeply mortified if she had lived to read the scathing and complete exposure of Keely's fraudulent demonstrations of his 'Inter-Etheric Force,' in her own magazine, the *Theosophical Review,* of May (1899), after what she had written about it in *The Secret Doctrine* (p 556-566, first ed.). She knew nothing personally about Keely, taking her impressions and facts at second hand from a friend in Philadelphia—a shareholder in Keely's original company, and from Mrs. Bloomfield-Moore, his enthusiastic disciple and backer; ...so, without stopping to test Keely's theories or verify Mrs. Moore's alleged facts, she flew off on a tangent into a most instructive essay on cosmic forces, and by her unguarded halfendorsement of the now-proven charlatan, exposed one more large joint in her armor to the shafts of the sneering enemies." And Olcott concluded: "But what does it matter after all?"[123]

It would take two years for another theosophist, A. Marques, to correct the errors in Olcott's apology, obviously made in a time when a direct answer was apparently needed, since Olcott could have found the answers that justified Blavatsky's opinion in *The Secret Doctrine*, Marques reasoned: "In the domain of Natural Philosophy, H.P.B.'s positive announcement has proved absolutely correct, namely, that in spite of all his genius, and in spite of his working on the most accurate basis, J.W. Keely, the discoverer of the 'Inter-Etheric Force and Forces,' would fail to make a success of his invention and discoveries. When she wrote, in 1888, the world, especially America, was anxiously awaiting the harnessing of a new power, the so-called 'dynaspheric force,' and its inventor was at the height of his most sanguine expectations for using psychic force and the latent faculties of the super-physical regions of the Ether. Yet H.P.B. boldly asserted that, although Keely was a natural-born occultist or magician, although his theory was perfectly correct and quite on occult lines, yet he would never be *allowed* (his italics) to perfect his discovery, though with no apparent reason for the failure except, as she stated, that 'the fifth and sixth planes of the Etheric, or Astral Force, will never be *permitted* (his italics) to serve for purposes of commerce and traffic' (S.D.I. 613), and because the discovery of this 'terrible sidereal force,' the 'Mash-mak' of the Atlanteans is 'by several thousand—or, shall we say—hundred thousand years too premature' (S.D.I., 615), and liable to bring disaster instead of help to humanity, while 'terrible secrets, untimely discoveries' are often due to the nefarious influence of the 'Brothers of the Shadow' (S.D.III, 488)...And truly, poor Keely, after many more years of unsuccessful efforts—towards the end of which he probably was driven to use trickery in order to make his financial backers wait patiently for the delusive accomplishment which he always felt just within his grasp—had to be buried, branded as an impostor or a fraud, while his only guilt was really to have been born ahead of his time.''

In direct reply to Olcott's apology, Marques wrote: ''But the curious part of the matter has been that when his failure became patent, and in harmony with H.P.B.'s prediction, her memory was assailed and she was taken to task for the very failure she had announced, while her pupils and friends could find only the lamest apology for her defense (*Theosophist,* XX, 687), taking her 'mistakes' and of her 'ignorance' of scientific discoveries, when in fact everything she said about Keely was quite correct—the only one who took up the proper justificative argument being Dr. Franz Hartmann (ibid., 764).''[124]

Franz Hartmann (1838-1912), theosophist and author of books on the Rosicrucians, occult symbolism and magic, and close associate of Blavatsky and acquaintance of Bloomfield-Moore, made several statements about Keely. He also met him at least twice. We've already seen Hartmann's sordid medical career, but in the fields of the occult he was more successful. He became president of the German branch of the Theosophical Society in 1896, headquartered in Berlin. In 1902, Rudolf Steiner would become its secretary. Hartmann also wrote *Magic, White and Black,* published in 1888, a long and curious digression on occult topics such as invisible beings, sound, the fourth dimension, music and harmonies, Akasha, the astral spheres and planes. Hartmann is also said to have founded a highly secret Rosicrucian Order in Switzerland.[125] Like Bulwer-Lytton, it is asserted that he was frightened with the prospect of premature burial.[126]

In 1905, Hartmann founded the Esoteric Order of the Rosicross together with German occultist and cabalist Leopold Engel. Leopold Engel (1858-1931), who called himself a ''nature doctor and magnetopath,'' was obsessed with the order of the Bavarian Illuminati and in 1897 founded his own Order of the Illuminati in the German town of Dresden. In 1906 he published his influential history of the Illuminati order, the *Geschichte des Illuminatenordens.*

Engel also wrote a very curious novel titled *Mallona,* the text of which he obtained through the remote viewing of a medium into the Akashic Records. The novel depicts life on the planet "Mallona," which was destroyed through the evil use of stupendous technology, and its remnants now form our asteroid belt. As a strange echo of Bulwer-Lytton's *The Coming Race* that Engel undoubtedly read—a German translation was available as early as 1874—flying devices are described, and also the usage of "ether-power." The concept of Akasha-chronicles, so widely dispersed through *The Secret Doctrine* that was published in a German translation in 1903, prompted Engel to write in *Mallona:* "It is a known fact that all things which have ever occurred, do not disappear without a trace, but instead are being photographed and kept in the universe. From every occurrence emanate light waves, which travel in the universe. Would one succeed to capture these light waves in another place and collect them in a suitable device, or to deliver them to a receiver, one would be able to recreate...the same image."[127]

In March 1888, Franz Hartmann had a long conversation with Keely and inspected his devices with great interest during one of his visits.[128] Allegedly Hartmann also made a sketch of one of the devices.[129]

Hartmann first heard of Keely years before he actually visited him, years before *The Secret Doctrine* was printed and years before Bloomfield-Moore tried to hire him to cure his daughter of mental illness. This could mean that Hartmann either learned about Keely through a contemporary newspaper or that, perhaps this is a glimpse of other uncharted channels, possibly of an occult nature through which news about Keely traveled. Hartmann in any case gives no clue, and with this it is unfortunately left open to conjecture.

Hartmann, however, wrote about Keely: "I have taken great interest in him ever since I first heard of him in 1882. I believe that the world is entering into a new era of existence, and will become spiritualized from top to bottom. As gaslight has driven away, in part, the smoky petroleum lamp, and is about to be displaced by electricity, which in the course of time may be supplanted by magnetism, and as the power of steam has caused muscular labor to disappear to a certain extent, and will itself give way before the new vibratory force of Keely, likewise the orthodox medical quackery that now prevails will be dethroned by the employment of the finer forces of nature, such as light, electricity, magnetism, etc."[130]

In the end Hartmann and Keely fell out, the former claiming that Keely would never be able to "utilize the force in mechanics," but that Keely's mission was "to spiritualize the world instead of advancing its material progress,"[131] which would help Marques in his arguments in justifying Blavatsky.

Burgoyne, who coincidentally came through Philadelphia—he had to check past the immigration authorities there while immigrating to America[132]—echoed Hartmann's earlier hopes, when he wrote that, "Startling discoveries in chemistry, electricity and all the physical sciences will be brought to light. Steam will be superseded by compressed air (gas), electro-magnetism (atomic power) as a motive power."[133]

Not only that, it also demonstrates, since the Theosophical Society and the Brotherhood of Luxor did not get along very well, that a deep-rooted interest in avant-garde technology is to be found across the whole occult spectrum, a topic to which we shall return in the following chapter. Burgoyne's book, which received favorable comments by Hardinge Britten who by now felt a deep hatred for theosophy,[134] was distributed in 1897 by a Chicago based esoteric organization, The Progressive Thinker Publishing House. Their advertisement slogan, which was published in the book *Ghost Land* edited by Hardinge Britten, reads: "Keep your brain vibrating!"[135]

Interestingly, there is another minor connection between the theosophists Olcott, Blavatsky and Keely, a minor triviality or a small coincidence so to speak. Olcott wrote that "One day in the month of July 1874, I was sitting in my law office...when it occurred to me that for years I paid no attention to the spiritualist movement. ...I went around the corner to a dealer's and bought a copy of *Banner of Light*.[136] Other persons in the Keely history who submitted their writings to this spiritualist periodical published in Boston were the freemasons and spiritualists John Ballou Newbrough and Keely's friend William Colville.

William Wilberforce Juvenal Colville (1859-1917) was a medium; of how he became aware of his talent, two versions exist. Colville claimed that in his very early childhood, his mediumship "originally declared itself." As a child he had a sensation of "information flowing into me. I can only liken my experience to some memorable statements of Swedenborg concerning influx of knowledge into the interiors of human understanding." Colville treated his mediumship as a talent that consisted of three features; one was clairvoyance, the second as "mental enlightenment" or "intellectual illumination" and the third being "the actual predicting of coming events." On May 24, 1874, Colville "experienced the first thrill of consciousness that it was my principal lifework to travel nearly all over the earth, guided by unseen but not unknown inspirers."[137]

History has a more sober opinion on the origins of Colville's mediumship and dryly states that his mediumistic talents were disclosed on May 24, 1874, coincidentally the year that marked Keely's appearance before a general public and just a month before Olcott would go around the corner to buy a copy of *Banner of Light*. That day, Colville became conscious of a spirit presence during a meeting. Until 1877, he was often found answering questions while unconscious, explaining his unawareness of his physical mediumship. At other times, he asserted that he heard every word he spoke as if it came from strange lips.[138]

Whatever the differences, 1877 would start Colville's career as a lecturer and medium. In the intervening years between 1874 and 1877, he had "many opportunities for witnessing extraordinary phenomena, as I became well acquainted with many prominent spiritualists. ...I had many opportunities for sitting in circles with Williams, Herne, Monck, Eglinton, and other extraordinary mediums, who, at about that time, were either in the inception or at the zenith of their fame." During his investigations, Colville was repeatedly being told that he was "a physical medium." Although he was not aware of this, he did admit that "planchette has worked for me repeatedly and automatic writing has been often with me quite an everyday experience." During the greater part of 1877-1878, Colville was "privileged to investigate the evidences of phenomenal Spiritualism all over England. The most private gatherings were open to me, and I was times without number privileged to sit with the most distinguished mediums."

His first performance as a medium took place in a Masonic hall on March 4, 1877. Apparently he did well, as the newspapers would herald him as "one of the marvels of the nineteenth century." As a consequence he started to tour as a lecturer for nineteen months between March 1877 and October 1878. He went to the United States towards the end of October 1878 and arrived in Boston. On reaching America he discovered that his arrival had been announced in the *Banner of Light,* "the oldest spiritualistic newspaper in the world" as Colville wrote.

After lectures in Boston, Colville would visit New York, Philadelphia, Chicago and other cities. Like Randolph, Newbrough and Leland, he travelled to various places and countries. In 1883, he would once again go to England, to return to the United States a year later. In 1885, he revisited England and a year later visited California for the first time. He joined the Theosophical Society in 1890.

In 1895 after a 10 year stay, he would once again visit England and Europe. In 1897, he held private midnight seances with a group of "several professionals." This group assembled twice a week at midnight. Their chief focus of attraction was, as that of Bulwer-Lytton and Disraeli once had been, a "huge crystal placed in the center of a large library table. The crystal was as large as an ordinary globe for containing goldfish, and into this brilliant object we all quietly but intently gazed."[139]

Colville not only co-edited *The Gnostic,* in which Dowd's "Rosicrucian Temple" was serialized, he also wrote profusely. His writings number over 125 publications, but with his *Dashed Against the Rock,* he left us his documentation of what must have been a series of remarkable conversations with Keely. When Keely died, Colville delivered a speech during the funeral services and gave a lecture titled: "In Memorial, John W. Keely." He also wrote the memorial address that was held on November 27, 1898, in Casino Hall, Thirteenth Street and Girard Avenue in Philadelphia, where he also gave his lecture on Keely the same month. The memorial address was published in 1899 by the Banner of Light Publishing Co. In it, Colville wrote that "Keely has been well looked upon as the fulfiller of many mysterious predictions. There are those among theosophists and others who have not hesitated to say that he was a soul embodied for a very special purpose; that he came to earth by direction of those mysterious masters who are called Mahatmas in the Sanskrit tongue, that he might give openly to this generation a secret which has been held in the keeping of a few especially illumined ones from times immemorial. It has also been said that, in consequence of the unpreparedness of the populace—in consequence of the lack of spirituality on the part of the great people everywhere, that obstacles have been thrown in Keely's way, even by the very spiritual messengers whose servant and representative he was."[140]

Unfortunately in both Colville's book and written speech there is a general lack of biographical details concerning their association. When, where and under what circumstances Colville met Keely is thus open to conjecture. Possibly Colville became acquainted with Keely as early as 1878, when he lectured in Boston, or around 1888, when Bloomfield-Moore's *Keely's Secrets* and Blavatsky's *The Secret Doctrine* were published, or in 1890 when he joined the Theosophical Society. The interest of a spiritualist and a medium in an inventor was possibly fueled by the ambiguous views of Keely that the theosophists shared. On one hand, Keely was considered as someone who rediscovered the fabulous powers and forces of the ancients - on the other hand, he was considered a person with special psychic capabilities.

Colville wrote profusely on a variety of subjects, such as astrology, spiritualism and reincarnation, fashionable subjects considering the occult milieus that he frequented. But his 1894 *Dashed Against the Rock*—in which we hear Keely speak— is a quite different book with futuristic ideas such as space travel, a theme to which, like Greg, Delisle Hay and Astor, he never returned to in his later writings. In Colville's case this may be explained by the fact that the book was, in all aspects, an account of Keely's visions. In Colville's autobiography, written in 1906, Keely is strangely absent, and Colville is more concerned with the depiction of his career as a medium and other psychic matters. The only impression that he wrote about his association with the inventor was found in his speech: "Keely was a man you could only know if you had a spiritual discernment; you could not get really acquainted with him simply by talking to him. The only way in which you could become familiar with him at all was by sitting down quietly with him and breathing in some of the mental atmosphere which he breathed, and feeling something of the spirit which animated him, then you might realize that you and he were spiritual neighbors."[141]

Undoubtedly, Cheiro felt the same fascination when he visited Keely in 1890. As we have seen, his memoirs are highly inaccurate. He could have conceived to do so either on the instigation of Stead or Blavatsky, whom he met once,[142] or indeed after visiting Bloomfield-Moore's London home. Perhaps a hint is to be found in one of his writings: "Is there a connection between music and communion with the dead?" he wonders.[143] Cheiro was a fanciful pseudonym appropriately chosen to name his profession; for the name was chosen by Count Louis Hamon (1866-1936), the most famous palmist or chiromancer of his time.[144]

Conventional history depicts Cheiro, who asserts that he never was "a member of any Spiritualistic Society, or any sect dealing with Psychic matters" and refused to become a member of the theosophical society when Blavatsky asked him to join,[145] more or less as a charlatan, although it is admitted that "it appears from the stories of many of his clients that his predictions were remarkably accurate."[146]

What Cheiro read in Keely's handpalm and what he concluded from what he detected is open to speculation. But his visit left a lasting impression on Cheiro, so much that he felt the need to record his experiences twice. His memoirs about his visit are lacking in accuracy. In another book, he wrote about his visit several times, but in heavy fictionalized form. In one of his short stories, Cheiro meets a person only vaguely described as "a London recluse," who explains to him that "he believed the law of vibration was the key to unlock the secrets of 'the beyond,'" but not only that: "In my investigations I have discovered certain chords of music that create the class of vibration necessary for the manifestation of still higher beings who inhabit what is miscalled the 'invisible world.'...I have discovered that certain chords create the class of vibration necessary for still more important manifestations." And echoing Blavatsky's statement with which chapter 1 opened: "the faintest chord of harmony produces its counter wave in endless space until like two affinities *they blend together and reappear* in greater strength (his italics)."[147]

Cheiro returned to this theme in another short story titled, "Turning Back the Clock of Time." Where Colville named Keely "Aldebaran," in Cheiro's short story Keely was now called "the Mystic" and the location was again shifted from Philadelphia to London. While certain passages such as those that describe the appliance of neon-lamps suggest that perhaps the Mystic was a fictional composite of both Keely and Tesla, we may deduct from passages in Cheiro's tale that he definitely meant Keely; "I had known from the first that he had mastered the most difficult problems of what may be called 'etheric electric waves,' and long before broadcasting receivers had been thought of, he had fitted up in his house a receptacle that collected both sound and speech from the ether, far beyond anything we have at the present day.[148] As a demonstration of the theory that the Law of Vibration was the key by which he could utilize some of the great forces of Nature, he could at any moment, by a chord of music on the organ, piano or harp, call whatever lamp he wished into action, or extinguish any or all of them in a similar manner. ...It is only the vibration and speed of the revolving molecules that hold the particles of iron, stone, or any other solid mass in a state of solidity. ...This man in his work had demonstrated that human life itself was only a matter of vibration. Thus, the Mystic was able to let a person leave his or her body 'for a certain time' by slowing down 'life's responsive throb to such an extent that the sub-conscious brain found its freedom.'"[149]

We further learn that the Mystic had "by another invention...utilized the electricity in the higher atmosphere, and collected it by storage batteries of enormous dimensions, and these, connected with vacuum glass tubes filled with 'neon' or some other gas, produced a light that appeared to flood every part of a room in which one of these lamps was placed."[150]

The Mystic was also able to control "electric forces that he was able not only to dissolve any metal, but he could direct the 'ions' of the dissolving metal into any organ or part of the body that he so desired."[151] Through this control the Mystic was able to cause electric currents to carry "ions" of copper, silver, gold or mercury into the body and cure many a disease. More wonders Cheiro encountered during a visit at the abode of the Mystic; a device called "register of Thought," the "delicately poised needle in this wonderfully constructed instrument" would be "affected by the aura or soul radiation of any person standing within a few feet of it and recorded by its movements the effect of thoughts passing through the brain."[152]

Cheiro also asserts that such a device was actually satisfactorily tested in England in August 1897. A fascinating prospect unfolds about one of Keely's devices being shipped to England, but we must not forget that Cheiro's memoirs lack in accuracy, and this time by choosing fiction as his form he presents us with an equally suspicious account dating it seven years after his visit with Keely. Very much in the realms of fiction is his description of a device built for the occasion, although one cannot help but think of the unexplained metal rod that was found dangling from the ceiling of Keely's workshop: "Near the window (was) a curious-looking couch covered with copper, insulated with glass feet...At the head of it lay a compass showing by the position of the needle that the couch lay north to south in a direct line with the magnetic current. On a table at the side I noticed a helmet of copper with a copper band so constructed to go down the spine, with two arms from it to go round the body and terminate in a twelve-pointed magnet on the solar plexus. Connected to the center of the helmet an insulated covered wire led through the open windows to a series of copper wires hanging from the edge of the high roof to a few feet from the ground. These in their turn were joined to an aerial of immense height over the house...the copper plate on which my feet would rest was connected with a wire which, passing through what he called 'a magnifier' at the other end of the room, terminated in an 'earth' zinc pole at the bottom of a well in the garden." In other words, "it was intended that my brain should be *exactly like the receiver in a 'wireless set'*" (his italics).[153]

Where Hartmann's occult companion Engel in his *Mallona* proposed the idea of capturing light waves that emanated "from every occurrence" in "a receiver" or "a suitable device" to "recreate the same image"—thus in fact watching into the Akasha chronicles— in Cheiro's tale the Mystic has actually built such a device. The apparatus, helmet, magnet and all is specifically made for that purpose the Mystic explains. "Science has proved that the light of some of our distant stars commenced its journey to the earth when it was 'without form and void.' ...That light reached this world, perhaps yesterday, after traveling thousands and thousands of years from some far-off star. What, then, if its photographic beams could be reversed and the scenes of long past ages could be reconstructed before our eyes. There is nothing lost—there is nothing impossible. Let us make the attempt."[154]

Cheiro bravely climbs on the couch, fastens the helmet on his head, places the curiously shaped magnet on his chest and "the Mystic" plays "a series of chords" on the organ. "The vacuum lamps responded, the lights changed and flooded the room with a pale gray shade like that of a ghostly dawn...The music ceased. I could see in the weird gray light the figure of the Mystic leaning over the keys, *waiting and listening* (his italics). A slight vibration came down the aerial leads, passed through my body and echoed back through the magnifier at the end of the room",[155] and the palmist is transported to the lost kingdom of Atlantis moments before its destruction. The Atlanteans take the destruction of their magnificent empire, the "highest pinnacle, greater intelligence can not be created, more knowledge cannot be attained," stoically, for "Men have become as gods, and being as gods, they bend their heads to Destiny."[156]

Hartmann's visit did not leave a lasting impression on him or Keely. We can only speculate about the nature of Hinton's aid. There were Colville's account of Keely's visionary ideas and Cheiro's distorted and fictionalized reminiscences. There was another occult contact however, who was relevant to Keely's research. This contact involved Seth Pancoast, one of the founding members of the Theosophical Society. Like Bloomfield-Moore, Pancoast is almost completely forgotten and is sadly absent in most studies of occult history. In contrast to Hartmann his knowledge of the occult was superior, as he was one of the most learned cabalists of his time.

In 1895, a contemporary newspaper remarked that Bloomfield-Moore made a statement that was "of interest as showing one of the sources of Keely's inspiration in the pursuit of his investigation."[157] At that time she confided to a reporter that, "To the fruits of the thirty years of research which our late townsman, Dr. Seth Pancoast, devoted to the study of the Hebrew Kabbala and to occult science, Mr. Keely is indebted for the instruction which has enabled him to fasten his machinery to the very wheelworks of nature, drawing from space a current of force and demonstrating by dynamic apparatus that it is the governing or controlling force of the universe."[158] To theosophists this was not exactly new. In 1888 Bloomfield-Moore was already writing about similarities in the lines of thought of Keely and Pancoast.[159]

Seth Pancoast (1823-1889) spent the first few years of his adult life in business, but when he was twenty-seven years, he began the study of medicine at the University of Pennsylvania. The year after his graduation as MD in 1852, he became professor of anatomy in the Female Medical College of Pennsylvania. At the end of the year he resigned to become professor of anatomy in the Pennsylvania Medical College. In 1859, he became professor emeritus. He held a private practice and taught at the college until 1865. At that time he interested himself in cabalistic literature. In this field, he became a noted scholar and built up what allegedly was to become the largest library of books dealing with the occult sciences ever assembled in America.

The ideas gleaned from his cabalistic studies curiously mingled with his medical and scientific knowledge, which led him to write a number of extraordinary books. The first of these was *The Kabbala; or the True Science of Light; an Introduction to the Philosophy and Theosophy of the Ancient Sages,* published in 1877, two years after his involvement in the foundation of the Theosophical Society. This book is said to be the first book ever written in the English language that tried to explain the Ten Sephiroth and give the mystical interpretation of the Holy Scriptures as contained therein.

Pancoast's treatise on the cabbala was republished the same year under the title *Blue and Red Light; or, Light and Its Rays as Medicine; Showing that Light is the Original and Sole Source of Life, as it is the Source of All the Physical and Vital Forces of Nature, and that Light is Nature's Own and Only Remedy for Disease, and Explaining How to Apply the Red and Blue Rays in curing the Sick and Feeble.* While its title suggests a new therapeutic idea, the book is a cabalistic writing in which mystery, science, religion, and medicine are blended.[160]

It is asserted that Pancoast was not only learned in the cabala and the theory of magic, but was also alchemically inclined and a practicing color therapist which is clearly demonstrated by the titles of his treatise.[161]

In 1875, before the founding of the Theosophical Society, Pancoast unsuccessfully tried to cure a leg injury that Blavatsky suffered while falling down the pavement.[162]

Pancoast's philosophies were in the same vein as those of E. Babbitt, who in 1878 privately published his classic study on the therapeutic value of colors and the etheric forces, titled *Principles of Light and Colour: Including the Harmonic Laws of the Universe, the Etheric—Atomic Philosophy of Force, Chromo Therapeutics & the General Philosophy of the Fine Forces, Together with Numer. Discoveries & Practical Applications.*[163]

There is some circumstantial evidence that the learned Pancoast wrote more than was every published. Some years after his death, his son wrote to Bloomfield-Moore. She stated that she received a letter from "the son of Dr. Seth Pancoast, author of the *True Science of Light,* whose manuscript works I tried to buy from his widow after the death of Dr. Pancoast in 1889. She refused to sell them then; but the son wrote to offer them to me in order to raise money for his collegiate course. I could not have encouraged him to hope from any assistance from me, in anyway, as I am not well enough to undertake the editing of them, but I at once referred him to Dr. Lounders after hearing from the doctor that he would revise and edit the work, if it proved to be what I had reason to think it was."[164]

It would be another highly influential occultist however, who publicly became involved with the Keely mystery long after those fateful events and long after most of these intriguing people faded from the pages of history. His statements provided alternative answers to the questions that have transformed the Keely history into the Keely mystery; questions such as why Keely's devices are now so strangely missing? What happened to his manuscripts? And why Keely was silenced and his inventions rigorously suppressed?

Rudolf Steiner (1861-1925) started his esoteric career in the Theosophical Society but referred to Keely more than once in his writings and lectures. After he joined the Theosophists, he drifted amongst others to the Ordo Templi Orientis for a number of years before he founded his Anthroposophical movement in 1912.

Steiner obviously learned about Keely through Blavatsky's writings, as so many did, or through Hartmann's statements whom he met.

The occultists and those in favor of Steiner considered—and still consider—him as a high initiate who was able to cross the barriers of space and time and could see the past, and even into the future. Steiner was aware of his capabilities early in his childhood. His Anthroposophy was to him the road to knowledge that would lead the spiritual part of mankind to the spiritual realms of the universe. The aim was to acknowledge its spiritual contents in the whole of creation.

Steiner also believed that man consists of four parts: as a physical body, as an ethereal being, as an astral body, and as ego; and that a close cooperation exists between the physical and the spiritual parts of both man and the universe.

Olcott hastily tried to rupture any connections between Keely and Blavatsky after Keely's alleged exposure, a situation that would be corrected several years later. Significantly Steiner instead began to lecture about Keely long after Keely's alleged exposure. The first recorded lecture in which he mentioned Keely was on March 30, 1905. He also referred to him during two lectures held in 1906, during three lectures in 1916 and again during a lecture in 1920.[165] It must be taken into account that Steiner held a great number of lectures, more than 6,000, which limits the times that he choose to speak about him to a rather insignificant proportion.

Moreover, the interpreters of Steiner's words on an occult level have satisfied themselves with the explanation that Steiner incorporated Keely in his grandiose vision on human evolution and the future of mankind—as Steiner did with Bulwer-Lytton—and that he wanted to say that the time was not right for Keely's inventions to be of benefit to mankind. If that is the case, Steiner merely echoed the opinions of Colville, Hartmann, Harte and Blavatsky. For Harte wrote in 1888, a short time before *The Secret Doctrine* was published, "Whether Keely's inventions will be a commercial success at present is another matter. The force, or, rather forces, which Keely handles, are the same as those known under other names in Occultism, and it is the belief of Occultists that these forces cannot be introduced into the practical life of men, or fully understood by the uninitiated, until the world is fit to receive them with benefit to itself—until the balance of the good and the evil they would work is decidedly on the side of the good. ...The discoveries of Keely have an occult side, which perhaps he himself may not fully perceive."[166]

Blavatsky would later write that "whenever such individuals as the discoverer of the eteric force—John Worrell Keely—men with peculiar psychic and mental capacities are born, they are generally and more frequently helped than allowed to go unassisted. ...Only they are helped on the condition that they should not become, whether consciously or unconsciously, an additional peril to their age."[167]

Notwithstanding his opinion of Keely, which is in part based upon the opinions of the theosophists,[168] Steiner spread the information about Keely in anthropsophical circles. Thus only recently references to Keely have been unearthed in the notes of Walter Johannes Stein,[169] a remarkable anthroposophist, writer of an erudite grail-study,[170] and the protagonist of Trevor Ravencroft's *Spear of Destiny,* written in the same manner as Bulwer-Lytton once wrote his *Zanoni* and *The Coming Race.* Stein also learned about the Keely motor from another source: "Mr. Dunlop told me that he saw the Keely motor, but he did not see it function." Interestingly, Stein remarks that this occurred before "the world at first learned more of this case through Madame Blavatsky."[171] Guenther Wachsmuth, who translated *The Coming Race,* was so taken by Steiner's remarks that he promptly wrote his book on the forces of the ether.[172]

Three of Steiner's accomplishments set himself distinctly apart from the theosophical thoughts on Keely. These accomplishments, most important in the Keely history, are the construction of what has been called "The Strader Instruments," of which a full exposé is given in chapter 11; the performance of Steiner's Mystery Plays in Munich in 1912 for which the Strader Instruments were meant; and a series of lectures during which he discussed Keely and his inventions. Steiner also modeled the figure of Dr. Strader partly on Keely in his *Four Mystery Plays,* published in 1912.[173] In the third part, Dr. Strader invented a device that operates on the fusion of vibrations. Through this device, mankind will be freed of toilsome labor and the use of expensive energy. Mankind, Steiner reasons in his *Four Mystery Plays,* would have all the time for self-tuition and spiritual development. But Strader was not to complete his remarkable device. In the fourth part of the *Four Mystery Plays* he dies while his device is still in the laboratory phase.

In 1918, Steiner gave six lectures, one of which is most interesting in connection to the Keely mystery. During this particular lecture, held on December 1, and titled "The Mechanical, Eugenic and Hygienic Point of View of the Future," Steiner explained that mankind would develop three new faculties during the next centuries. This would happen in the same natural way as mankind's mental faculties had been developed during the past. These three new faculties would be the new mechanical faculty, the new hygienic faculty and the new eugenic faculty. Steiner believed that especially the English and the Americans would develop the new mechanical faculty. New machines and mechanistic devices would be the result. These would operate through "the laws of the fusion of vibrations."[174]

He also confided to his audience that indeed there existed secret societies. But he did not mean orders such as the Golden Dawn, the Hermetic Brotherhood of Luxor, the Freemasons, the Rosicrucians, the Ordo Templi Orientis or the Theosophists. These orders, while being secretive and surrounding themselves with an aura of mystique, were never secret in a strict sense.

Steiner made it quite clear that he did not mean any of these occult orders. He had been initiated in most of these orders and he was honest enough to admit that this secretiveness that was also apparent with his Anthroposophical Society was nothing more than a relic from the early days of theosophy. At that time it merely served to provide the society with a "special distinction."[175]

We could easily dismiss Steiner's statements as pure allegations typical of an occultist, eager to clothe himself with a supposedly profound occult knowledge. But the fact that Steiner saw most of the 19th century occult orders from the inside gives food for thought. Was he simply trying to impress his audiences? This we will probably never know, because as Disraeli in his novels and speeches and Bulwer-Lytton in his letter to Hargrave Jennings before him, Steiner never really bothered to identify these mysterious secret societies. During his lecture he only confided that these societies existed in the English-speaking countries and were "sources through which, by certain methods of which I perhaps one day will speak...truths may be begotten according to which one may rule things politically."[176]

He also stated that, "In the denial to others of a certain kind of occult knowledge, that is being cherished in these centers, lies an immense power."[177] This knowledge was strictly reserved only for those mysterious secret circles, for according to Steiner this was "the only way by which it is possible to attain world domination."[178]

These secret societies plotted, according to Steiner, to conquer the entire east "that begins with the river Rhine and continues further east to Asia," and to establish a caste of rulers in the West and a caste of slaves in the East,[179] thereby morbidly foreshadowing Hitler's own plans, partially based on occult concepts, that would plunge the world in a black abyss of horror, pain and sorrow.

He also told that the initiates of "English secret societies" very well knew that "through the use of certain abilities, which are until now hidden from mankind, but which develop themselves through the law of resonance, machines and other mechanical contrivances could be set in motion." Then Steiner exposed a bit of the deeper meaning of his *Mystery Plays:* "You can find a hint in that which I have brought in connection with the person of Strader in my *Mystery Plays.*" With this Steiner of course meant a device that he had developed and that is known as his mysterious "Strader Instruments."

But he also delivered a most puzzling statement: "These things are today in development." What did Steiner mean? Was he just adding a little atmosphere, an aura of mystique to his lecture, was he drawing on the reservoir of occult legend and hearsay, or was he hinting at the fact that somewhere, in total secrecy an unknown group was constructing avant-garde technology along the lines of Keely's discoveries?

Calling the fact that so much of Keely's legacy is now missing, perhaps what Steiner further said to his audience obtains an ominous meaning. Echoing Blavatsky who wrote that, "It is just because Keely's discovery would lead to a knowledge of one of the most occult secrets, a secret which can never be allowed to fall into the hands of the masses,"[180] Steiner continued with a grave warning: "These affairs are being guarded as a secret in those circles on the subject of material occultism. There are engines possible, that, because one knows its vibrational curve, can be set in motion through a very small human influence. ...But they will, when that which I call mechanical occultism will be put in practice, which is an ideal of these secret circles, deliver not only an equivalent of five- or six hundred million in human labor, but they will achieve about one thousand million in human labor. Therefore nine tenths of the human labor of the English speaking populace will be superfluous. But mechanical occultism will not only make nine tenths of the labors...superfluous, it will also make it possible to paralyze every rebellion of the unsatisfied masses. The ability to set engines in motion according to the law of resonance with harmonious vibrations, will develop itself with the English speaking populace. Of this, those secret societies are well aware. On this they count when they will attain the dominance over the entire population of the earth."[181]

Steiner's allegations may have the ringing of some actual, unnerving truth. Perhaps for this he had his reasons to warn, yet to speak in vague allusions and riddles. For only after closer examination some people who were involved in the Keely history reveal their shadowy alliance or association with what might be just another college fraternity, one is obliged to join out of social habit and tradition when undertaking a study at a university, but of which it is asserted to be one of the most powerful, dangerous secret societies of the last century: the chapter of the order Skull & Bones. This society does not seem to have overt occult or esoteric overtones, but instead seems to be a breeding ground for illuminated politics.

The society was founded at Yale University in 1833 by General William Huntington Russell and Alphonso Taft, after Russell obtained a charter to do so from an unnamed German Society. The chapter exists even today and is outwardly modeled on a Masonic institution; thus it uses the number 322 to denote itself. It uses the symbol of the skull and crossbones and its members undergo initiatory rituals.[182] Possibly the chapter of the Skull & Bones order was a continuation of the Pi Beta Kappa lodge that is said to have been imported by Benjamin Franklin from France in 1776.[183]

Dr. Brinton, who at one time witnessed one of Keely's experiments and who translated Keely's terms in a comprehensible list, was a member of the Scroll & Key society, a group that revolved around the chapter of the order Skull & Bones, and was founded at Yale a few years after the Skull & Bones chapter was founded. Wayne MacVeigh, at one time Keely's lawyer, whose "dexterous move" was instrumental in releasing Keely from jail in 1888, studied at Yale. It was he who made "a brilliant speech...for the resuscitation of the Brothers and Linonia."[184] The English equivalent of the Skull & Bones is referred to as "The Group," and although John Jacob Astor's name does not crop up, it is asserted that "The Astor name is prominent in 'The Group' in England, but not in the Order in the U.S."[185]

The most unsettling thing is that outwardly these persons seemed to have aided Keely. The question thus remains open as to whether or not these persons, members of either sinister groups—or in fact innocent college fraternities—revolving around or having to do with the Skull & Bones, were indeed involved in a complex scheme or shadowy machinations, and whether or not they were in league with, or instead quite against the unknowing Keely.

Dale Pond's replica of Keely's Musical Dynasphere (front and side view)

Provisional Engine

The globe and drum revolve in opposite directions through the action of etheric force which is transmitted via a wire of platinum and silver.

Keely in his workshop

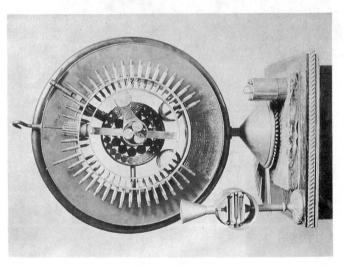

Compound Disintegrator

Vibratory Indicator, Attractive Disks, Musical Globe, and Medium for testing centripetal force.

Keely's Provisional Engine

Keely with Vibrodyne Motor

Nikola Tesla (1856-1943)
Brilliant inventor who knew of Keely's discoveries.

Nikola Tesla's laboratory in New York, which mysteriously burned
to the ground in March, 1895.

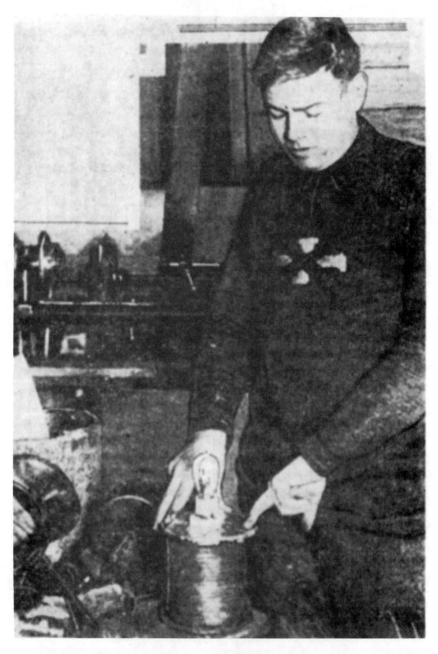

In 1919, Alfred Hubbard built and demonstrated his first device at the age of 19. He claimed that his device could take energy out of the air.

When Hubbard's device was successfully tested, the local newspapers reported the news on the front page.

Rudolph Steiner (1861-1925)
Occultist and one-time member of the Ordo Templi Orientis. Founded
the anthroposophical movement. Wrote the *Mystery Plays* and developed
the enigmatic Strader Instruments. He lectured on Keely and told his
audience that certain English secret circles were developing technology
based on Keely's discoveries.

Steiner's first Goetheaneum, built of the same wood used in the construction of violins to ensure proper resonance and vibration. Mysteriously burned down in 1922.

Steiner's mysterious Strader Apparatus. One of its parts consisted of a metal which had not yet been discovered.

The Strader Apparatus with its accompanying devices

Three devices accompanying Steiner's Strader Apparatus

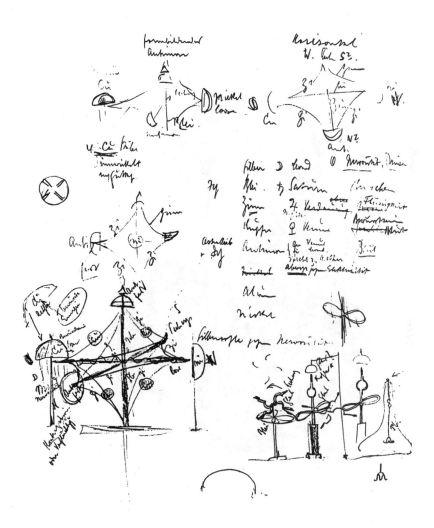

Sketches and working drawings of the Strader Apparatus

Saint Yves d'Alveydre (1824-1909)
French occultist, kabbalist, author and inventor of the Archeometer.

L'ARCHEOMÈTRE

Rapporteur synthétique des Hautes Études

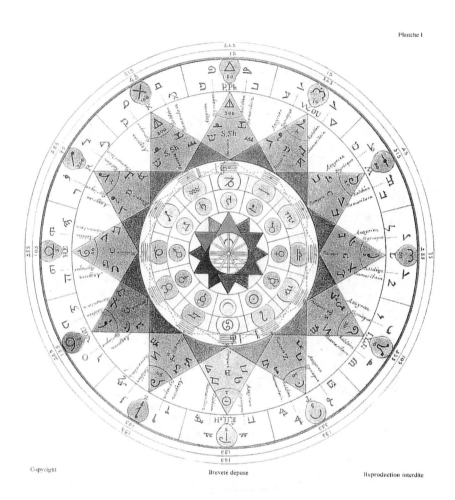

Les III lettres de construction : A, s, Th.

Saint Yves de'Alveydre's Archeometer that "translated into the material the word, form, color, smell, sound, and taste, the key to all religions and all the sciences of antiquity."

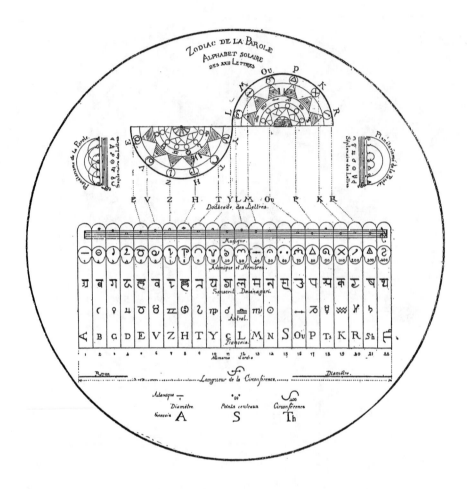

Schematic depiction of the correspondences on the Archeometer between numbers, letters, colors and musical notes, the sign of the Zodiac and the planets.

Joseph Maria Hoëne-Wronski (1776-1853)
Inventor of the Prognometer. Brilliant mathematician and initiate of
the highest order.

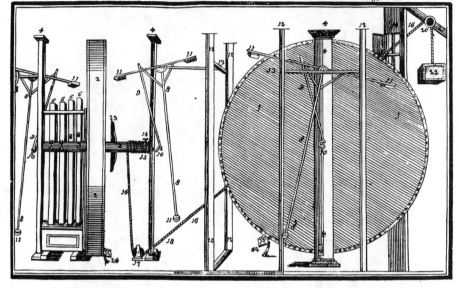

Only surviving sketch of Wronski's Prognometer. It was scrapped before his death, and later discovered by Eliphas Levi in a Parisian junkshop.

The self-moving wheel that Orffyreus exhibited at Merseburg, Hesse-Cassel in 1715.

Johan Ernst Bessler, alias Orffyreus (1680-1745)
Invented a machine that allegedly operated independent of any known
source of power.

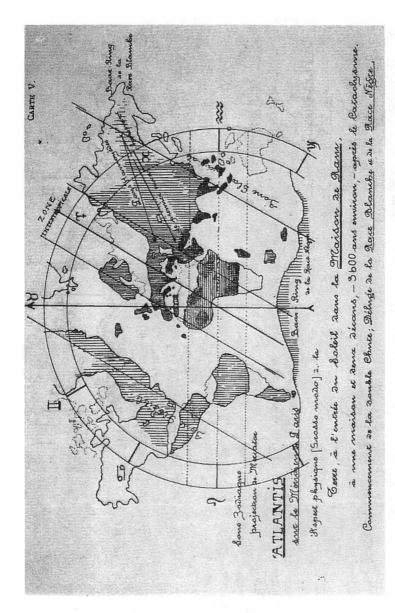

Map from the Appendix of the book *Western Occultism and Eastern Esoterism* by Auguste Van Dekerkove, alias Zanne (1838-1923), showing the Zodiac and Atlantis on the Parisian meridian.

10

The Secret Tradition
Occult Technology and Free-Energy

"...transporting systems, and the lighting of cities, and the operation of factories may someday be the outcome of...what I'd call mechanical witchcraft."

Charles Fort, *Wild Talents*, 1932

"The technology of the future will be magic. But magic is: technology with other means."

Eugen Georg, *Verschollene Kulturen*, 1930

There was some interrelationship between Keely and the academic sciences, but his entanglement with the occult communities of the time was considerable. There was already a strange technical dimension to the occult underground that was partly connected to the rapid scientific developments of the 19th century. This strange dimension helps to explain the immense success that *The Coming Race* enjoyed in the occult circles, for his novel struck a deep note of recognition.

These circles cherished the occult concept of the will of the magician, the adept, the initiated. With this will, the mind-force, and its proper application, anything would be possible. This concept was used by the early alchemists and Rosicrucians, by Mesmer and Von Reichenbach, and further developed by Bulwer-Lytton in *The Coming Race*. Deviating from this concept was the idea that Keely's devices were propelled by his very willpower, the waves of the brain, as set forth by Blavatsky, later Steiner and ultimately Fort.

It was the occult concept of the will that was an important ingredient of the occult technology that in that respect was a prophetic forerunner of today's cybernetics, virtual reality and most innovative science. For not all was pentagrams, old grimoires and the muttering of evocations when the stars were right.

That strange technological dimension is only mentioned in several studies of the occult, but never has been the subject of a study in itself.[1] It is in this exclusive perception of reality that we encounter wonderful tales of homunculi and golems, of automatic people and artificial heads that communicate. There the skies are filled with the fabled vimanaas of the Great White Brotherhood. In this section of reality, the spiritist uses devices such as the Ouija board as a rudimentary interface, or the secluded environment of the Psychomanteion for dialogues with other-dimensional inhabitants.

It is also in this hazy area that the irrationalist whose mind is forever tuned to grandiose vistas of the universe discovers new rays, weird magnetism and strange powers. Through assiduous study the very foundations of cosmic forces and the laws of the universe are uncovered. It is in the darkest and most unexplored regions of the occult that the willpower of the initiate is linked to strange devices, to achieve a magical transformation of the universe itself, an aim that Rudolf Steiner hinted at discussing certain secret societies.

It is therefore quite possible that somewhere deep in the occult underground, Keely's work was continued after he died, his devices carefully hidden and his manuscript diligently studied precisely with this purpose in mind. Based on the existing documentation this is impossible to verify, let alone prove. We know that circles surrounding the Golden Dawn were very much aware of Astor's futuristic writings, of Bulwer-Lytton's novels, of Jules Verne's futuristic tales but also of such strange tales of fiction as Bradshaw's hollow earth novel *The Goddess of Atvatabar*. But whether or not the members of these orders were actually involved in the construction and use of a variety of techno-magical instruments that went beyond the traditional wand and other ritualistic appliances is now matter for conjecture. Contemporary studies make no mention of such endeavors, although, as we will see, there is documentation that in a number of instances several members of the Ordo Templi Orientis were actually either interested or involved in the construction and use of techno-magical devices.

Whether or not Steiner took precautions of not becoming more explicit about the nature of the secret circles that were engaged in the development of avant-garde technology, or simply dismissed the whole idea, is equally open to conjecture. Nowadays we only have Steiner's unsettling remarks.

However, what makes this possibility likely is that a strain of influence of Keely certainly is discernible in more than one instance. There was also a secret tradition that was nurtured by a number of occult orders, esoteric societies and lone individuals that possessed a deep interest for avant-garde scientific discoveries.

In the process, the philosophies of many of these occult orders would be mixed with futuristic concepts and alternative sciences, as would more generally accepted occult doctrine. There is also some documentation that mysterious devices were constructed, for many of which now only a rudimentary description, a rumor, a hint or a whisper remains.

A good example of this unique blend of occultism and strange science is the case of the German occultist and writer Ferdinand Maack (1861-1930), who founded a Rosicrucian order in Hamburg. In 1897, Maack devoted himself to the study of his newly discovered rays and his dynamosophic science. Maack also wrote lengthy essays that would be essential for a direction in magic that was called mathemagia, a blend of mathematics and magic that would center around the appliance of numbers for magical and ritual ends.[2]

In the meantime, orthodox science drove the seekers of perpetual motion, the renegade scientists with their self-made jargon and the garage inventors who were hunting for free-energy into the arms of the occult world, and it was inevitable that the two would meet. Thus, the German Baron Von Reichenbach (1788-1869), who discovered what he termed the "odylic force," is today only mentioned in the occult textbooks. While experimenting, Von Reichenbach noticed that sensitive persons could perceive radiation that others could not see. Von Reichenbach termed these rays the cosmic Dynamid and named these after the Sanskrit word Od.

In the cases of Mesmer's animal magnetism and the findings of the pioneers of radiesthesia, one is now equally forced to turn to the history of the occult. The early history of free-energy is therefore also very much a part of the occult; in its pages are found various tales of forgotten 18th and 19th century free-energy inventors who, while themselves not directly related to occult doctrine, at least found a more welcome home there after their rejection by the orthodoxy. But long before the 18th and 19th centuries, rumors and legends of strange machines and unheard-of technologies surfaced from time to time in the occult communities.

The famous French magus Levi, for instance, records the story of Rabbi Jechiele, who was an adviser at the court of the French king Louis XI in the 13th century. Jechiele possessed a "brilliant lamp that lighted itself." "What one said about his lamp and its magical clue proves that he had discovered electricity, or at least that he knew how to make use of its principles; for this knowledge, as old as magic, was transmitted as one of the keys of the higher initiation," Levi explains.

Levi notes that the lamp had neither oil nor a wick, but those who went by his home at night would see "a brilliant star in Jechiele's house; the light was so bright, that one could not stare into it, and it projected rays in the colors of the rainbow." Another device, used to protect himself, emitted electric shocks: the rabbi "touched a nail driven into the wall of his study, and a crackling, bluish spark immediately leapt forth. Woe to anyone who touched the iron nail at that moment: he would bend double, scream as if he had been burned, after which he would run away as fast as he could."

Levi also mused on the tale of Jechiele's contemporary, alchemist and occultist Albertus Magnus (1193-1280): "At the same time lived Albert the Great, who was considered by the people to be the grand master of all magicians. The chroniclers assure that he possessed the secret of the Philosopher's Stone and that he had solved, after thirty years of labor, the problem of the android; that is to say that he constructed an artificial man, living, talking and answering all the questions with such a precision and subtlety, that Thomas Aquino, annoyed for it not keeping its silence, struck it on its head."[3]

This iron man purportedly served Albertus Magnus for years, opening doors for guests, asking what they wanted and deciding whether or not they could see his master. Details about the appearance of this iron man or principles of operation lack completely, and the scarce bits that are available do not correspond to each other. Generally it is told that this remarkable automaton was made out of "wood, metal, wax and leather." At other instances Albertus Magnus is credited with the construction of only a head that could answer questions.[4]

Another variation of this legend recounts that when Thomas Aquino destroyed the artificial being, Magnus cried: "Thomas, Thomas! Thirty years of labor thou hast destroyed with one stroke!"[5]

There is some confusion, even among high occultists, about the meaning of this tale. Levi searches for a symbolical meaning on an initiatory level and warns that, "This is the popular fable; let us see what it means," after which he explains that, "The mystery of the creation of man and his first appearance on earth has always preoccupied the curious who searched for the secrets of nature. ...The name Adam in the Hebrew language signifies the red earth; what then is this red earth? It is what the alchemists search for, hence, the great work is not the transmutation of metals. ...it is the universal secret of life."[6]

The earliest records, although made a long time after Albertus Magnus died, contain a clear coherence with astrology and with magic: the body parts of his android were constructed according to the position of the planets, the eyes for instance were constructed in accordance with a certain solar constellation, a curious fusion of the concepts of technology and the arcane. What must be taken into account was that Albertus Magnus, although today chiefly remembered as an occultist, was also one of the most important scientists of his day, a veritable homo universalis who mastered botanics, physics, astronomy and mechanics.

In Magnus' tale it is hinted that the legendary technology of building artificial people, or crude robotic beings had a much longer history attached to it. He wrote in *De Anima Libri Tres:* "It is said that Daedalus made a statue of Minerva out of wood that was moveable in all its limbs and, through the movement of the tongue, appeared to sing. This movement explained itself thus: in the inner of the statue were organs of mercury inserted and they appeared to move after the movement of the statue."[7] Possibly this tale was a distortion of the very earliest oriental designs of a perpetuum mobile with hollow wooden stakes filled with mercury.

While much of this curious fusion of technology and occultism was lost or erased from the pages of history, it was said that the ever-elusive, mysterious Rosicrucians were the keepers of parts of this knowledge. It comes as no surprise, then, that Bloomfield-Moore wrote about Keely's vibratory force in connection with that mystical organization of the Rosicrucians, and reminded her readers that: "...Everybody knows that a note struck upon an instrument will produce sound in a correspondingly attuned instrument in its vicinity. If connected with a tuning fork, it will produce a corresponding sound in the latter; and if connected with a thousand such tuning forks, it will make all the thousand sound, and produce a noise far greater than the original sound, without the latter becoming any weaker for it. Here then, is an augmentation or multiplication of power, as it has been called by the ancient Rosicrucians, while modern scientists have called it the law of induction. Modern science heretofore only knew the law of the conservation of energy; while to the scientist of the future the law of the augmentation of energy which was known to the Rosicrucians will be unveiled."[8]

This was not the only secret to have been guarded by the Rosicrucians. It was also claimed that they possessed the secret of the "everburning lamps." These lamps were capable of burning for extremely long periods of time without a known means. Such a lamp was allegedly found, still burning inside a tomb dating back to Roman times as recently as the 1930s, during the construction of a road near Budapest.[9]

Hargrave Jennings, the 19th century occultist who knew Randolph of the Hermetic Brotherhood of Luxor and who corresponded with Golden Dawn initiate Arthur Machen and Bulwer-Lytton, collected a number of tales about these legendary mysterious lamps: "...Rosicrucius, say his disciples, made use of this method to show the world that he had reinvented the ever-burning lamps of the ancients, though he was resolved no one should reap any advantage from the discovery." The tomb of Christian Rosencreutz, the mythical founding father of the Rosicrucian Order, was said to have been illumined by these everburning lamps.[10] Naturally these tales would influence the direction that some writers took with their works of fiction. A perpetual lamp that is being fueled by carbonized diamonds was featured in a 1899 tale titled *The Master of the Octopus.*[11]

Long before the invention of the telegraph, the telephone and other means of communication, it was believed that certain initiates had technical means to transmit messages over great distances. Allegedly Johan Trithemius (1462-1516)—who coincidentally is also credited with having the secret of the ever-burning lamps—described a device that resembles the current radio.[12] In a book that is attributed to him, Albertus Magnus gives an elaborate and modern sounding description of a strange telegraph in a passage entitled, "The wonderful secret of how to build the sympathetic plate of numbers, with which one can communicate in an instant with distant friends." The telegraph consisted of two boxes made out of "fine steel," constructed "in the same manner as the cases for sea compasses" and having "the same weight, dimensions and appearance." These boxes had "reasonably big" surrounding rims on which were engraved the letters of the alphabet. On the bottom of these boxes a small appendage was fixed on which some sort of needle was fastened. These needles were made out of a carefully selected magnetic stone.[13] The legendary Rosicrucians are said to have built such devices.[14]

Another means of communication, but with totally different beings, was provided by John Dee (1527-1608), the brilliant magus, mathematician, alleged Rosicrucian, astrologer royal to the court of Queen Elisabeth I, and some say spy. Dee constructed a complex device to communicate with what he termed "angelic beings." Influenced by Trithemius, Dee reasoned that intercourse with "spiritual creatures," was one of the highest ends that man could aim at. At this point, magic would enter Dee's life and work. The exactness of mathematics, together with the philosophies of the Hebrew cabbalah and the Arabic magico-alchemical works, formed the foundations on which Dee built his system of magic in a scientific manner.

In 1581, he began with the establishing of contacts with the "angelic beings." At first he relied to a certain extent on equipment, such as wax tablets, a skrying table, a gold lamen and several shewstones of obsidian and rock crystal. This eventually evolved into a sophisticated system which involved setting the skrying stone, of which several were used including a crystal ball and a black obsidian mirror, upon an elaborately engraved sigil called the "Sigillum Aemeth." This Sigillum Aemeth in turn was placed upon a special table that was inscribed with a hexagram, enclosed within a frame of Enochian letters, supporting seven specially designed talismans. The whole was insulated from the floor by a further four wax tablets, which were miniature versions of the Sigillum Aemeth.[15] The results obtained through this method, and what would become Dee's legacy, was a strange language called the Enochian language which would be studied by countless esoteric and occult societies, among them the Golden Dawn.

Another striking example of strange, occult technology is found in a rare 18th century hermetico-alchemical book, which is worth a lengthy quotation. In it a friend of Benjamin Jesse (1642-1730) who was a cabalist and Rosicrucian from Hamburg, writes of the wonders that befell him while inspecting Jesse's workshop. Jesse died a year before, and the author states that he writes the account "so that one may imagine a bit what sort of physics of the ancients in the golden age will come to light."

The wonders were all to be found in Jesse's "prayer room in which never before a living human had entered or knocked on its door." This "prayer room" had a highly unusual safeguard to begin with: "...he took me to this prayer room and coated the joints and seams of the door with a translucent crystal substance, which he used as if it had been wax. After this he pressed his seal upon it, which was crafted out of gold, so that in this substance the seal of his signetring was imprinted, that immediately hardened, so that the seal would have been broken in two if the door would have been moved just slightly." The key to the prayer room was inserted in a little box that was treated with the same translucent crystal substance and his signetring. After Jesse did this, he threw his signetring in the crystal substance which then "dissolved like a piece of ice in hot water, and it fell to the bottom of the glass as a white powder, and the crystal substance colored red. After this he also melted the glass with the crystal substance and he gave me such a glass with the keys."

Jesse died, and as he requested, the writer gave the keys and the box to his relatives. With the help of the crystal substance, also termed "crystal water," the seals on the box melted away, the key to the prayer room was taken and the seal on the door of the prayer room was also melted. We now only have the detailed account of the writer of the strange devices that he saw in the room, which must have been extraordinary indeed: "In the middle of this prayer room a table of ebony stood, its plate was round, its rim coated with pure gold. For this table was a small chair for kneeling down. In the middle of this table stood an instrument of wonderful design. Its lower part, or the base, was round and made out of pure gold. Its middle part was made out of translucent, shining crystal, in which the everlasting fire was enclosed that emitted shining rays: its upper part was made out of pure gold and in the shape of a bowl. Immediately above this instrument hung a crystal on a gold chain, fashioned in the shape of an egg, so that the everlasting fire was enclosed if it emitted its rays. At the right side of this table I saw a golden box and a small spoon, in the box was a balm of reddish color. At the left of this table was a small chair made out of pure gold, on which a book lay with twelve pages, also made out of pure hammered gold which were so flexible, as if it were paper."

A bit of the reddish balm was placed in the bowl of the strange instrument on the table, and "immediately a pleasant smoke went up that refreshed the senses, and, what was even more miraculous, in its ascend the smoke touched the fire that hung over it in the crystal egg in such a way that it emitted terrible rays like the thunder and the stars."

The writer witnessed more strange wonders; they also found a "small box, made out of ebony but coated on the inside with pure gold. In it were twelve instruments made out of pure gold, wonderfully wound and crafted, around them engraved with symbols and letters. We went to the next box, which was larger. In it were twelve mirrors, not made out of glass, but out of an unknown substance, very neat and clean; in the center of these mirrors were odd symbols: their rims were fastened in golden frames. ...Then we proceeded to a larger room; in it was a very large mirror that was called 'Salomon's mirror' by Jesse and the wonder of the world, in which mirror he could join all images, every one of the entire world. Lastly I saw an ebony cabinet in which a globe was made out of a very odd substance. Jesse said that in it the fire and the soul of this world were also enclosed, and because of that it moved all by itself, in the same way as our world. I also saw, hanging above this cabinet, another cabinet. It was a cabinet with a special instrument, in the way as a timepiece, that had a roadmark or a pointer, but instead of the indications of the hours it had letters. Jesse said that this instrument moved in the same manner as the one he had in Switzerland. ...At that time in that prayer room I witnessed incredibly wonderful things through the movements and the use of these wisdom instruments, things that I can impossibly write down, nor am I allowed to, but this I would like to share with you...more I cannot do."[16]

Obviously Bulwer-Lytton studied these strange accounts well or perhaps he drew upon his own experiences, as he describes similar strange techno-magical devices in his haunting tale *The House and the Brain*: "...our main discovery was in a kind of iron safe fixed to the wall. ...In this safe were three shelves and two small drawers. Ranged on the shelves were several small bottles of crystal, hermetically stopped. They contained colorless volatile essences, of what nature

I shall say no more than that they were not poisons—phosphor and ammonia entered into some of them. There were also some very curious glass tubes, and a small pointed rod of iron, with a large lump of rock-crystal, and another of amber also a lodestone of great power. ...We found a very singular apparatus in the nicest order. Upon a small thin book, or rather tablet, was placed a saucer of crystal; this saucer was filled with a clear liquid—on that liquid floated a kind of compass, with a needle shifting rapidly round, but instead of the usual points of a compass were seven strange characters, not unlike those used by astrologers to denote the planets. A very peculiar, but not strong nor displeasing odor came from this drawer."[17]

Then there was the sidereal pendulum which hung in a glass encasing. With a conventional code, one obtained messages through the striking of the pendulum against the glass encasing. These sidereal pendulums were found all over Europe in the 18th century and as late as the 19th century.[18] And indeed, in a letter to Edison, Olcott remarks that, "you told me that you were making some experiments with a pendulum, to test the dynamism of the will: since then I have seen many such tests successfully made."[19]

Not only the original and in many aspects legendary Rosicrucians were allegedly involved in technological endeavors. These were to be found across the whole occult spectrum. One of the very first recorded instances, in which we find no occult influence at the surface, was the invention of Johan Ernest Elias Bessler (1680-1745). In Germany, he began to be known around 1712 as the constructor of various self-moving wheels. These he exhibited freely, but always with their mechanism concealed by casings forming part of the wheel and revolving with it. He exhibited his largest wheel at Hesse-Cassel in 1717. Gould writes that Orffyreus' wheel is "the only instance on record of a machine, capable of doing external work and yet apparently independent of any external or known source of power, having been exhibited in public and subjected to official tests."[20]

While there is no occult influence at the surface, the way Bessler obtained his pseudonym gives food for thought. Bessler constructed his pseudonym "Orffyreus" by placing the letters of his name in a circle. By choosing the opposite letters he obtained his pseudonym. This not only brings the countless revolving wheels in Trithemius' book *Polygraphia* to mind, it could very well be that here we have an atrophied account of a device not dissimilar as Magnus' or Jesse's apparatus or one that Bulwer-Lytton described.

In a pamphlet written by Bessler and published in 1715, he gives an outline and principles of the machine, "but that explanation is at variance with all modern ideas of mechanics. If, as he claimed, he had discovered a new source of power, he was either unable or unwilling to give a correct description of it," remarks Gould.[21] Bessler, or Orffyreus as he chose to name himself, destroyed his wheel in a fit of rage, without ever disclosing how it worked. When he died, he took the secret of his invention with him to dissolve forever in the mist of history.

In order to preserve another secret, towards the end of the 18th century, avant-garde science mingled for a while with freemasonry. Around 1780, Franz

Anton Mesmer (ca. 1734-1815), who was to become a source of inspiration for Bulwer-Lytton, swept Europe with his mesmerism. Mesmerism was a curious fusion of avant-garde science and occult preconception, and in the process Mesmer would coin the term "animal magnetism," which he saw as the soul of all that breathes, not unlike the ideas that Keely gravitated towards while describing the source of his discovered force more than a century later. Conveniently, at Mesmer's times French High-Grade masonry was at its peak, and it is sometimes alleged that it was freemasonry that introduced Mesmer in the better-situated Parisian circles.[22]

Mesmer directed fluids by the movement of his hands and directed these through tubes and bathtubs or in a glass of water. By means of his animal magnetism he obtained strange results. It is said that people thus treated were often given to weep or to sleepwalk, and he cured many persons afflicted with fits of all kinds. No wonder then, that many freemasons were trying to obtain Mesmer's secret. However, it was established that only "sensitized" persons could procure the same results, and from this sprang the notion that an order was needed, to "sensitize" those persons.[23]

Since Mesmer was a French High-Grade mason himself, a grade was established to promote and preserve the art and the secrets of animal magnetism. This little-known chapter in the complex history of freemasonry has become known as "magnetic masonry." The plan led to the founding of the Order of Universal Harmony in Paris in 1782. The purpose of these magnetic masons was also to create a healing center, which might radiate over the vast circle of initiates. For this, the initiates had to be ritually purified before they could conduct the process of magnetic healing themselves, somehow an occult forerunner of the process in which Keely had to teach certain persons to sensitize his equipment.

In 1784, lodges of this magnetic masonry were established in several French cities, including Versailles, Lyons, Bordeaux, Grenoble, Nancy and Marseilles. When the French revolution broke out in 1789, the order either dissolved or went underground. When Mesmer died, he died almost forgotten.[24] His influence on freemasonry is to be found in a direction that is also called "Mesmerian Masonry," in which this spiritual current of the 18th century was connected to freemasonry.[25]

Perhaps this Mesmerian current influenced a German group that is surrounded by mystery; the Free-Masonic Order of the Golden Centurion. This order supposedly was founded in Munich in 1840 by a number of rich German industrialists and well-to-do citizens[26] and is described as "One of the most important, and certainly the most diabolically mysterious." Around the 1840s, this order used a strange device called the Tepaphone. But contrary to Mesmer's endeavors decades before, the Tepaphone was used for an infinitely more sinister purpose. Allegedly the Tepaphone was a "machine which, when coupled with the will of a magician, could kill anyone no matter were they were."[27]

While various theories abounded upon its exact working, it was believed that the device could load or unload a person with the vital odic force. The Tepaphone was being described in one instance as made out of "multiple optic lenses and a copper spiral consisting of twenty-four coils in the center of which was a copper plate." An image of a person could be placed beneath the lenses and in the stream of electrical current that ran through the instrument. In this manner the person of the image would be affected positively or negatively. The spirals were tools for engaging the concentrated mental force of the operator or operators of the instrument, in order to guide its effects.[28]

Of John Murray Spear (1804-1887), the American spiritualist preacher and founder of the forgotten "new motive power movement," it too can be said that he had one foot in the territories of the occult, while at the same time his restless mind created other wonders. Like Mesmer, it was to benefit mankind. In 1854 he constructed a motor at High Rock in Lynn in Massachusetts, which he called the New Motor. He intended it to be self-generative. The same year and across the ocean a strange book titled *Der Organismus des Weltalls,* or *The Organism of the Universe,* by German occult philosopher U. Milankowitsch was published.

In it he remarked that, "When I want to build a machine...that wants to move and that shall root in the earth, I at first think of that machine. ...I design a blueprint. This blueprint, this idea originated in my mind, according to which...this machine should be built, and when this machine...is built, the idea that was located in my mind, proceeded in objective reality, is realized in nature, and has become one with...the machine. Thus the idea was the primal image, its form and law according to its rules...the machine is built. When we go from the workshop of the mechanic into nature, we will find an immense, boundless and endless building, the house of God, consisting of countless parts, or a moving, living, immense World Machine, consisting of a countless number of revolving and revolvable wheels. Now, the blueprint of this building of nature, this living World Machine, is the idea itself that is objectified and melted together with the same. The blueprint of the whole of this building of nature is therefore the primal idea, the absolute idea, the idea of ideas."[29]

It is doubtful if, with his enigmatic statements, the German occultist somehow foresaw what Spear was constructing at the same time. It is equally doubtful if Spear read the philosophies of the German occultist and his musings on the nature of the World Machine, but it provides a partial insight into the motives of a priest and spiritualist to construct a curious machine-like device. It also reflects the exalted spirit of the times, as experienced by the esoterically inclined.

Spear, much like Newbrough—who was also a spiritualist—claimed that he did so at the instigation of one of the groups of spirits by whom he was controlled. He had been active in the antislavery, peace and temperance movements, and became a medium in March, 1852. He claimed that his book *Messages from the Superior State* was dictated by the spirit of John Murray, the founder of the sect of Universalism. Whatever its causes or literary origins, it heralded his first public appearance as a medium. Spear was also in the habit of journeying all over the country as the spirit moved him, "at the command or direction of spirits to whom he professed himself willing a childlike and unquestioning obedience."[30]

A year later, Spear confided to a Boston newspaper that his spirits made "important declarations" to him as he visited Niagara Falls. Forty years later, this place would see a very different kind of magic, but this time in the form of one of Tesla's visionary ideas. Spear's spirits declared that they had formed various associations. One of these was the "Association of Electricizers."[31]

Before the construction of his New Motor that led to a whole movement, "the new motive power movement," Spear experimented with mineral and vital electricity as a means of developing the latent powers of mediumship. He also sought to promote the influence and control of spirits through the aid of copper and zinc batteries, "so arranged about the person as to form an armor, from which he expected the most phenomenal results." However, an experiment tried in St. Louis "proved, so far as external effects were concerned, a complete failure."[32]

In the fashion of Levi's symbolical explanations and of Albertus Magnus' construction of an android, Spear too had other things on his mind; he "had long indulged the idea of embodying in some tangible form the crude conceptions of certain minds (not limited to the earth spheres alone), who have labored to discover and scientifically control the mystery of the life principle." Eventually Spear and his array of invisible spirit counselors thought that they had made this discovery, and a Boston spiritual periodical, the *New Era,* declared that "the association of Electricizers in the spheres were preparing to reveal to mankind a 'new motive power,' God's last, best gift to man," a work that was "destined to revolutionize the whole world" and "infuse new life and vitality into all things, animate and inanimate." From time to time, the Boston periodical would drop mysterious hints concerning Spear's discovery, which was "to awaken the world to wonder," but finally it announced in its pages that "high spiritual intelligences, through the organism of Mr. John M. Spear, had given directions for the construction of a living machine," termed "a new motor."

Consequently, strange reports began to circulate in spiritualist circles. In one of these, a Boston woman, also a spiritualist, was named as "the mother of the new motor," and "absurd and impossible stories were bruited about concerning the practices by which 'the life principle' had been infused into its organism."[33]

Nevertheless, the *New Era* soon printed an article headlined, "The New Motive Power, or Electrical Motor, otherwise called 'Perpetual Motion'—The Great Spiritual Revelation of the Age." In it, its editor who was Spear's friend but not a spiritualist himself, proudly announced that "after about nine months of almost incessant labor, oftentimes under the greatest difficulties, we are prepared to announce to the world, first that spirits have revealed a wholly new motive power to take the place of all other motive powers. And second, that this revelation has been embodied in a model machine by human cooperation with the powers above." The last statement was the vague utterance that the results thus far obtained, were "satisfactory to its warmest friends."[34]

Spear's "electric motor" or "The New Motor" was designed to "correspond to the human organism," it had "a brain, heart, lungs etc.," and it should "perform the functions of a living being." The queer device also had "some little balls, connected with the machine," which "for some months have given evidence of motion." The device also was equipped with a large wheel, the "grand revolver," upon which "all the executive power is made dependent." Wiring of some sort covered the apparatus, "Each wire is precious, sacred, as a spiritual verse. Each plate of zinc and copper is clothed with symbolized meanings, corresponding throughout with the principles and parts involved in the living human organism. ...The various parts of this mechanism, both the wood work and the metallic, are extremely accurate, and so mathematically arranged with reference to some ulterior result or effect." Poles and magnets were also arranged in a specific way in the device.

The statements of the famous spiritualist Andrew Jackson Davis - who went to investigate Spear's remarkable machine, which he described as a "peculiar construction" - give some insight in the proposed working of Spear's New Motor: "The philosophy given through Mr. Spear, upon which the mechanism is predicated, is this: First, that there is a universal electricity. Second, that this electricity has never been naturally incorporated with mineral and other forms of matter. Third, that the human organism is the most superior, natural, efficient type of mechanism known on the earth. Fourth, that all merely scientific developments of electricity as a motive power are superficial, and therefore useless or impracticable. Fifth, that the construction of a mechanism on the laws of man's material physiology, and fed by atmospheric electricity obtained by absorption and condensation, and not by friction or galvanic action, will constitute a new revelation of scientific and spiritual truths, because the plan is wholly dissimilar to every human use of electricity."[35]

The New Machine was to derive its motive power from the magnetic store of nature, of creation itself that was defined by Milankowitsch as an incredible World Machine. Since Spear's New Motor derived its power in such a way, it was to be as independent of artificial sources of energy as was the human body.[36]

When a woman, obeying a vision that she had, went to the High Rock at Lynn where the New Motor was displayed, she suffered "birth-pangs" for two hours. From this possibly epileptic seizure, she judged that the essence of her spiritual being was imparted to the machine. At the end of that time, it was averred that "pulsations" were apparent in the motor.[37]

The New Machine failed to work, or at least its actions remained inconclusive and unsatisfactory, and even his fellow spiritualists didn't think much of the device. Eventually the machine was brought to the village of Randolph in Massachusetts and housed in a "temporary building." There the New Machine, costing Spear nearly $2,000 to construct, was destroyed by a mob of superstitious villagers.[38] Saddened, Spear, who wanted to give mankind "God's last and best gift," disappeared from the pages of history, except for an occasional reference.[39]

What it demonstrates, aside from the superficial similarities between Spear's New Motor and Keely's disintegrator, is the fact that Spear's case was the reason that certain spiritualists were interested in Keely's discoveries in the first place.

Spear's story, strange and intriguing as it may be, was not an entirely isolated one. In 1850, just a few years before Spear was visited, young farmer Jonathan Koons, who lived on an isolated farm in Milfield Township in Athens County, Ohio, had his encounter with spirits. His neighbor, John Tippie, also lived on a farm some two or three miles away. On the farms of each they erected a log cabin, consisting of a single room about 15 by 12 feet. The buildings were built by Koons and Tippie under the direction of "what claimed to be the spirits of human beings," as "circle rooms, and fitted up with instruments, etc., from written plans and diagrams drawn out by the spirits with their own hands."

In each room was a "spirit machine," and although there is no definitive time period in which young Koons and Tippie were engaged in their construction, they were probably built around the time that Spear was busily assembling his New Motor. The spirit machines consisted of "a somewhat complex arrangement of zinc and copper," with "two drums...fastened with copper wires upon wooden supporters" at the top of a table. The table was intersected with copper wires wrapped with zinc. "On the upper cross wire hang some copper plates, cut in the form of doves, to which are suspended a number of bells, which the spirits sometimes ring." The spirits alleged that the spirit machines served the purpose of collecting and focalizing the magnetic aura used in the manifestations, also described as the "collecting and retaining the electricity of the circle and it is charged before giving any demonstrations, at every sitting." The charging of the singular devices was accompanied by a "startling noise." The devices, also described as "novel batteries" were placed upon a long wooden table, by the side of which lay several music instruments, provided "according to direction."

Visitors witnessed strange doings; "spirits" would play the musical instruments, loud knockings were heard and sometimes the log cabins shaked on their foundations. These spirits identified themselves as "of the most ancient and primal order of man." They spoke of the biblical Adam as of a "comparatively modern date," and indicated that they were "by no means the first of earth's inhabitants," though antedating the biblical Adam by "thousands of years."[40]

In the year that Spear had his revelation at Niagara Falls that would lead to the construction of his New Motor, Koons entrusted to paper a religious philosophy given to him by his visiting spirits. He wrote that the spirits declared that, "the electric element forms the various paths in which planets and all other known bodies in space travel and move in their respective orbits, but that nothing...can penetrate the realms of the 'subtler fluid,' yet it divides and permeates all space, and seems to hold in control the infinite realms of the electric element. ...There is a grand central territory in the universe. ...It embraces illimitable though unknown realms; yet its position as a vast central point is defined, from the fact that from thence, and to thence, seem to tend all the illimitable lines of attraction, gravitation and force, which connect terrestrial bodies, and link together firmaments teeming with lives and systems."[41]

Koons' spirits even imparted to him a philosophy regarding the working of their spirit machines. From it, we learn that one of its purposes was to overcome gravity. This detail alone may very well class Koons' and Tippie's spirit machines as two of the very first recorded instances of antigravity devices that were actually built with that purpose in mind.

"It is said that spirits, in their communion with earth, manifest through two primitive elements; first, an electro-magnetic element of which the spiritual body is composed; next, a physical aura, which emanates from the medium, or can be collected from material substances, analogous, it is supposed, to the element of 'vitality.' ...From the combinations of these two, namely the emanations of the spirit and the medium, a third or composite is formed, which is affected by the atmosphere and human emanations. From the preponderance of the electro-magnetic or spiritual element, the laws of cohesion and gravitation can be overcome, and through this spirits are enabled to dissolve and recompose substances with great rapidity, heave up and carry material bodies through the air, and causing them to float or sink in proportion to the strength of the battery formed."[42]

Thus, visitors of those strange log cabins would witness the musical instruments floating in the air, suspended from gravity while a horn, "a tin trumpet of two feet long," would produce "speaking, whistling, singing and addresses." Each time it would rise into the air, produce the sounds after which it would "fall to the table." The phenomena seemed to begin each time with "tremendous blows on the table, ceiling and walls."[43]

In 1854, Koons' and Tippie's log cabins were still shaking "like a tree in a gale of wind," the spirits exhibiting "extraordinary pyrotechnics, seemingly to consist of luminous bodies flying about with the swiftness of insects." Koons would pick up a violin and drew a bow across it; "immediately" another violin was sounded. Tippie's cabin also exhibited the same phenomena; although persons there "neither saw writing nor a spirit hand." The music in Tippie's cabin, however, was "all produced by spirits," and was "more varied and interesting than Koon's."[44]

Numerous people visited the strange cabins where gravity seemed to exist no more. Affidavits were signed. The cabins were searched, but nothing of a fraudulant nature was found. Not all was well though; Koons' and Tippie's houses were attacked by mobs, their barns and crops destroyed by fire, "their children set upon and ill-treated." The phenomena seemed to have waned in later years and would ultimately come to a complete standstill; it was suggested that the reason for this was that Koons and Tippie lost their mediumistic gifts: "it must be remembered that...the presence of a large family of highly mediumistic children and the electrical nature of the locality where the circles were held must be taken into account."[45] Koons and Tippie split up, and the ultimate fates of those wonderful spirit machines which according to witnesses were able to suspend gravity, is unknown.

Emma Hardinge, in her time a well-known medium and one of the founding members of the Theosophical Society, and of whom we are indebted to for the previous passages about Spear, Koons and Tippie, cites an instance in which we traverse the boundaries of the actual and historical. In a strange book published under her auspices, the anonymous narrator of the allegedly autobiographical sketches received an initiation into the mysterious Ellora Brotherhood while in India. The subterranean temple of Ellora—which indeed very much exists—where the initiation took place was in fact a strange device of gargantuan dimensions: "The whole temple was furnished with fine metallic lines, every one of which converged to six powerful galvanic batteries attached to the silver thrones by six of the adepts. These persons, adepts in the loftiest sense and most significant sense of the term, received their inspiration from the occupant of the seventh throne, a being who, though always present, was not always visible."[46]

Through the electrical system of this elaborate battery, the positive pole being the seven hierophants, and the negative being the assembly of neophytes, the narrator was mentally impressed with fantastic images of cosmic events,[47] much like the visions of Koons and Newbrough's *Oasphe*.

In 1873 in Paris, Eliphas Levi, the great French magus and friend of Bulwer-Lytton, was quite happy to stumble upon the metal parts of a device called the Prognometer in a junk shop. The proprietor bought the dismantled device at an auction of the collector of handwritings and curios, Valette, who was executed during the revolt of the Parisian Commune. The proprietor paid 500 francs for it, but he sold the device to Levi for one fifth of the original price.[48]

The Prognometer, built by the Polish mathematician Hoëne-Wronski, was a mechanical device that could predict certain trends in the future history of mankind, or so it was claimed. "While the master lived, the disciple had not been allowed to see the Prognometer. Now he bought it, recognizing Wronski's writing in the mathematical symbols which covered the contrivance," writes Webb.[49] "Wronski trusted his secret only to Marquis Sarrazin de Montferrier, whose son-in-law was the last grandmaster of the Templars. I heard about this secret wonder, that Wronski had guarded as jealously as Menelaous had Helena, but I had some doubt as to the reality of its existence. Besides, I knew that before Wronski died, he had dismantled all his machines and had sold the copper to people from the region of Auvergne," explained Levi,[50] and once again, as in the case of Mesmer, we see an order of definite esoteric and secretive nature involved in the preservation of occult technology.

Joseph Maria Hoëne-Wronski (1776-1853) was a remarkable man, and it is with him that the revival of 19th century occultism on the continent indirectly begins. It was Wronski who initiated Levi between 1850 and 1853, who then became known as the sole instrument of the 19th century revival of the occult sciences. After a career that had led Wronski from the Polish rebellion of 1794 in the Russian army which he left with the rank of major in 1797, he studied philosophy in Germany, enlisted once again in the Polish army, and worked in 1810 in the Observatory of Marseilles. Around this year, Wronski had his illumination, and as a consequence, he claimed to have discovered the Absolute. The Absolute is the knowledge of truth, which may be reached through the human reason. While Wronski claimed that this was achieved through rational thought, it nevertheless is almost impossible to understand him, since he wrote his theories down in dense, mathematical terms. Another important theory of Wronski was that man could create reality from the total of the impressions of his senses.[51]

Brilliant as Wronski may have been, he was not able to communicate or depart with his ideas, which were steeped in occult lore. Of this, Webb writes that "Wronski was knowledgeable in Cabalistical matters as was obvious to early commentators on his work. He also knew Boehme and was familiar with Gnostic teachings. ...Wronski maintained that the goal of man was to become god-like; like other occultists, he veiled his meaning with an impenetrable curtain of jargon. His teachings were not for the vulgar but only for those who would make the effort to penetrate his mathematics."[52]

The further career of Wronski has some resemblances with that of Keely, although Wronski had a much, much more poverty-stricken existence. As in Keely's life, Wronski's only child died young. As with Keely, business teamed up with intellect, but to no effect. In Wronski's case, this came in the form of a French businessman, Pierre Arson. He met Wronski in 1812 and proceeded to take a course of instruction from him. He also agreed to subsidize the publication of Wronski's oeuvre. Around 1831, however, the two fell apart; Arson seems to have been disappointed in the final revelation of the Absolute, and so he not only revoked his agreement, but also published a hostile broadside.[53]

Wronski replied and soon the two would find themselves in court. Curiously, Arson never ceased to regard Wronski as a genius. Another never properly explained matter was the fact that during their mutual recriminations in pamphlet form, Arson attracted the attention of a secret society about whom he never was certain whether they were in league with or against Wronski.[54]

There are speculations as to the identity of this secret society; while Webb thinks it likely that it involved a revived group of French Martinists, another suggestion is that it was one of the Saturnian Brotherhoods, said to have been revived in Warsaw by Wronski, with outer courts in Krakow, Poison and Thorn, although these lodges were ultimately destroyed by various wars.[55]

The occultist-mathematician kept turning his mind on things technical; in 1833, he developed a system for steam locomotives which he called a Dynamo-genic System that allowed the engine to dispense with rails. Wronski also invented a tracked vehicle, and the contract that he signed in 1833 with the Messageries Générales de France probably was for its prototype. He should have been able to live comfortably from the monies given to him by the company; yet when he perceived further mechanical principles from his invention, and he decided that it was his duty to publish these, the company withdrew all support. The remaining years of his life were spent in absolute poverty, his only feat being Levi's initiation in the last three years of Wronski's life.[56]

It is through Levi, that a description of Wronski's Prognometer has survived at all, and Levi hints that this was not the only device that he built: Wronski "dared to involve himself with inventions, he constructed mathematical machines, revolving axes which were put together in an admirable fashion. ...Only his machines would not work because the copper and bronze of his devices would not acknowledge the truth of his philosophies. ...His most fevered and most kept secret investigation was the invention of a divinating machine, also called prognoscope, that calculated all probabilities and drafted equations of occurrence that had happened in the past, happened now and would happen in the future, in order to establish all possible values."[57]

One day Wronski found out how to do this, and "he asked workmen to come over and ordered them to the most extreme secrecy. Nobody saw the design of his machine, but he let the workmen construct the machine in bits and pieces, and being a bit of a mechanic, put the parts together himself. All was immensely complicated but as harmonious as the universe itself. The construction of the Prognometer cost enormous amounts of money."[58]

It consisted of two globes riveted together, which, by two crossing axles, turned in a large immovable circle. Two small pyramids which alternately opened and closed themselves contained the principles of all the sciences. The summary of all these sciences was engraved, corresponding to their analogies, on the two globes that revolved around the two axles. According to Levi, the machine resembled the heavenly globe, covered with polished bismuth, mounted on a carriage of gilded copper.

"There were two smaller globes on the top of which were mounted two three-sided pyramids. One symbolized godly knowledge, while the other symbolized human knowledge. These always revolved contrary to each other, so that the harmony was the result of the analogy of its counter parts," Levi writes.

One of the small globes, the one that symbolized godly knowledge, was also called "a polarisator" and it was adorned with a lightning flash and a polished compass. Around this globe four letters, A, B, X and Z, were fixed, corresponding with its Hebraic equivalents, that had the value of "other Hebraic letters." From this globe two branches pointed, both having small compasses that indicated the proportion of "what was above and what was below." The other globe, representing human knowledge, carried "the flamboyant star of the magus," which Levi also calls "the sign of Salomo." This sign was viewable from "two sides," and the flash of lightning, pointing towards the globe, terminated in a five-pointed star, symbolizing "human initiative and human autonomy." On the circle were the signs of the zodiac and apertures which "opened and closed at will." On the ports of these apertures, the names of the sciences were written; under the ports, Wronski wrote the names of the "fundamental axiomata." There were 32 ports, on each of which were written "the names of three sciences." "The axiomatas were engraved with the greatest precision, but with a handwriting so small that even with a magnifying glass these were hardly discernible," Levi writes.[59]

On the inside of the big globe, which was part dark and part light, Wronski wrote equations of the comparative sciences, and on the big motionless circle he wrote the fundamental principles. "All sciences are the degrees of a circle revolving around the same axle," and elsewhere stood, "The future is contained in the past but is not wholly contained in the present," and "the rays of the

Prognometer represent the summary of all knowledge." When Levi touched one of its parts, "the globe made out of bismuth opened itself and revealed on the inside another globe that was also covered with mathematical equations."[60]

It is not known what happened to Wronski's Prognometer upon Levi's death two years after he found the device. Except for one etching, no picture of the Prognometer exists. It was Wonski's Prognometer, the "extraordinary calculating machine" that is considered a possible inspiration[61] for another strange device, invented by French occultist Saint-Yves d'Alveydre (1842-1909). His fame in occult circles is derived mainly from his book *Mission de l'inde en Europe,* in which the 19th century esoterists learned more about the subterranean realm of Agarttha.[62] But Saint-Yves d'Alveydre did not consider this or his other esoteric books to be his greatest works, nor his ideas on the use of seaweed as a means of nurture. In the last period of his life Saint-Yves d'Alveydre was totally devoted to his great invention which he called the Archeometer. He even obtained a patent, no.333.393, dated June 26, 1903,[63] for this invention.

The Archeometer was a grandiose machine, which "translated into the material the word, form, color, smell, sound and taste,"[64] it was the "key to all religions and all the sciences of antiquity," and consisted of a disc or discs of colored cardboard with some very complex diagrammatic arrangements.[65]

In the pamphlet *Archéomètre. Brevet d'invention no 333.393* that was printed in 1903,[66] Saint-Yves d'Alveydre wrote in eight pages everything that he wanted the public to know about his mysterious device. From this we learn more about the design of his Archeometer. The disc was divided in concentric zones. These zones or divisions contained the correspondences that existed between numbers, letters, colors and musical notes, the signs of the zodiac and of the planets.

On it was also found the invaluable alphabet of Watan that, according to Saint-Yves, had been used by the Atlantean race. These letters were held of the utmost importance since through these one could rediscover the elements of the symbolic and figurative signs developed in antiquity and the meaning of which had been lost since time immemorial. Also included in the Archeometer was a metric system, destined to reform sonometry, that could be used in the determination of the proportions of all the graphical constructions.[67]

While it is suggested that Wronski's Prognometer was perhaps the inspiration of the Archeometer,[68] Saint-Yves could equally have been inspired by Keely's musical charts. Upon examination, one is stricken with the uncanny resemblance that Keely's charts have with the discs of Saint-Yves Archeometer.

Saint-Yves was probably influenced by Keely's proposition to show the various colors of sound on a disc, each note having its color in order to demonstrate that the same laws which develop musical harmonies develop the universe. After all, when Saint-Yves invented his Archeometer, the Prognometer was already half a century old and perhaps a legend for 30 years.

Saint-Yves terminology as exemplified in the parlance of the "red race" points towards Theosophical influences, which are also clearly visible in his esoteric oeuvre. And theosophy was the largest and most easy accessible channel through which Keely became known in the occult communities of the 19th and early 20th centuries in America, England, Holland, and in France.

When Keely began to receive attention from the press and the general public, several other inventors came forth to tell of their inventions, and news of their activities reached the surface.[69] Often these stories are irritatingly vague. Inventors sometimes visited Keely and it is from this that we very briefly learn of their existence, but unfortunately nothing of their inventions. When he was in jail, a young inventor who had "been waiting for eight years for a patent from Washington," tried to see him, without success.[70] Mr. Browne, an inventor from Brooklyn, visited Keely in his laboratory and claimed that "He could do all that Keely could do."[71] Dr. Dupuy of New York was "for years experimenting in this field without meeting with Keely's progressive successes,"[72] and there was of course Kinraide, who experimented in his fully equipped cave with magnetic needles.

Sometimes, more information is available and tantalizing glimpses on the inventions themselves are given. In 1875, A. Arnold confided to a newspaper that he had "received by mail from an unknown source a pamphlet entitled 'The Keely Motor.'[73] The first paragraph of this pamphlet is a quotation from a paper published by me some two years ago, in a New York scientific journal. I know nothing of the Keely motor nor any of the persons connected with it. But for the benefit of the public I will say that I have in practical operation, capable of doing useful work, a simple apparatus which generates a constant pressure of fifteen pounds per square inch, without heat, chemical or electrical action, or the employment of any other materials than a small quantity of air and water. I use this low pressure only for convenience, but can at pleasure increase it indefinitely in the same apparatus. I can assure the public that producing a pressure only limited by the strength of a generator, from no other source of power than a small quantity of cold water, air moving slowly therein, is no 'humbug,' nor new discovery, but is a fact, long shown to scientists and mathematicians. And, furthermore, the apparatus I have for effecting it cannot be patented by other parties. Respectfully yours, A. Arnold. Tenafly, N.J., July 4, 1875."[74]

Two weeks later, a notice appeared in a newspaper of a new motor in operation in Hamilton, Ontario that was said to rival the Keely motor. "The motive power consists of nine parts of air and one of coal-gas, which operates as an explosive. The inventor challenges the Keely motor to a trial."[75] Although there was considerable interest in such inventions, as is clearly demonstrated by all the publicity that surrounded Keely, nothing more was heard of these inventors.

Four years later in 1879, a Chinese sub-prefect, Tung Yu-ch'i, claimed to have invented "a machine which would go of its own accord and generate sufficient power to propel a steamer of the largest size." The enterprising sub-prefect went to Shanghai with a credit of 3000 taels, given to him by the governor-general, and "plans and specifications of the most mysterious nature, in which clogged wheels, tubes, and other contrivances were inextricably jumbled together." A working model of this device was constructed in "the native and foreign workshops" of Shanghai, which "would not work."

The next year would find the persistent inventor in Peking, where he obtained a royal edict: "An Imperial Decree—The censorate has memorialized us to the effect that Tung Yu-ch'i, an expectant sub-prefect in the province of Anhwui, proposes to construct a steamboat to be impelled by a cold vapor generated without the use of fire, which shall supplant the one using fire. Its construction is already well-nigh completed, and it is estimated that 3000 taels will suffice to finish it. A diagram, with illustrations of the invention, has been presented to the memorialists for their inspection. Should the steamer invented by the officer in question be found capable of quick motion and adapted to practical use, it will of course be proper to adopt it."[76]

Nothing more was heard of Tung Yu-ch'i or the steamer that was nearly complete. In the same year that he was constructing a working model in the workshops of Shanghai, the discovery by Wesley Gary (1845-1875) of a neutral point between the poles of a magnet, allowing for nearly perpetual motion, was printed in the respectable and bourgeois *Harper's Magazine* of March 1879. His discovery also drew the attention of the occult communities. It was noted in *The Theosophist* for November of the same year.[77]

Steiner also learned about Keely through his affiliation with the Theosophical Society, took his idea on a threefold society from Saint-Yves d'Alveydre,[78] and would become a major player on the occult scene.[79] He lectured on several occasions about Keely and his discoveries, and like Blavatsky he was a wide-reaching and influential channel through which much information about Keely traveled in the occult underground. In some occult quarters this was not exactly new, since Blavatsky, Hartmann and others had preceded Steiner in that respect.

Steiner, an inhabitant of the esoteric substratum of that time, not only knew Hartmann, but also knew a host of other major occult figures. While he was general secretary of the Berlin branch of the Theosophical Society in Germany in 1906, he obtained a charter from Theodor Reuss to found a lodge of the Ordo Templi Orientis in Berlin. Before that, he was involved in a Viennese esoteric society, the Guido von List Gesellschaft that was founded in 1905 to honor that dark Viennese initiate Guido von List. List's notes and references betray the enormous influence that theosophy would have on his writings. After all, Blavatsky's *The Secret Doctrine*, that seminal work that would deliver Keely to the occult circles anywhere where Blavatsky would be studied, was translated into German and published in 1903. A German edition of Scott Eliott's *The Lost Lemuria* appeared in 1905, and both works found their way into List's writings. Thus he also mentioned the airships and the cyclopean structures of Atlantis, already written about by Scott-Eliott.[80]

The Guido von List Society also numbered amongst its members his friend Jörg lanz von Liebenfels. Also ever present was Franz Hartmann who is described by von List's biographer—himself a prominent theosophist—as "very honored among us."[81]

Liebenfels, also heavily influenced by theosophy, would found the esoteric templarist order Ordo Novi Templi, with as its innermost and secret core the "Lumen Club."[82] He held the most singular ideas; in his *Theozoologie oder die Kunde von den Sodoms-Äfflingen und dem Götter-Elektron,* published a year after *The Secret Doctrine* and two years after Fjodorow died, Liebenfels painted his bizarre vision of a new race of man that would be once again equipped with the same electro-magnetic-radiological organs as the ancient races or gods. According to Liebenfels, these extraordinary sensory organs were for the reception and transmission of electrical signals and gave the owners telepathic powers and omniscience.[83] Liebenfels blended his extraordinary ideas on religious topics with recent discoveries in the fields of electronics and radiology, such as the N-rays discovered by Blondlot in 1887, Röntgens discovery of the X-rays a few years later, the discovery of radioactivity by the Curies in 1898 and the application of radio communication by Marconi and Herz between 1898 and 1904.[84]

From that strange blend of occult thought and avant-garde technology that clearly obsessed him we may assume that he had studied Blavatsky's writings on Keely well, but it also had a more practical side to it. Liebenfels also successfully applied for several patents, one for some kind of propulsion system for ships, submarines and aeroplanes. The U.S. Naval office took these patents during World War II.[85] Liebenfels alleged that he corresponded with the most singular persons; if true, we may glimpse a deeper layer through which Steiner's warnings cannot be simply brushed away, for early subscribers to Liebenfels' ariosophic magazine *Ostara* were, amongst others, August Strindberg, Lord Herbert Kitchener and Lenin, who met Liebenfels in Switzerland.[86]

Hartmann, who had closely rubbed shoulders with Liebenfels in the Guido von List Society, had also been working with Reuss around 1902 in an occult group called the Ancient and Primitive Rite of Memphis and Misraim.[87] It is alleged that Steiner joined this order in the same year and worked with Reuss until 1914.[88] Reuss, who had been a member of the Theosophical Society since 1885,[89] obtained the charters for this order from John Yarker, who also sold Blavatsky a Masonic Diploma.[90]

Hartmann knew Karl Kellner, the founder of the Ordo Templi Orientis.[91] It is asserted that Kellner derived much of his teachings through a Paris branch of the Hermetic Brotherhood of Luxor, the order of the American mystic Paschal Beverly Randolph.[92] Burgoyne, who is also briefly referred to in the previous chapter, was the secretary of the order.[93] Beside Hartmann and Reuss, other prominent occultists connected with the O.T.O. in its infant or later stages were John Yarker, Golden Dawn founders William Wynn Westcott and MacGregor Mathers, Papus and Aleister Crowley.[94]

Hartmann met Keely around 1887, had a lengthy conversation with him and investigated his devices with the greatest interest.[95] In connection with Keely, Hartmann also referred to a strange invention made by O.T.O. founder Kellner, an invention of which we know very little. Hartmann's short reference though, provides another glimpse deep into the occult territory, where strange technology is being developed.

That Kellner (1850-1905) was the person to involve himself in such an undertaking is not that surprising, considering he was already technically inclined. He studied natural sciences with a special emphasis on chemistry in Paris, where he may have met Robert Donston Stephenson, who studied chemistry in Munich and medicine in Paris and who pursued occult studies under Bulwer-Lytton. While in Paris, Kellner could equally have met Parisian alchemists such as the mysterious and reclusive Louis Lucas, or even the legendary Fulcanelli. Although no documentary evidence of this exists, it is alleged that in Paris, he was involved with at least one occult order - so the possibility remains.

On his return to Vienna while still a young man, he erected a laboratory for industrial consultation. He invented a procedure for the manufacture of paper, which was highly successful. The following years saw more of his inventions, and in 1895 he held 63 patents. When he died, he had more than 80 patents, among them one for light bulbs and a 1897 patent for color photography.[96]

According to Hartmann, the instrument invented by Kellner "collects and produces electricity directly from the ether of the atmosphere without any friction of solid corporeal substances and without any chemical agency."[97] Elsewhere Hartmann writes that, "Mr. Keely's power seems to be derived by changing the vibrations of cosmic ether. The machine which my friend Dr. Kellner has invented seems to be based upon the same principle, only, while Mr. Keely transforms these vibrations into some force connected with sound, Dr. Kellner's machine transforms them into electricity."[98]

There are more fascinating cases, but quite isolated from the witches' cauldron of esoteric orders and societies and not always with apparent occult flirtations on the surface. In 1888, *Scientific American* published a short announcement with the suggestive title "Keely Outdone." In it, a certain William Timmins, an English mechanic residing in Pittsburgh, claimed to have invented a machine by which "untold motive power can be stored or used without the expenditure of fuel." The article stated that "...he has been engaged for two years in perfecting the invention, and is now negotiating with the governments of England, Russia and the United States for the sale of the right to use his discovery." Interestingly, although *Scientific American* brushed Timmins' device off with the remark that, "The machine seems to be simply an air compressor of the simplest sort," it also noted that part of the device consisted of "two layers of bars containing eleven different minerals, the magnetic influence of which is the secret of the inventor."

Timmins called his device a "Pneumatic Generator" and stated that it could be used to propel the largest ocean steamer afloat or to move 80 laden freight cars in one train. It would be an ideal engine in warships, but also "can be used as a defense against hostile attacks by means of air chambers placed behind the armor plating,"[99] but in the coming years, no trains or ships were manned with his invention.

Another highly intriguing case is that of Nathan B. Stubblefield. In the 1880s, Stubblefield of Murray, Kentucky, claimed to have invented a wireless telephone with which he could "transmit vocal messages with clarity through the ground."[100] According to Gerry Vassilatos, Stubblefield "developed an extraordinary receiver of ground electricity (which produced great quantities of electric power) and numerous 'vibrating telephones' which were used by local residents in 1887. The telephonic devices were patented in 1888 and represent the first commercial wireless telephones, using the ground as the transmission medium. ...We have photographs of his telephone sets. These reveal small, ruggedly built wooden cases which are surmounted by conventional transmitter-receiver sets. Heavy insulated cables run to the outer ground from his apparatus. Stubblefield developed an 'enunciator' (horn loudspeaker) which amplified the voice of callers. These sets of his appeared in numerous demonstrations on the East Coast, from New York to Delaware. ...Numerous private and public demonstrations of this first system were made in Murray, Kentucky (1886-1892), where the mysterious black boxes were seen. Two metal rods were stuck in the ground a few feet apart from each distantly placed set. Speech between the two sets was clear, loud, and startling despite distances of 3,500 to 6,000 feet."[101]

Stubblefield conducted a series of tests for the U.S. Navy, which was favorably impressed. In 1898, he published a brochure to attract investors. In this brochure, he "insisted that power for his device was not generated in the cell. He calmly stated that the cell received energy from the earth." Vassilatos also writes about Stubblefield's intriguing motor, "The device features several mobile pithballs around a compass-like perimeter, resembling the equally mysterious electrostatic hoop telegraphs of the 1700s." According to Vassilatos, Subblefield's last claim made to a neighbor two weeks before he died was that, "The past is nothing. I have perfected now the greatest invention the world has ever known. ...I have taken light from the air and the earth...as I did sound."[102] Stubblefield was not the only one to transmit the spoken word in this manner. In 1893, it was written that the Englishman W. H. Preece, chief engineer and electrician to the Post Office, had "put up a wire a mile long on the coast near Lavernock, and a shorter wire on Flatholm, a little island three miles off in the Bristol Channel. He fitted the latter wire with a 'sounder' to receive messages, and sent a message through the former from a powerful telephonic generator. That message on the mainland was distinctly heard on the island, though nothing connected the two, or, in other words, the possibility of a telephone between places unconnected by wire was conclusively established." And referring to the 1893 experiment on Lake Michigan to send signals to Mars by means of two long crossed lines with lights, the author muses that, "There is a possibility here of inter-planetary communication, a good deal more worthy attention than any scheme for making gigantic electric flashes." The writer then proposes the idea that the ether is the medium through which these messages are carried.[103]

There were of course others who experimented with the etheric force that seemed to have a use to everyone's liking. In 1891, Prof. Oskar Korshelt while living in Leipzig, invented, built and experimented with a device which he called a Sonnenäther-Strahlapparat, or Solar-ether ray apparatus. As Keely before him, Korschelt thought that the ether could be used for the treatment of various diseases, and his Solar-ether ray apparatus was meant as a healing device. Korschelt also published a book about his invention and the use of ether.[104]

The Solar-ether ray apparatus collected from sunlight "the living force of the ether," that could be "radiated parallel in any preferred direction or from such a collecting point evenly in all parts of a room." The Solar-ether ray apparatus also worked on humans, animals, plants and even crystals, "very much like the sun itself, but without the harmful influences of heat and light." Korschelt had succeeded "with this apparatus, without any intermediating human nerve-system, only with instruments, to establish the working of the force which Reichenbach called 'the Od,' and which is generally called since Mesmer organic (animal) or healing magnetism."[105]

Korschelt stated that his device "after countless tests" had finally solved the riddle of healing magnetism or mesmerism; "the ether particles, collected and radiated from the fingertips of a mesmerist or through the ejecting of the body molecules, especially the skin, prove that there are two kinds of healing magnetism; that which functions with a foreign force and that which functions with its own force. ...The first kind is not different from the working of the Solar-ether ray apparatus."[106]

Wilhelm Hübbe-Schleiden, chairman of the German branch of the Theosophical Society, proudly announced a few months later that, "the researches continue with the same quietude, and are now expanded on plants in the living room and in the fields," and persons who had acquired Korshelt's device were asked to communicate their experiences.[107]

Korschelt kept on improving his device, and from these reports we learn something of the appearance of his Solar-ether ray apparatus. During 1893, he developed a "new wire disk (nr.7), that in the same manner, but much more powerful, has strange windings." The disk was 13 cm in diameter, and consisted of two disks, the one on the bottom made of zinc, the one on top made of copper. This disk was "bored through with spiral windings," and through these openings a "wire cable made out of five wires of copper, of which each one was gilded, one was silvered, one nickeled and one was covered with zinc, the fifth however being without a covering." The disk contained another small appendage or spire, around which "another spiral wire enveloped itself around copper pins, which are turned upwards with the spire." Both sides of the device could be used, during a period of 10 to 30 minutes. Korschelt believed that he was successful, as "very sensitive persons had seen the ether rotate around the wire disk." These sensitive persons also saw another etheric ray rotating counterclockwise with the first ray, and slightly above the device a "conus formed," in which "various sparks of different colors in their confusion form all the colors of the rainbow."[108]

That Korschelt's devices, long after their initial conception, enjoyed some success in those quarters where such strange inventions were appreciated we may surmise from the writings of 1920s German occultist and neo-Rosicrucian G.W. Surya. Surya, of whom more is written in the next chapter, relates that he witnessed how "again and again under the Korschelt devices which were put on a tripod, sick persons would sweat profusely for fifteen minutes" and he refers to Korschelt's devices in his writings. According to Surya, Hartmann also "observed the healing of an old woman through these devices."[109]

A year later, in 1894, while in America Colville's and Astor's books were published, the Golden Dawn opened a temple in Paris, and French journalist magnetizer Auguste Van Dekerkove suddenly decided to change his name to "Zanne." This peculiar name was given to him by his mysterious "spiritual masters." Zanne at once started to write his occult history of the world, a book that would eventually number more than 2,000 pages.[110] His great interest was to find a means to reconstruct the primordial language of all mankind. His interest was similar to that of Saint-Yves d'Alveydre, who tried to reconstruct the true meaning of the symbolic signs with the help of the language of the Atlantean red race, the alphabet of which he had placed on his Archeometer. Both the interests of the French occultists were similar to the pursuits of J.K. Rensburg, the friend of Belinfante.

Also in 1894, Belinfante started to lecture in Holland about the forces of the ether in connection with gravity,[111] and Dutch inventor Robert Pape, while residing in the Frisian province, claimed to have discovered a ray in which life could be prolonged. Pape built a device which he called a "Life Wave Generator." Unfortunately neither the device itself, nor a coherent description, has survived. Pathogenic and dangerous microbes placed in the field of the ray showed no sign of growth, Pape claimed, and it is alleged that his device was looked at and even tested by the scientific community.[112]

But the rewriting of history was one thing, as was the construction of the primal mother tongue. With the ultimate aim of a modification of the surrounding universe in mind, man would recreate creation or at least redefine reality. In 1896, a Ulysses Grant Morrow, ardent follower of Cyrus Reed Teed's hollow earth doctrine, invented and completed work on a measuring device that would prove that the earth was not only hollow but that it reclined upwards, since the doctrine of Cyrus Reed Teed, also known as Koresh, held that mankind was living on the inside of this hollow sphere. The device that was actually built was called the Rectilineator.[113]

In 1902, Clemente Figueras claimed to have built a device that could generate electricity without the use of fuel. Tesla was aware of Figueras' claims, for he sent a clipping to his friend Robert U. Johnson, the editor of *Century Magazine*.[114] In Russia, Michail Filippow (1858-1903) claimed to have invented a method for the wireless transmission of the effect of an explosion from one place to another.[115] Filippow belonged to that part of the Russian intellectual coterie that blended metaphysics, esoterism, science, technology, mysticism and religion into a curious whole. The central figure of these circles that included amongst others Tolstoi, who was mentioned in connection with Keely in one of Steiner's lectures,

Dostojevski, and Gorki who admired Filippow, was the orthodox-Christian author Nikolaj Fjodorow.

Fjodorow is a remarkable case of a highly religious and mystically inclined person who saw avant-garde technology as a means to achieve an incredible spiritual end, much as the 19th century occultists did with their passion for technology, but with lesser ambitions. His case serves as an illustration of where such a road of thought might ultimately lead. It would be interesting to trace certain elements of Fjodorow's philosophy in some of Tesla's experiments, such as the never properly explained Wardenclyffe Tower, and in some theories of the brilliant Constantin Tsjolkovski.

Fjodorow, who died in 1903, led a humble and ascetic life. The librarian with the long beard never slept more than five hours a day on a hard wooden board and sustained himself only on tea, cheese and salted fish. He was a staunch believer in judgment day, but in his opinion this could only be achieved through man himself. One thought apparently obsessed him: to find a way to raise the dead. If only the ancestors would be given a new life, man would be truly equal to Christ. The first step to reach this would be the construction of a gigantic electrical ring around the earth that would obtain the energy from the sun and the magnetism of the earth. With this ring of electrical force man would thus be able to create artificial earthquakes and floods, so that the dispersed molecules of our ancestors would be lifted out of the soil. In order to recognize these particles, man would grow special organs, so sensitive that man would perceive "the growing of the grass and the movement of all the molecules of the universe." Then the particles of our ancestors would be "sowed together."[116]

In his remarkable visions that preceded those of Liebenfels and traveled in many aspects far beyond those of the German ariosophist, Fjodorow expected that the earth was far too small to contain all our ancestors; however, man would be completely "biologically transformed" and would re-create his "own organism" into a "perfect work of art." Mankind would be thus transformed into a "new, cosmic and immortal being" that could take "any form" and would "colonize the entire universe." The entire universe, all its suns and planets, would be used as raw material for the construction of this "new being." Universe would be liberated of the force of gravity and would become a continuation of mankind—in fact, one giant conscious being. And, as a brilliant echo or evocation of the age-old occult doctrine that was assiduously studied by its initiates, new man would be equal to God.[117]

11

Vril from Atlantis
Keely's Legacy

"Just imagine a machine that is constructed in such a way, that it does not operate by steam or electricity, but by those waves that man generates in his tone, in his speech. Just imagine such a motor that one may operate by those waves or perhaps by the generation of his spiritual life. It was still an ideal. Thank god that it was an ideal at that time, because what would have become of this war when this Keely-ideal had become a reality in those days!"

Rudolf Steiner
Weltwesen und Ichheit, lecture held in Berlin, June 20, 1916

When Steiner lectured about Keely on several occasions between 1905 and 1920, he found an occult landscape that was already riddled with experiments with the ether and strange devices that were sometimes commercially exploited, as in the case of Korschelt's apparatus. In all probability some of Steiner's occult associates learned about Keely as early as 1882 after Hartmann's visit or perhaps by reading *The Secret Doctrine,* which was published in 1888. It must be taken into account that the occult underground with all its connections, alliances and liaisons was hardly a coherent whole, aiming to achieve unison through one thing although often its inhabitants strived at the same idea.[1]

There is a mere glimpse of what Steiner meant and enclosed in a riddle. He made elaborate drawings and sketches of mysterious devices called the Strader Apparatus and its three accompanying devices. Although as we have seen, there were many more devices sprinkled across the occult spectrum, Steiner's Strader Apparatus has attained somewhat mythical proportions. However strange these devices were, they were meant to be seen; the four devices were to be used in Steiner's theater plays, known as the *Mystery Plays.* The original models of the devices disappeared from the first Goetheaneum that burned down. The origin of the fire that destroyed the building has sometimes been explained as the doings of Nazis who hated Steiner, or as the result of faulty wiring. But we might also speculate that the fire may actually have been started by the highly unusual experiments involving the research of the forces of the ether which were conducted there and which might have gone wrong, experiments of which years later the experimenter was still not allowed to talk about.[2]

The first Goetheaneum building was built out of the same wood as that used in the construction of violins, thus ensuring the large domed hall capabilities of resonance and vibration. Steiner's devices did survive the fire that struck the building though, since so conveniently "for some kind of reason they were not present during this night."[3] An explanation for the disappearance of the original models was that they were not cared for because the significance of them was not recognized at the time.[4]

Likewise, little documentary information survived. Although the Strader Apparatus was the main device, it was accompanied by three other, smaller and equally enigmatic devices. Steiner made three sketches of the main device and its three accompanying devices, which were used as an ensemble in his fourth part of the third of the *Mystery Plays,* titled *Der Hüter an der Schwelle,* or *The Guardian at the Threshold,* its title a veiled tribute to *Zanoni.*

There are also a number of drawings made by Oskar Schmiedel, who built the Strader devices to Steiner's specifications, in addition to a number of unidentified sketches. Schmiedel claims that Steiner gave him elaborate and very detailed descriptions for the construction of the models,[5] which he and some technicians built in the winter of 1912-1913,[6] coincidentally the same period that Keely's "secrets" were allegedly sent to Sweden.

What function the Strader Apparatus and its three accompanying devices exactly had is not clear. Steiner's orders for the staging of the devices in the fourth part are only minimal: "On his (Strader's) table are to be found models of mechanisms." Walter Kugler remarks that, "it could be of significance for a further exposition of the Strader Apparatus that there is no mention of an 'apparatus,' but of 'models of apparatus,' which are the first prototype, and still need a further development."[7]

Schmiedel nevertheless speaks of how "impressive" Steiner's specifications were: "One part was even to consist of a metal or a substance, that was not yet discovered. It is difficult to say with any certainty what the means of these devices were. The main device appeared to me to be something like a condensator for rays and workings that would emanate from the cosmos, possibly also a transformator for these. Various metals: antimonium, copper, nickel and also uranium pitchblende were used; besides a surrogate for the above referred to and not yet discovered substance that was to be painted blue. Except from this main device, a number of others were indicated. On the wall hung a half globe of copper. Its inner side was pointed towards the main device. Another device represented perhaps some kind of measuring device. ...Once he also mentioned when this invention of Strader would become reality. It would be in a not too distant future. Unfortunately I do not recall these dates."[8]

When the play *Der Hüter an der Schwelle* was performed in 1912 in Munich, there had been so little time that only a contrivance was ready,—a shadow of what was to be—built on Steiner's directions by Imme von Eckartstein. During the winter more time was available to build an exact model by using different metals for the half globes; one made out of antimonium, one made of nickel, the fourth half globe, or shell, was half made of copper, the other half made out of the metal "which was not discovered yet." Under this half globe hung the thinnest "feeler organs," made out of very thin goldwire. On the fourth side of a lead cross was to be assembled a spire of uranium pitchblende. The connections of the six spires were partly copper and partly tin. From one shell to another opposite led a spiral glass tube. But "even more mysterious were the three accompanying devices, one of which representing a glass bowl with a platinum wire, molten or hung in it, the second was a flat lying figure eight made of glass tubing, adorned on the top with a spire of coal, with a small shell of copper on it. The third contrivance was to be again equipped with four uranium spires in horizontal fashion. The shape of this device suggests that it could rotate. Electricity should be kept away from it."[9] The main device was to be put on a table: somewhat in the distance stood the other three devices, also on a table. The copper bowl adorned a wall.[10]

Owing to its mysterious nature, the Strader device invited several occult explanations, the most interesting and complex being that it could be seen as "the inversion of Steiner's design of his Heavenly Jerusalem" of 1908. From there a deep plunge is taken into the fourth dimension, cabalistical doctrine, the tarot, Gnosis, heat and light as more-dimensional phenomena, related to movement and vibration, Atlantean energy procedés, and crystals as the "substitute for the perpetuum mobile." The Strader device, we learn from all this, was the pre-design of an ether converter, which, "contrary to Keely's converters, had almost no moving parts (transformator part) that lead to the increase of power."[11]

With the ideas behind his *Mystery Plays* and the construction of the curious Strader devices, Steiner found himself in very good company. A year later in 1913 in France, Albert Caillet published his massive bibliography on the occult sciences: a staggering compendium of the occult, hermetic, alchemical and Rosicrucian publications from the earliest days of printing until that time, a work so complete that it is in use even today and has not been surpassed. But in Germany, other things were taking shape. The occult was rapidly transfiguring itself into a new order. A whole occult movement blossomed feverishly, deeply embedded in the German social strata, with as its most important philosophical cornerstones Mesmer, Von Reichenbach, Bulwer-Lytton and ultimately Keely, who in the eyes of many a theosophist and anthroposophist had discovered a force that was best kept hidden from mankind. And they had good reasons for thinking so; Europe was creeping towards what would be the First World War, and a 1913 wave of unexplained aerial objects equipped with powerful lights that struck parts of England, Holland, France and Germany was explained as the doings of "spy ships of the enemy." The mysterious objects were accordingly nicknamed in the newspapers as "scareships."

Fevered and strange the times must have been, just a year away from the horrors of the First World War. But in the meantime, Steiner had conceived his devices, and experiments were made by others, such as the photography of thoughts—called psychography—in which Friedrich Feerhow claimed that he had succeeded.[12] Feerhow also published his theories on the human aura, subtitled "a new contribution to the problem of the radioactivity of man,"[13] and also his "influence of the earth magnetic zones on man," in which he considered the aurora borealis "an odic phenomenon," in accordance with Von Reichenbach, who also thought so.[14] Feerhow mused on the Od-rays and the N-rays as discovered by the French scientist Blondot, "named N-Rays because the French researchers had no knowledge of the results of their German predecessor."[15] Pamphlets, booklets, books and learned treatises were issued in which not only Blondot's N-rays, but also Dubois' "organic fluorescens," Prof. Jäger's "Anthropin," Justinus Kerner's "Nervengeist" or "Nervespirit," Dr. Baraduc's "Biod," Ziegler's "Pouviour irritant," Darget's "V-rays," Korschelt's "Dark suns (etheric) Rays," of course Mesmer's "Animal magnetism," Luys' and Rochas' "Effluviums," Dr. Narkiewicz-Jodkos' "Electrographics" and a multitude of other discoveries of new forms of force and energy were discussed.[16]

The very borders of conventional science were not only traversed, but also severely questioned. Means that once were considered pure magic were dusted off and transformed; thus bewitching and casting spells were a "form of telepathy," the moving tables of the spiritists surely were the result of as-yet undiscovered energies and forces. The negation of gravity was discussed,[17] as were telesthesia, telenergy, mental suggestion and magical thought transfer.[18] The magnetic healing of wounds, the fluidal body of man and the doctrine of thought waves, "the understanding of thoughts as psychosphical energy...already formulated by the Russian researcher Dr. Naum Kotik," was explained in yet another brochure.[19] Magic was considered to be a form of "nature science" and a doctor, Adam Voll, wrote upon the dowsing equipment and the sidereal pendulum,[20] while Robert Blum confided his theories to anybody who would read his book that was published in three tomes suggestively called *The Fourth Dimension*. In the first part, Blum followed Keely's doctrine in propagating a threefold order to be found in all nature, while the third part was suggestively entitled *In the Realm of Vibrations*.[21] Interestingly enough, a German translation of Shelley's *Frankenstein, or the Modern Prometheus* also was reprinted during that period.[22] Much of the above was published by theosophical publisher Max Altmann. His *Zentralblatt für den Okkultismus,* published in 1908, was to be the first monthly occult periodical in Germany.[23] Altmann also published *Moderne Rosenkreuzer* by occultist and neo-Rosicrucian G.W. Surya in 1907. The book that was in its seventh edition in 1930 treated much of the above and recommended many of these brochures. Meanwhile from the theosophical quarters, authors such as

Leadbetter and Besant had already rained down a torrent of books and pamphlets on thoughtforms, the ether, the Akasha, chakras and the human aura.

Two years after the construction of Steiner's first Strader instrumentarium, which was only a tip of the German occult iceberg, in Paris the French alchemical circle around the fabled alchemist Fulcanelli was deeply involved in the construction of a "Turbo Propulseur," or a "curious aerial sled," that was first conceived by Jean-Julien Champagne in 1911. We learn nothing of its appearance, its propulsion or in fact the underlying motivation for an alchemist to conceive of such a device, save a sketchy description. The vehicle was powered by a motor-driven propeller, and was said to be "quite suitable for travel on roads."[24] What the actual motor consisted of was not mentioned. Elsewhere it is being described as a "polar vehicle," and its propeller was still in construction in 1914.[25] It was eventually presented to Tsar Nicholas II, allegedly through the French occultist Papus, who resided at his court around that time.[26]

In 1914, while the French alchemical circles were still busy with the construction of the propeller of their strange vehicle and shortly before the outbreak of the First World War, a German named Franz Philipp purportedly succeeded in lifting his "Sonnenkraft-Triebwerk," or solar-powered motor, into the air in Germany.[27] Perhaps the experiment was the result of the earlier-mentioned widely held discussions in Germany on the negation of gravity, in matter as well as the human body, but we do not know, for the only thing that Philipp wrote considering the nature of his invention was that he discovered its principles with a device of his making for the study of the solar protuberances during a total solar eclipse. Based on what he then discovered, he would construct several aerial craft and eventually spaceships with "ever better propulsion systems." According to Philipp, all this would lead to space travel well before World War II.

An impenetrable and suspicious curtain of vagueness surrounds his claim. Studying his bizarre book, one is quick to dismiss the writings of Philipp as those of a lunatic, or as a parody on UFOs and ufology in general. To make matters worse, nowhere in his book does he lift the veil on the exact nature of his supposed solar-powered motor, while on the other hand the pages are filled with wild and unproven allegations without a shred of evidence that any of it was true.

The case however obtains a sinister and ominous dimension when we read elsewhere that Philipp was found dead in his Berlin apartment in the late 1970s, after having laid there for two weeks. It is also stated that at one time the CIA tried to "assassinate Philipp at least once." Then there is mention of his connections with self-styled counter Pope Clemens XV, who was the head of the Order of Saint Michael and the founder of the "New Church" located in the east of France and in 1970 was said to have numbered some 50,000 members in Germany, France and Italy. This New Church believed in the salvation of mankind with the help of extraterrestrials.[28] The fate of his solar-powered motor, if it ever existed outside of his imagination, is now unknown, but perhaps the fact that the memory of Philipp is again cherished in certain contemporary German occultist underground circles where his ideas are currently linked with what has become a whole ideology—or a religion in the making—of vril, speaks for itself.

In this strange history, we again encounter isolated cases. A year after Philipp's discovery and a continent away, an electrical engineer named C.E. Ammann, residing in Denver, started working on something of a more tangible nature and what he eventually called "the atmospheric generator." A Denver newspaper reported seven years later that the device, according to Ammann, could "draw energy out of the air." The atmospheric generator was described as a "compact cylindrical object with two small brass spheres protruding from the top." Inside was to be found "an arrangement of steel wires and minerals so fixed as to draw the energy from the air," somehow echoing Timmins' pneumatic generator a quarter of a century before. The device was demonstrated in 1921; it was attached to an automobile which was then driven around the city. The device was inspected; it was agreed that Ammann, who was then 28 years of age, had "made an invention that will revolutionize power," and that "we have long known that certain minerals exist which, if properly arranged together, would furnish power." Ammann, we learn, was planning to go to Washington for a patent that very same week in 1921, but we hear no more of him or his device.[29]

The air was filled with energy in those days. In 1916, Harry E. Perrigo invented a converter to extract electricity out of the air. He worked on his device between 1916 and 1927. He discovered that it produced more electricity when a breeze was allowed to circulate through the room or when a warm body stood close to the antenna of the device.[30] A great invention, but of which we also hear nothing more.

Perrigo was followed in 1917, just a year later, by an Armenian immigrant named Garabed T.K. Giragossian, who lived in Boston. Giragossian claimed to have discovered an inexhaustible source of free-energy, which he called The Garabed. He stated that his discovery would make the steam engine a thing of the past. His source of free-energy would power ships, airships and locomotives. Although he was vague about the exact nature of his discovery, prominent Bostonians would vouch for his honesty. Giragossian apparently convinced Congress, for in 1918 President Woodrow Wilson signed a resolution to protect him and his discovery. Interestingly, he did not want money, but protection. Four scientists and an engineer saw his device on June 29, 1918, and this commission delivered its report, just one paragraph long, the same day. Its conclusion was that Giragossian's claims were false.

The invention seems to have been a massive flywheel that, once set in motion, would be kept turning by a miniature 1/25 horsepower electric motor. When linked to a brake-type dynamometer, the spinning wheel would briefly produce up to 10 horsepower before it stopped. Giragossian and friends insisted that the commission did not understand his invention, and Congressional supporters held more hearings in 1923 and 1924. Although he maintained that his Garabed would "reshape the destiny of mankind, creating an age of reason, an everlasting happiness," Congress voted no further action. Resolutions on Giragossian's behalf were proposed almost annually until 1930.[31]

In 1919, two years after Giragossian's first public appearance, 19-year old Alfred Hubbard built and demonstrated his first device. Like Perrigo, Hubbard also claimed that "he was getting energy out of the air." One experiment made with the Hubbard transformer was the propelling of an 18-foot boat around Potage Bay near Seattle. A 35 horsepower electric motor was linked to a Hubbard transformer that measured 11 inches in diameter and 14 inches in length. It furnished enough energy to drive the boat and a pilot at a reasonable speed around the bay. Later Hubbard alleged that his transformer was powered with radioactive substances. Hubbard admitted that he had used the idea of power from the air to protect his real idea for a patent, and that his machine created electrical energy directly from radioactive materials, which he did not name. As far as can be determined, no U.S. patents were ever issued to Hubbard concerning this device, and Hubbard, like Ammann and all those before him, disappeared from the pages of history.[32]

The same can be said of Lester Jennings Hendershot, who developed a device that he always maintained worked on the force that pulls around the needle of a compass. Hendershot claimed that during his experiments, he learned that by cutting the lines of the magnetic north and south he had an indication of the true north, which in his opinion was not the magnetic north of an ordinary compass. By cutting the magnetic field east and west, he found he could obtain a rotary motion, in which we find an echo of Keely's statements and of the very early designers of magnetic perpetuum mobiles. Hendershot built his first unit in the 1920s, but nothing seems to have come from his invention. There are vague tales that he went to Washington for a patent, but somehow this city does not favor young inventors, for as far as can be determined no U.S. patents were ever issued to him, and his doings in later years are largely unknown.[33]

The same, or a possible grimmer fate awaited young inventor John Huston of Prineville, Oregon. Around 1920, he claimed to have invented a way to take "heat out of the air" with condensers. According to an eyewitness, the first poorly insulated rig boiled water in 20 minutes. The device was claimed to replace fuels, to be good for household heating or refrigeration and to be able to run railroad engines and steamboats. Huston and his father formed a company of 20 stock-holders and he built an up-to-date model of his device. The model apparently worked better than expected, and Huston and his father took it to San Francisco to demonstrate it, since they hoped to interest manufacturers in building the device on a royalty basis. According to the eyewitness who talked to Huston after his return, Huston told him that, "The machine can be made so hot that it will destroy itself. Reverse the machine, and the temperature will go as low as 250 below zero." According to Huston, manufacturers in San Francisco "refused to build the machine because it would throw too many men out of work. It would also kill the sale of fuels, the major cargo of steamships at that time." He also stated that, even though the U.S. refused to patent the device, Canada and England did patent his invention. However, his remarkable invention disappeared, and he died a young man at the age of 22 in 1920 or 1921.[34]

Although the original Strader instrumentarium also disappeared when Steiner's Goetheaneum burned down on December 31, 1922, that strange blend of technology and occultism did not disappear with it. It was perhaps this current, possibly experienced during his stay in Berlin between 1921 and 1922,[35] that prompted Georgi Ivanovitch Gurdjieff to write about fantastic "tri-cerebral beings," the "all-pervading Okidanokh" and the "tri-centric beings of Saturn," who possessed a bizarre device or installation, called the "khrakhartsakha," that was invented to study hitherto-unknown aspects of this cosmic substance.[36]

What very much existed outside the borders of such an exercise in the fantastic was the German magical order Fraternitas Saturni, founded in 1928 in Berlin. The order had its roots in an occult society called "the Pansophical Lodge," or "Pansophia," originally founded as a loosely organized study group in Berlin shortly after the First World War. The group became more organized and changed its name "the Grand Pansophical Lodge of Germany, Orient Berlin" in 1923. After a visit by Aleister Crowley, a schism occurred; the Pansophia Lodge ceased to exist in 1926 and the Fraternitas Saturni came into existence.[37] The Pansophia lodge, whose founder knew Ordo Templi Orientis cofounder Theodor Reuss quite well, was already bent on the curious fusion of technology and the arcane. Not much is known of the activities of the Berlin Pansophia lodge. However, in one of its publications one finds references to Babitt's *Principles of Light and Colour* and German occult philosopher Milankowitsch, who so elegantly formulated his theories on creation as a giant "world-machine" at the time that Spear was constructing his New Motor.

The publication also promised a training to any initiate of the first grade in which one would learn such matters as "unexplained miracles of nature," dowsing, the art of the pendulum, the od-force and magnetism, hypnosis and telepathy, matters that had set the German occult substrata a decade earlier in such an exalted and expectant state. Moreover, the initiate of the Pansophia lodge was required to experiment with "Od-rays" and the pendulum.[38]

Beside studies in cabalistic doctrine, sidereal astrology and sympathetic magic, the initiate would learn during alchemical courses the secret of the Aurum Magicum: "This magical gold of fluidal and fiery appearance is also called the ointment of the wise, with which the true magus may establish the greatest miracles: 1. the magical Perpetuum mobile of the sidereal heavens in eternal circulation etc., and 2. With it the eternally burning lamps of the ancients can be lit."[39]

The mind of the neophyte is carefully prepared for a journey into the spheres of the arcane. In a long and curious digression on alchemy, the initiate is immersed in a vast sea of hermetical and technical metaphors. Thus we read of "the spirit of imagination," of "love that mingled with wisdom," of "light-life" and of the dead, the "old Adamic man" that may rise through "the tincture (cosmic electricity, auric force, electro-)," of "etheric regions" and the "alchemical tincture."[40] It is hinted that "alchemy has only one universal, not two and more. Yet this one can manifest itself in seven different rays, forces, energies, vibrations in various spheres or planes of existence."[41]

But we also read that, "In the new coming times we will refute the hypotheses which were acknowledged until then, about the origin, rotation and evolution of the stars, suns and planetary systems. The human race will be magnificent and will be sensibly organized, will evoke a new meaning of life, with the help of which a totally new knowledge of nature's secrets will blossom. A new science will develop, in the direction of transcendental physiology and psychology. ...The world ether, these days once again rejected by science, will no longer be a vague hypothesis, when mankind will unriddle the secret of its existence through a variety of inventions and discoveries, through which those mechanic-material, chemical-technical conjunctions will disappear into nothingness. New problems will be undertaken by the human brain, which then has to torment itself with until now unknown and because of that rejected cosmic forces and energy quantums of unknown dimensions. The astral regions will once again be accessible for everyone, and new elements and ferments, a new biogenesis on sidereal principles, therefore a new heavens with a new astronomy and astrology, and a new earth with a new geosophy and tellurism will expect a new type of man, after the downfall of the old man. ...But the universal brotherhood knows by looking at the giant world clock, the sidereal zodiac, almost precisely the right spiritual moments coming in a few decades...the most favorable projection of a new race of man, a higher developed type of man."[42]

Unfortunately the exact nature of the "variety of inventions and discoveries" is not further explained or described, although we are left to assume that these were based on other principles than the known mechanical and empirical ones. And leaving that puzzling vista of magico-spiritual eugenics behind, which—as is clear from other passages not cited here—was influenced by Steiner and Blavatsky, but possibly also by Bulwer-Lytton, the writer leaps into cosmic distances. There, we learn, the sun and the planets are parts of the same thing; point and counterpoint; "one and the same ether, one of it has become positive— and is named the sun, the negative one is called the planet. Both are but one etheric globe, of which its center is the sun, its periphery the planet is called. ...The sun cannot exist in the absolute middle of the solar system, because of the opposition of the planets, who likewise want to be in its center. Since the universe can only exist in bi-central form, there is also no universal central body. It is there, but in the form of the sun and the planet. Only God is monocentral—the world is the bicentral god. God is the monocentral world. The behavior of the planets to the sun is one of a polar attraction and repulsion, power of the primal law in the solar system, power of light, of radiation. This attraction and repulsion is only possible, because the planet of its own force when it approaches the sun too close, in itself tilts from the negative to the positive pole—becoming the sun—and when it distances itself from the sun, it once again tilts the positive pole and nourished the negative pole in itself. And that is the cosmic motor!"[43] the Pansophia lodge jubilantly declares.

With this occult vision of the solar system as one giant cosmic motor, the Pansophia lodge adhered itself to Keely's views on the earth's poles as being negative and positive, the polar and depolar force, the "interchange of polar and depolar sympathy" and "the sympathetic attractive, the force that draws the planets together," which was "the same that regulates the motion of the planets in their recession from each other." Aside from the uncanny similarity, we are left to muse—by absence of any reference to Keely in the publication of the Pansophia lodge—as to what degree the techno-magicians were aware of Keely's endeavors or in precisely what way they put this knowledge into practice.

It was the Fraternitas Saturni that evolved out of the Pansophia lodge, and which was an occult society that followed Crowley's doctrine to some extent, that definitely conducted experiments along those lines where technology and the occult meet. Although it is undisputed and historical fact that the order did so, very little is known about these researches. "Certainly one of the most unique aspects of Fraternitas Saturni magical technology," writes Flowers, "is (or was) its involvement with electrical instruments to enhance or to effect magical ends. This was part of a general field of interest among initiates of the Fraternitas Saturni, a field which included the study of, and experimentation with, the magical effects of high-frequency sound, electromagnetic fields, so-called 'Tesla energy,' ozoniation of the atmosphere, ultraviolet light etc." These experiments were put in a theoretical framework of teachings concerning "aethric waves," or the chakra system, thereby providing these with a magical basis.[44]

As is clear from the writings of the Pansophia lodge, the foundation was already in place, and in that respect the research of the Fraternitas Saturni may be seen as a continuation of that tradition. Certainly, Germany had somehow survived the First World War, although not under the most favorable circumstances. But with its survival and the ensuing economic disorder, the occult substrata blossomed again into strange foliage.

That curious fusion of technology and occult lore also infected the German film industry and in some instances the fringes of fantastic literature. In 1916, during the height of the devastating world war, a six-part serial, *Homunculus,* was released, its theme relying heavily on Paracelsus' concepts of creating an artificial man and on Shelley's *Frankenstein,* which had received a reprint three years earlier by Altmann's publishing house. Coincidentally the same year that *Frankenstein* received its reprint, German filmer Paul Wegener came across that other ancient legend of the creation of an artificial man; the legend of the Golem, which led him to film the tale a number of times, in 1914, 1917 and in 1920.

The year 1919 saw the appearance of *The Cabinet of Dr. Caligari,* in which hypnosis, Mesmerism and somnambulism, so assiduously studied by the German occult milieus, played an integral part. A year later the Austrian magazine solely dedicated to fantastic and horror literature and one of the very first in the world, *Der Orchideengarten* or *The Garden of Orchids,* published a special issue consecrated to "the world call of the German electronic industry" as it stated on its first page. It did this in its own unique way. The special issue was titled *Elektrodämonen* or *Electrodemons* and it featured such tales as H.G. Wells, "Lord of the Dynamos," Hans Reisiger's "Elektrischer Sabbat" or "Electrical Sabbath" and Alf von Czibulka's "Das Wiehern in den Transformatoren" or "The Whirling in the Transformators." The tales were accompanied by weird etchings of machines and grotesque demons, which were melted into a bizarre configuration. A portfolio of drawings of Otto Muck, titled "Maschinen," was advertised in which he rendered machines and engines into a hybrid of biological and mechanical creatures, gifted with a strange semblance of life.[45] The founder of one of the most striking magazines of the fantastic ever produced and that lasted only three years was Karl Hans Strobl, a then well-known writer of dark, unusual tales who had studied at the University of Bingen at the same time that Hugo Gernsback, who would become Tesla's friend, studied there.[46]

Also in 1920, the German film *Algol* was released in which an evil extraterrestrial entity from Algol, in occult lore known as the demon star, donates superior technology to an earthling, a machine with which one can become master of the world. In 1922, *Nosferatu* appeared in the German cinemas. The film superbly transformed *Dracula,* written by Bram Stoker, into a haunting play of light and shadow. The sets of *Nosferatu* were designed by Albin Grau, a leading member of the Fraternitas Saturni.[47] And in 1926, the same year that the demise of the Pansophia lodge from the occult stage made way for its successor the Fraternitas Saturni, the magnificent *Metropolis* was released. In one of its more disturbing scenes, ancient magic and avant-garde science in unison create an artificial human in a laboratory filled with sparkling technology, but also a clearly visible upside-down giant pentagram.[48]

Although not much is known about the researches of the Fraternitas Saturni, some light can be shed upon this matter by studying the beautifully printed issues of the official organ, the *Saturn Gnosis*. Its pages are riddled with terms as "electromagnetic fields" and "planetary spheres," "etheric planetary forces," "cosmic will" and "ether movements."[49] We read of the etherwaves of the Weltgeist, a concept not unlike Mesmer's animal magnetism, which also appears in the writings of Bloomfield-Moore: "The word ether is from 'aitho,' to light up or kindle. According to Pythagoras and all the oldest philosophers, it was viewed as a divine luminous principle or substance, which permeates all things, and, at the same time, contains all things. They called it the astral light. The Germans call it the 'Weltgeist,' the breath of the father, the Holy Ghost, the life-principle."[50]

Expressionist trance studies of Atlantic and Saturn demons, an exposition of which was held in Berlin in 1931,[51] stand side by side to references to the Akasha chronicles, pictures and articles on by-now familiar figures as Wronski, Levi, Paracelsus and Agrippa and comments on the writings of Jennings, Hartmann, Maack and Rudolf von Sebottendorff, the founder of the infamous Thule Society, which branched off the Germanen Order whose spiritual fathers were such men as Guido von List and Lanz von Liebenfels.[52]

The initiates of the Fraternitas Saturni perceived man as an "electro-magnetic power plant," and electrolysis, electricity, ultraviolet, infrared, röntgen rays and radio activity are being discussed together with the magical concept of "radioactive electromagnetic space-force fields," which "through certain centers in man's etheric body and in the nerve system" make an exchange of "cosmic and human rays" possible.[53]

The Fraternitas Saturni conducted its magico-technical rituals in order to create with their will "impulse waves, which reach into the cosmos. In their rhythmic change these will reach the electro magnetic force fields, not only reach, but in their radiation create from them."[54] There are veiled allusions as to the magical technology involved in all this: "One can also see sacred cult ritual as cosmic physicism. First a concentration of energies in a suitable and chosen matter, such as in the most used cult devices, led, copper, silver, gold, which represent the chiefly magical metals. By word or vibration the atomic structure of these metals is set in motion. By the magnetic wailing the etheric radiation of the force motion is centralized, directed in a given direction, to achieve a binding to astral matter."[55]

There are hints of the devices used in their curious techno-magical rituals and experiments through which "etheric rays may be assimilated and transformed." "Through the use of antennas built of copper, silver and gold—because of the motions of their electrons have a bigger absorption for cosmic rays—which are switched on through persons with a certain state of mind and through auto-suggestion."[56]

Glimpses are given of other devices such as Korshelts' Solar ether Waves-Apparatus of which much of the technology apparently was derived, judging from the above and following descriptions, and a certain Dr. Eckhoff who invented devices which "through their spiral working of to opposite placed spirals," made out of an "extraordinary metal," provided the human body with "etheric rays," a process that he called "telesion."[57]

The author also tells how, "In our circles an engineer has constructed a copper spiral with 27 windings, the center of which is a copper bowl. An antenna is created thus, which is able to receive ethereal waves." The device apparently worked better when a person with "strong magnetic powers would radiate on it," and it is commented that "similar methods, although of a different kind, have given likewise results through the appliance of high frequency electricity, that go back to the research results of Nikola Tesla." It is also stated that "modern high frequency devices are being used."[58] The concepts of "fields with crystalline structures that are short circuited whirlfields," "Euclidian-Newtonian-Galilean-Space-Time worlds," and "the giant organ of our starry worlds" are discussed, together with the "power currents of cosmic and mathematical structural laws, built on sound systems (Klangsysteme)" and "gravity potentials of electromagnetic and mechanical laws."[59]

We learn of "primary suns and their force fields," the energy of which is "ultra energy in a total vacuum," and we are confided that "this central or primary solar field has a cellular structure with the possibilities of aggregation of the space-substance in its highest form." Then it is stated that, "The condition of this space-substance as fleeting electricity has, changes itself in the central solar field in a plus condition and—Vril."[60]

And with this statement from the pages of the official organ of the Fraternitas Saturni the wheel has turned full cycle. Keely's visions of obtaining a force through the ether from the "very wheelworks of nature,"—an idea derived from the learned cabalist Pancoast and Nathan Loomis, linked to Bulwer-Lytton's vril-force—had traveled a long way, had crossed the fin de siècle with theosophy and had survived a devastating world war with anthroposophy. The original concepts had transmutated on each occasion when occult groups applied their wonderful ideas for their own quests. They surfaced again in Berlin of the interbellum, where the German Vril Society, another occult order or group obsessed with strange technology, fully allied itself with Bulwer-Lytton's vril ideology and Keely's discoveries and took its tenets to the very limit. Although the existence of a society by that name never was,[61] the group as such did exist, but under a totally different name, and totally different from what modern mythology would want us to believe.[62]

In 1930, the year that Surya's *Moderne Rosenkreuzer* went into its seventh edition, the last of the magical issues of the Fraternitas Saturni was issued and the magical order went underground, ultimately to be banned by the Nazi regime three years later, a little booklet of a mere 60 pages appeared under the imprint of the "Astrologische Verlag Wilhelm Becker," located at Schlosstrasse 59 in Berlin-Steglitz. This booklet bore the appropriate title *Vril. Die Kosmische Urkraft. Wiedergeburt von Atlantis.* The author chose to hide himself or herself under the pseudonym "Johannes Täufer." The publisher was an organization that named itself "Reichsarbeitsgemeinschaft Das Kommende Deutschland," whose headquarters were also located in Berlin, at the Pallastrasse 7.[63]

A second booklet appeared the same year, also by the "Reichsarbeitsgemeinschaft" and titled *Weltdynamismus,* but issued by a different publishing house, Otto Wilhelm Barth Verlag. Both booklets referred to each other. It was this Reichsarbeitsgemeinschaft which would be the model for the legendary Vril Society. Peter Bahn, who can be credited with the discovery of the true identity of the Vril Society, found no clues as to the nature or origin of the group itself, and we learn little of the Reichsarbeitsgemeinschaft from their own description in their "Vril" booklet. But it is likely that this group was somehow a continuation of the Pansophia lodge or the Fratres Saturni. Since the Reichsarbeitsgemeinschaft left no clues it is now open to conjecture, but we may safely assume that at one time members of this group not only shared the same occult substratum in Berlin but also shared the same ideas on the fusion of technology and the arcane. A passage on Atlantis in the booklet of the Reichsarbeitsgemeinschaft that describes the Atlanteans as "psychophysical dynamotechnicians and not mechano-machinists"[64] involuntary evokes the earlier cited shimmering visions of the Pansophia lodge: "...mankind will unriddle the secret of its existence through a variety of inventions and discoveries, through which those mechanic material, chemical-technical conjunctions will disappear into nothingness."

Not only that, they evoke the same passion for a search for alternative technology in the light of occult doctrine as demonstrated by a book that was also published in 1930, the same year that the booklets of the Reichsarbeitsgemein-schaft were published and Surya's book was reprinted. As Surya's book, the book was issued in Leipzig, the place where occult publisher Altmann resided. It was titled *Verschollene Kulturen,* or *Hidden Cultures* and written by Eugen Georg. In the ponderous tome Georg digressed on topics such as magic, the occult, Atlantis, Lemuria and Hörbigers' Welteislehre, but also on the "cosmic propelling forces" and the proper use of solar, wind, tide, thermal and earth rotational energy. Explaining these topics, Georg confronts the reader with a multitude of terms as "metamechanical-organic," "energetic-metatechnical," "organic-technical" and "machine technical," "magia-technology" and "meta-technology." Amongst others, Georg cites as his sources the writings of Hartmann, Maack, Liebenfels, List, Levi, theosophist Scott-Eliott whose remarks on Keely we have read, and Surya's *Moderne Rosenkreuzer.* Georg also mentions Schappeller's space-force engine and vril several times in connection with Atlantis, like the Reichsarbeitsgemeinschaft would do the same year. One of Georg's conclusions was that to him magic is in fact technology, but with other means.[65]

Here it is important to stop for a moment and study Surya's most puzzling book *Moderne Rosenkreuzer* or *Modern Rosicrucians* more deeply. For while its title might suggest a study on Rosicrucian topics of his day, in fact the book is quite different and, as is the case with Georg's *Hidden Cultures,* its contents clearly help define the occult surroundings in which the mysterious Reichsar-beitsgemeinschaft came into being.

The name G.W. Surya was a pseudonym; behind it hid the person Demeter Georgiewitz Weitzer (1873-1949). Weitzer knew Hartmann and wrote about Kellner's alchemical work in his "well equipped alchemical laboratory."[66] Weitzer, or Surya as he choose to name himself, also belonged to the resurrected Rosicrucian occult sub-scene of his day and published a number of treatises on alternative medical science, astrology and Paracelsus.[67] But it is with him that we may see how the German occultist-technicians were influenced by Keely.

In the pages of his book, Surya mentions *Zanoni* on several occasions, quotes Hartmann, discusses thought photography, Korschelt's devices, the Od-rays and the ether while using a single term that also appears in the writings of the Pansophia lodge and the Reichsarbeitsgemeinschaft: "Electricity is but a special form of the general cosmic energy. By and by science has arrived at Dynamism...for the ether is...the foundation of all electrical manifestations, but also matter!"[68]

But the most curious and interesting part, presented in such a way that one is not sure whether the anecdote concerns actual incidents or a fictional account, involves the ship of a certain Lord E., called the *Sirius*. The ship, traveling more than twice as fast as "the best Lloyd steamboat" was "a hundred meters long and had four mighty funnels placed behind each other, which through their intervals, indicated that the boiler-rooms occupied one third of the length of the ship," Surya writes. The *Sirius* had double lining, watertight compartments and was equipped with two screws[69] and a smaller rescue-vessel aboard, called *Dependance of Sirius*. His vessel has "its own machine," a powerful petroleum engine of 300 horsepower, but also a mysterious "additional machine," that Lord E. invented and similar to one that also propels the *Sirius*.[70]

Mysterious as the additional machines may be, Lord E. nevertheless wishes to explain their working: "...five years ago I made wonderful discoveries during my experimental studies of the ether. You know my friend, that I have set it as my task to fathom the double sphinx of force and matter. ...The speed that my additional machines, built on the ground of my discoveries and inventions, grant my *Sirius*...was for me a sublime moment, when my *Sirius* cleaved the waves, propelled by 'Ether Force' ...Never must mankind, in its current stage of development, obtain these terrible forces that I have discovered."[71]

Here Surya momentarily interrupts his account to insert a large note that occupies most of the page. "Compare *The Secret Doctrine*... The coming force," Surya hints in the note. And it is a hinting that he employs, for while he cites several lines from her chapter that delivered Keely's discoveries to the occult world, Surya, further digressing on the nature of the ether and also citing Tesla does not mention Keely, the central character of that particular chapter in *The Secret Doctrine*, although Surya would do so in another book.[72]

We further learn from the mysterious Lord E. that, "In 1877 a German scientist had already penetrated the physics of the ether or in the secret of matter...[73] and that two of the four funnels of his *Sirius* are camouflage, 'to divert curiosity,' and that his 'additional machines' are also called 'etheric machines'...So you see that the name 'additional machine' is also a kind of mask, so that I won't betray my principles during a conversation."[74]

Finally a mere glimpse is offered on those etheric machines: "just somewhat above the water...two short bronze tubes with horizontal axles showed, similar to torpedo launching tubes, each with a diameter of 60 centimeters. From these a hurricane-like constant jet of air ejected." The bow of the ship had the same two tube-like appendages. "When I want to obtain a counter motion, the forward tubes will start working and the backward ones stop, that is, they will not obtain ether power anymore. Diagonal tubes, simultaneously switched on, turn the boat. ...This is everything that I will explain ...How I obtain this enormous ether wave is my secret and it must stay that way. In order to obtain power I proportionally dematerialize a small amount of matter; 4-5 milligrams pro second for the two tubes. In reality a thousand explosions pro second take place, and this seems to us a constant stream."[75]

What is left for Surya to remark is that Stefan Brandt, the protagonist through whom we learn of the existence of Lord E. and his ether-driven vessel, mutters that "the Atlanteans must have built their airships the same way." Indeed Lord E. confidently admits that this must be the case, and he in turn muses, "It would be easy to build a veritable airship with these forces. ...To build an airship of 25,000 kilos one only needs a tube with half the diameter of one of the reaction-tubes of my *Sirius*."[76] But Lord E. leaves it at that, since "One who has traveled so far to self knowledge, will not be smitten down by arrogance, pride, the will to rule, just to play 'Master of the World' for a short time."[77]

We will leave Surya's account of modern rosicrucian Lord E. here, and we are left to ponder upon his expression "Master of the World." For Jules Verne wrote a tale involving a powerful airship having that very title, and thus we are left in uncertainty if with this particular expression Surya provided a clue, meaning that he merely employed Vernean trickery, a fictionalized account, or wrote about a real incident involving a real person and having to do with real etheric engines. And although as we will see in the last chapter, the very mention of Verne might yield another explanation, we must also take into account that Surya's most singular book featured on its title page the subscript "an occult-scientific romance." Thus Surya more or less admitted to having employed the same literary procedé as Bulwer-Lytton had done with his *Zanoni* and *The Coming Race*, that is, to divulge to the reader certain information, carefully hidden in what outwardly seems a fanciful tale, a work of fiction. As an afterthought we might remark that Surya's descriptions of the reaction-tubes of the *Sirius* somewhat resemble a certain passage in Greg's *Across the Zodiac*: "...I went on deck...to examine the construction of the vessel. ...Her electric machinery drew in and drove out with great force currents of water which propelled her with a speed greater than that afforded by the most powerful paddles."[78]

Surya's book also featured an advertisement on one of the last pages with the slogan "fight materialism!" that rallied any interested person to join a group called "Der Deutsche Neugeistbund," claiming that the group was "a member of the enormous chain of new spiritual movements which nowadays circumvent the globe," the Tibetan equivalent called "Swastika." Mention is made of "Ortsgruppen," or "local chapters" and the "Neugeist Centrale" or "New spirit central point." From the advertisement it is clear that some sort of neo-Rosicrucian movement is meant: "Germany can only then be saved and made ready for a new rise when the world view of the Rosicrucians is lit over the soul of this country!" it exclaims.

Stripped of its choosing occult sides, it sounds similar to the ambitious plans of the 18th century Parisian magnetic freemasons to create "radiant healing centers" to promote Mesmer's techniques, and to the statements of the Reichsarbeitsgemeinschaft that labeled itself as "not political," claiming that "in every German city units are created" and stating that "knowing ones point the roads to the practical education of the Uranian radiant people." The question then is, did the Reichsarbeitsgemeinschaft belong to a hidden or largely kept underground tradition, having as its more important philosophical stations Mesmer, Reichenbach, the Rosicrucians, Bulwer-Lytton and Keely? And were the plans of the

"Deutsche Neugeistbund" and the "Reichsarbeitsgemeinschaft Das Kommende Deutschland" part of a grandiose national or perhaps ultranational scheme or on the contrary quite competing in nature? Again, when we return to the writings of the "Reichsarbeitsgemeinschaft Das Kommende Deutschland," nothing in that respect is found. What the writings of the Pansophia lodge, the Fraternitas Saturni, Surya's and Georg's books do demonstrate, however, are the occult surroundings in which these philosophies matured and that gave birth to the elusive Reichsarbeitsgemeinschaft.

About the booklets of the Reichsarbeitsgemeinschaft, Bahn aptly remarks that "not one word is being said about Bulwer-Lytton and his novel *The Coming Race*, although the connection of the term 'Übermensch' and the display of vril as a powerful energy reminds one about the description of the Vril-ya in Bulwer-Lytton's novel." We may add such an omission also applies for the curious vision of the coming new race of man that the Pansophia lodge had printed five years before in the same city. Bahn also points out that "in the writings of the Reichsarbeitsgemeinschaft remarks on such other societies as the Thule society lack completely, and the group seems to have had a positive attitude towards Christianity, a clear difference from the most völkische-esoteric movements with affinities to National Socialism,"[79] neither are such references found in the writings of Pansophia, Fratres Saturni or Surya.

We can further conclude that, aside from having seen how Surya hinted at Keely by pointing towards Blavatsky's chapter, nowhere in the writings of the Reichsarbeitsgemeinschaft is Keely or his discoveries mentioned. There is a connection between the latter two, however frail. Not surprisingly, this link is to be found in early theosophy. The German publisher, Wilhelm Becker—who issued the first booklet of the mysterious German group—belonged to the leading German astrological scene long before World War I. Becker stayed in London for several years as a student of Alan Leo (1860-1917), one of the most important astrologists of late Victorian England. Initiated and prepared by Leo, Becker opened up a flourishing astrologer's shop in Berlin in 1910.[80]

Leo was introduced to theosophical gatherings led by Blavatsky in London in 1889 by Sepharial, yet another influential astrologer. The following May, Leo formally joined the Theosophical Society and remained a devoted member for the rest of his life. His membership overlapped briefly with that of William Butler Yeats, who resigned the same year to become a member of the Golden Dawn. The co-founder of the Golden Dawn, William Wynn Westcott, was a colleague of Sepharial in Blavatsky's Inner Circle that only counted 12 members.[81] Of Leo it was said that he was "a serious and hard working man," who never read many books except *Raphael's Guides* and Blavatsky's *Isis Unveiled* and *The Secret Doctrine* with all the references to Keely. These books he studied "seriously."[82] There is circumstantial evidence that shows something of the range of his occult interests; in the writings of his wife, Bessie Leo, one encounters the terms Mesmerism and animal magnetism.[83] There are other connections; in an occult magazine published by Otto Wilhelm Barth who also published the second booklet of the mysterious Reichsarbeitsgemeinschaft, the official organ of the Pansophia lodge was announced.[84]

The Reichsarbeitsgemeinschaft, who conveniently chose to publish their booklet with Becker's and Barth's publishing houses, saw the earth as having "the same structure as an apple sliced vertically in two halves." As Keely saw half a century before them, and what was termed the cosmic motor by the Pansophia lodge in accordance with Keely's views, the group saw in the north pole the anode, or positive, and in the south pole of the magnetic axis the cathode, or negative. From this, the Reichsarbeitsgemeinschaft drew certain technical and physical conclusions for the usage of vril-energy, which it also called "the all-force of the forces of nature." Certain devices, described as "ball shaped power generators" would "channel the constant flow of free radiant energy between outer space and the earth" and would enable a "specific use" of this energy.[85]

These ball-shaped devices, in which two bar-shaped magnets stuck, were conceived as small "world globes." In the interior of such a device was the "electrovital filling mass," and through an impulse a "grounded earth element" would be activated and electrically charged. The energy thus created could be tapped and could be used to drive engines. The used energy would be renewed through the grounding of the earth force field, the voltage in the ball aggregate would remain the same, and Bahn concludes that "we deal here with a sort of converter for radiation energy." The impulse however was given through radio technical means by a so-called "primal machine." Only one on the whole of the earth was needed. Here the Reichsarbeitsgemeinschaft becomes suitably vague in its description; the "primal machine" is described as a configuration of seven ball-shaped cells, but nothing more is said about its exact function.[86]

How then does all of this tally with Keely's visionary concepts? In order to establish this and by any absence of reference to Keely—except for Surya's hint—we must follow strange trails, and take into account that the techno-magicians of the Reichsarbeitsgemeinschaft possibly meant that the description of the primal machine was to be taken as a metaphor, or else their ideas on technology were so avant-garde in concept that these traversed the boundaries of the material and space-time. For as an echo of that age-old Indian cyclical doctrine that would lead to the perpetuum mobile, 19th century theosophical literature of influence speaks of "seven globes" that "form the field of our system of development. Our earth is the fourth globe in this chain. Around this chain of worlds the life-wave travels seven times and this traverse is called a Manvantara."[87]

These and other theosophical concepts proved so powerful that a contemporary of the Reichsarbeitsgemeinschaft, theosophist and astrologer A.C. Libra saw the planetary system as one "etheric body," in which the sun, the planets, the moons and the comets are "the material organs or the centers of power."[88] That idea sounds remarkably similar to that of the Pansophia lodge who saw the solar system as "one ether globe," and the occult order of the Fratres Saturni who held it that "our starry worlds" are in fact "one giant organ." Libra's books were published by the same company that issued the book by Dutch astrologer and theosophist Thierens, whose thoughts on Keely's supposed psychic ability we encountered in chapter 7.

Libra was well read in Germany[89] and could therefore be another influence to the Reichsarbeitsgemeinschaft or perhaps the techno-alchemists and occultists of the Pansophia lodge and the Fratres Saturni. Also, we must not forget how one of the booklets of the Reichsarbeitsgemeinschaft was issued by publisher Becker who specialized in astrology, and the other by publisher Barth, who in his occult magazine devoted a laudatory article to Libra's books.[90]

Much in the style of the Pansophia lodge, Libra also explains that according to him, the sun emitted ether rays, "called prâna by the old Indian adepts," which were transformed on the planets into light and heat.[91] The driving force of the cosmos is "the centrifugal electric solar force," which is in balance with the resistance of the center-searching magnetic currents of space.[92] "In the same manner," Libra continues, "the heart of every micro-chasm is the sun of this world in miniature." The heart is, according to Libra, the cave of Bethlehem in which the godly spark, Christ, is being born.[93]

Steiner, too, undoubtedly exerted his influence on the doctrine of the Reichsarbeitsgemeinschaft. In 1908, he gave a cycle of 12 lectures about the apocalypse of St. John. The number 12 was significant because it was also the number of the houses of the zodiac. In the third and fourth lectures—which added together form the number seven—Steiner digressed on "John's letter to the seven communities in Asia," and on "the unveiling of the seven seals." If all this number dissecting sounds as if too much is being made of mere coincidences, that indeed might be the case. On the other hand it is a well-documented fact that to a learned occultist, a high initiate, there is no such thing as coincidence and the slightest detail might be the cache that holds an important message. After all, the symbolism and meaning of numbers and measures are of the greatest importance to occult doctrine. Steiner also made it clear to his audiences that "in the occult...one never has expressed himself clearly, but has created something that should conceal the true nature."[94]

Thus, according to Steiner, the contents of an initiation and the prophecy of the cycles of the development of mankind in which he had adorned Keely with a brilliant role are carefully hidden in the apocalypse of St. John. The importance that Steiner saw in the apocalypse of St. John, and Libra's writings on the godly spark, help to explain the pseudonym of "Johannes Täufer" or "John Baptist"— the biblical figure who baptized Jesus—that the author of the booklet *Vril. Die Kosmische Urkraft* chose.

Reading how Libra informs us of the essence of "cosmic man," with "seven nerve-centers that correspond with certain zodiacal signs," sheds perhaps some light on the hidden meaning of the phrase "Uranian radiant people."[95] Also, a year after Steiner's lectures, a New York theosophical publisher issued a book that purports to be the real message according to St. John. In it we read that the term "radiance" denotes "the luminous cloud or aura enveloping the purified man or Initiate, and which is visible only to the inner sight."[96]

Perhaps amidst all this we glimpse the real symbolism of the primal machine of the Reichsarbeitsgemeinschaft: the primal machine is man itself. Man that has evolved and has become truly the fulfillment of itself, only one on the face of the earth is needed, man as a brilliant embodiment of the cosmic motor. Possibly with this metaphorical interpretation we also perceive what Keely ultimately foresaw with his line of research that would enable him to link his devices to the waves of the brain, what Steiner was driving at in his 1918 lecture, or how it came to be that Blavatsky explained Keely as one who was paranormally gifted. We may even venture so far as to think twice of the apparent symbolism of the street number of the address of the Reichsarbeitsgemeinschaft in Berlin, which was seven. There is another slight reference in Libra's writing that helps us to follow the hidden trail and chart the occult substrata; while digressing on the meaning of the zodiac, Libra refers to Burgoyne's *The Light of Egypt*.

Burgoyne, who we met in the previous chapter, was heavily influenced by Bulwer-Lytton. Bulwer-Lytton's intriguing concept of vril was a major influence on Blavatsky, who in turn became a friend of Bloomfield-Moore and one of the most prominent occultists to write about Keely. Steiner was also impressed by the writings of learned initiate Bulwer-Lytton and had studied Blavatsky's writings. He lectured on Keely and even stated what he held as the real nature and significance in his *Mystery Plays*. In his third lecture on the apocalypse of St. John, entitled "a letter to the seven communities," Steiner alleged that a certain community is a representative of the sixth time-period (Steiner thought there were seven), a time-period in which mankind's spiritual life is being prepared. Amazingly, or perhaps not, since all could be a matter of coincidence, the name of this community is Philadelphia, the place where Keely lived.[97]

Bulwer-Lytton's concept of vril was also a major influence on the philosophies of the Fratres Saturni, and it is almost too convenient that the year that the Fraternitas Saturni went underground, the Reichsarbeitsgemeinschaft made itself publicly known. The former, evolving from the Pansophia lodge, could therefore very well have provided the latter with a legacy of inspiration, sharing as they did the same geographic locale and in all probability the same occult substratum. Bahn however points towards another and almost certain source of the ideas of the Reichsarbeitsgemeinschaft. "Everything that the Reichsarbeitsgemeinschaft stated in 1930 on the subject of 'primal force,' the 'primal machine' and 'ball-shaped aggregates' stated, was already published two years before in a book about the 'Raumkraft' or space force-theory by the Austrian inventor Karl Schappeller."[98]

Schappeller (1875-1947) is certainly one of the most mysterious persons of the 20th century free-energy scene. It is alleged that his work on free-energy philosophy started around 1894. This coincidentally was the year that both Colville's and Astor's books were published in America, Pape discovered his life-wave in Holland, in France the Golden Dawn opened a temple, and Zanne started to rewrite history to fit his occult insight.

Information about Schappeller's early years is hard—if not impossible—to obtain, and we can only speculate to what degree he was influenced by Keely. It is possible that a certain connection existed between Keely and Schappeller; Schappeller stayed in Vienna in the 1920s where he undoubtedly would have mingled with the esoteric coterie surrounding occult neo-templarist Liebenfels, who in turn was connected to Hartmann. And three decades before Schappeller's arrival in Vienna, a Viennese paper published several articles about Keely, and Bloomfield-Moore visited that city and was questioned about her support of Keely. Whatever the influence of ideas that might have resulted from such connections, Shappeller's ideas themselves show a remarkable similarity to those that Keely had half a century before him.

In Vienna, Schappeller collected a menage, many of whom where young engineers. He developed new ideas on the suppliance of energy and found financial backers, even in industrial and clerical circles. In 1925, he bought an old castle in his hometown of Aurolzmünster, where he led a luxurious lifestyle that often brought him in financial difficulties. While this led to a negative image, his labors at the castle were of a different nature. There he had Franz Wetzel, one of the leading radiestesists of his time, amongst his co-operators. Schappeller also received financial backings from industrial companies such as the Siegerländer industrialist Fritz Klein, and through middlemen from the German emperor Wilhelm II, living in exile in Holland.

An English shipping company was interested in Schappellers "Raumkraft" aggregates, and at one time conducted serious negotiations about a ship's engine that was to be derived from his "Raumkraft" aggregates, evoking Surya's anecdote of Lord E.'s ether-driven vessels, the *Sirius* and the *Dependence of Sirius*. Schappeller's wanderings during the Second World War are uncertain, and indications on the use of his research are completely lacking. When he died, all of his research papers and documents disappeared,[99] and it is asserted that, as a strange and haunting echo of Steiner's remarks, certain interested parties and international control aspects made every effort to banish Schappeller's ideas and plans on his energy-converter from the pages of history.[100]

Schappeller considered the primal force as "that which holds the earth in its inner together,"[101] a statement remarkably similar to Keely's, and Newbrough's for that matter. Schappeller stated that "in the whole of nature, there is no nothing, no useless space. Where there is no matter, there is energy, a so called empty space is therefore a space filled with force. ...Space controlling is energy, space filling is matter. Because the cosmos is a closed vacuum, it is an immense space of energy."[102] This line of thought is analog to that of Belinfante when he wrote 23 years before Schappeller that "space between the heavenly bodies is filled with ether,"[103] or in fact any other ether-theorist.

Schappeller's aim was to create a "constant discharge" between the cosmos and earth, which he considered to be a reservoir of force, and the cosmos. Since the word "plasma" was still unknown at the time that he conducted his experiments and formulated his theories, he called a similar phenomenon "electrical vapor," "luminous magnetism" or "luminous ether." The use of the dynamic principles of the ether led to the construction of a ball-shaped device in which the "luminous magnetism" was created that could be held permanently and was used as the conductor between the earth and the cosmos.

Schappeller's device resembled a miniature earth and was built from two precisely calculated half globes with the hulls consisting of magnetic parts with an inner room built of a nonmagnetic diaphragm. Inside this globe were two magnetic poles of a "certain shape." Connected to these poles were thin tubes, filled with an "electrical mass." The tubes, separated from the inner hull of the globe through an isolating substance, were configured in a number of closely wound spirals. The "electrical mass" with which the tubes were filled consisted of a substance that was partly wax. These tubes were then "permanently electrified" and were connected with a pole to the grounding.

The second pole originated in the middle of the globe, which was filled with an "electrostatical mass" that was Schappeller's secret. When energy was taken from the globe it recharged itself continuously in the same quantities. The globe was fixed to some kind of "magnetic arm" that was called the "rotor," while the globe itself was called the "stator," although this unusual device had no moving parts. When the globe was charged and switched on, a magnetic needle reacted on the north and south pole similar to the actual north and south poles of the earth. Switched off, the globe was magnetically neutral. It is alleged that a globe of only 15 cm diameter delivered an astounding high number of kilowatts. The "luminous ether" regenerated itself and only dissolved when the globe was opened. The filling of the globe was done by a special "filling machine" of which unfortunately no technical drawings or specific descriptions exist.[104]

Interestingly, a 1930s German esoteric magazine with ariosophic leanings published an article by Schappeller and referred to Liebenfels in connection with Schappeller, thus giving more substance to the possibility that Schapeller indeed rubbed shoulders with the coterie surrounding this dark initiate while in Vienna, or that both parties were at least aware of each other. Liebenfels studied *The Secret Doctrine* very well, and perhaps thus a trail leads back through time and geography to John Keely in Philadelphia.

The article did not mention Keely but lamented that Schappeller and Liebenfels were neglected researchers, as was Frenzolf Schmid, it pointed out. Schmid succeeded, so the article goes, with the help of a device of his making to divide the cosmic rays in its three components. These were the "primal rays" or "death rays," the "pure rays" or "healing rays," and "primal additional rays" or "indifferent rays."[105] In 1929, Schmid published two booklets about his invention and claimed that with his device he was able to use any of these rays as he liked, for instance as a healing method, like Keely had claimed almost half a century before him.[106]

Liebenfels was quite aware of Schmid's exploits, as he wrote on a number of occasions about Schmid and his curious invention. They also added more weight to his own strange philosophy: "Add to this as another important fact the discovery of Frenzolf Schmid, according to which light is broken down in three rays, primal or death rays, life rays, indifferent rays (or, as I call them, carrier or isolating rays). ...Frenzolf Schmid has, with this discovery, established scientifically my thesis on the trinity. The primal rays correspond with the 'father,' the healing rays with the 'son,' the indifferent rays with the 'holy spirit,' that by means of isolation and direction steer the other rays."[107] And elsewhere Liebenfels extols the virtues of a treatment with one of Schmid's devices for the healing of disease, while treating one of Schmid's publications on the subject.[108]

Clearly, as so many of the 1920s European occult milieus, Liebenfels was quite taken with inventions and discoveries that involved cosmic rays and stupendous energies. In his writings the terms "Urkraft" or "Primal Force" and ether crop up, again suggesting at least a common origin with Schappellers' inventions and Keely's discoveries.

Liebenfels also uses such metaphors as "the radio broadcast station" for God, evoking the ideas of Dutch grail seeker Rensburg. Rensburg, who like Liebenfels admitted his indebtedness to *The Secret Doctrine*, wrote on the communication of man with godlike beings as "inter-astral telepathy, that is, marconigraphy of an organical nature, from their nerve system directly to our nerve system...," further explaining that "as long as there existed religion on earth, meaning the connection with the gods, this was never anything else than inter-astral telepathy."[109] But with gods, Rensburg does not mean the same gods as those that feature in Liebenfels' dark blend of Wotanism and Christianity: Rensburg muses on "organical marconigraphy from star to star" and on "material beings on other stars that exert their influence on us through telepathy, soon amplified by means of inter-astral radiography."[110]

Enough is said here of Rensburg's highly original but today totally forgotten writings, in which he refers to Golden Dawn initiate Waite, to Edgar Rice Burroughs' Mars Novels and to Jules Verne, and who ponders on the possibility of life on the sun and of metals, of which "it is proven that they have feelings," in turn evoking Keely's ideas on "a crazy piece of metal." As brilliant and unusual a thinker as Rensburg was, he would pursue his ideas further and thereby also quote the writings of the dark Viennese initiates List and Liebenfels. But as a horrible irony of history, Rensburg, who was Jewish, died in 1943 in a Nazi deathcamp, such focal points of unimaginable suffering being partly the result and the culmination of the racist ideas of the ariosophical initiates Liebenfels and List.

Returning to Liebenfels' writings, we find citations taken from *Zanoni*, and his thoughts on inventions such as the "Atomedium," "a sort of automatic dowser (or pendulum) that partly corroborates the discoveries of Reichenbach and de Rochas"[111] or of the discovery that "the human blood emits rays that is capable to dissolve hydrogensuperoxyd."[112] But whereas in these writings of Liebenfels the name Keely is nowhere mentioned, it is safe to assume that the dark initiate from Vienna also had been keenly aware of Keely's discoveries. Liebenfels was heavily influenced by *The Secret Doctrine*, and he knew Hartmann, who visited Keely on a number of occasions, moreover his particular frame of mind where avant-garde technology and the occult sciences meet makes an interest in Keely's discoveries quite obvious.

To what extent Liebenfels was involved in the case of Schappeller remains uncertain, as in the case of another contemporary Austrian inventor, Victor Schauberger. Around the time that Schappeller was working on his space-force machines, Schauberger claimed to have developed an astounding device that he named the Implosion Turbine. Water was pumped in a spiral-shaped tube in which it then imploded. This drove the pump and converted it into an electrical motor. Schauberger mentioned the existence of an "all pervading fifth element, Aether." Although much has come to light in recent years concerning Schauberger, as with Keely's and Schapeller's devices the whereabouts of Schauberger's machines— which were shipped to the United States after the Second World War—still remain unknown and he too seems to have been the victim of an interested party with international control aspects.[113] In recent years, researchers have noted the similarity between the theories of Keely and Schauberger.[114]

We have tasted something of the very troubling and bothersome aspect that plagued those early occult pursuers of unorthodox rays, Keely, Schappeller, Schauberger and other free-energy inventors: that uncertain and dark corner which perhaps was the same that Steiner alluded to in his lecture, and has been named "an interested party with international control aspects" by one author. It is obvious that this unidentified current was much more powerful than any of the occult and esoteric societies that surrounded these free-energy inventors. Its frightening aspects led Jacques Bergier to theorize about a secret organization that he suggestively called "the men in black," that has always accompanied every progress of mankind during the ages and therefore also the endeavors of those free-energy inventors, something that the theosophists and again Steiner already warned about. Bahn speculates on the absence of publications of the mysterious Reichsarbeitsgemeinschaft after 1930—that perhaps this group in itself was meant as a front organization, as, except from their identity, they never really cared much for secrecy: "clearly the Reichsarbeitsgemeinschaft Das Kommende Deutschland was not a secret society. It published its philosophies with two relatively known publishing companies and it recruited for members in its pamphlets with an address that anybody could find."[115]

But of the Reichsarbeitsgemeinschaft as being a front organization, Bahn states that it is "Not to be excluded, although on the other hand not proven...the Reichsarbeitsgemeinschaft could have been a front organization of an occult circle, whatever it called itself, that really operated in hiding and observed and filtered potentially interested by means of the Reichsarbeitsgemeinschaft, before they were admitted to the inner core. ...The division between the 'secret order' and front organization possibly created a double protection; against unsavory characters...and against repression from the state, that was taken on by the front organization and thus diverted from the inner core."[116] That after 1930 no other publications were issued by the Reichsarbeitsgemeinschaft could mean, according to Bahn, that "the group did not attain the resonance that they hoped for."[117] Could the Reichsarbeitsgemeinschaft have been the last brilliant ploy to either promote or employ the same principles that Keely discovered more than half a century before? Was Steiner's first Goetheaneum at one time the place where certain initiates conducted strange experiments in the fashion of Keely's discoveries? Unfortunately, such lines of thought and the questions that they yield are until this point quite unverifiable.

Whatever the causes, the reasons or the motives, nowhere does one come across so many instances of patents, papers, documents, technical drawings, devices and inventions that have gone mysteriously missing after the often sudden, unexplainable and sinister deaths or disappearances of persons than in the history of the search for alternative energy sources. That, too, is a historical fact, even though a sad and frightening one, even if one wants to ignore or downplay this, depending on the point of view taken on history in general.

Relatively recently a warning as to this frightening aspect was issued,[118] and elsewhere two examples of inventors who disappeared under never-resolved circumstances are offered. There is the case of John Andruss, who claimed that he found a method of making petrol from water by adding a special liquid to it. His disappearance has never been solved. Apparently both the British and the U.S. governments were sufficiently impressed by his claims and even set up a joint panel of experts to study his work. "Finally, the stage was reached when a tank of a motor-boat was filled with water, and Andruss poured into it a glass of his secret composition. Selected men were in the boat and it is stated that it roared out over the waters of a lake near New York. A final test was to be made on the speedway at Indianapolis; but on the morning of the (successful) test, Andruss did not turn up. From that day in 1925, until this, not a trace of him has ever been found."[119]

The second example is the case of German mechanical genius Rudolf Diesel, who "on the night of 29 December, 1913, eight months before the outbreak of the First World War, sailed with the Harwich steamer from Antwerp." He never reached Harwich, and his nonarrival and what happened to him remains a mystery to this day.[120]

Webb states that, if there is any truth in the dictum that occultists have a special relationship with the imagination in their pursuit of other realities, we might find an extraordinary amount of creative work in, for instance, the realms of mechanical invention.[121] He includes such people as Hugo Gernsback who coined the word "television" and the term "science fiction" and Austrian Hans Hörbiger, the strange prophet of the Welteislehre that became the official cosmogony in Nazi Germany.

But as we have seen in many instances, this creativity went far beyond common mechanical invention and stretched itself into those planes of existence where they felt that like Keely they could touch and measure the very stars, where wonderful rays, cosmic energies and primal forces were the rule: where man is not simply made out of flesh and blood alone but is also a creature that is, like the surrounding universe, a source of radiant energy, a power plant of unimaginable possibilities who holds the promise of becoming godlike in the end.

Clara Bloomfield-Moore (1824-1899)
An extraordinary person with a deep interest in science and the occult.
She supported Keely for 15 years, and remained his friend until his death.

Helena Blavatsky (1831-1891)
Founder of the Theosophical Society and author of *Isis Unveiled* and
The Secret Doctrine, in which she devoted a chapter to Keely and his
discoveries.

Henry Steel Olcott (1832-1907)
Co-founder of the Theosophical Society. Lawyer, early spiritualist
and editor of *The Theosophist*.

Wilhelm Hübbe-Schleiden (1846-1916)
Founded the Theosophical Society and the German occult periodical,
The Sphinx. Wrote about Korschelt's mysterious solar ether ray devices.

A rare 1920s photo of the underground temple of Ellora in India, place of initiation by the mysterious Ellora Brotherhood. According to Emma Hardinge Britten, the place was a strange device of gargantuan dimensions.

Emma Hardinge Britten (1823-1899)
Medium, spiritualist and founding member of the Theosophical Society.
Published the strange account of the underground temple of the Ellora
Brotherhood.

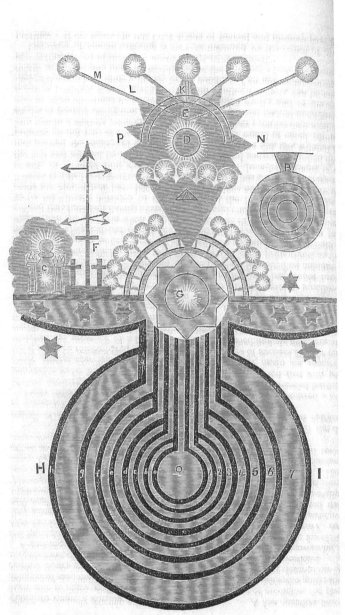

Diagram of the Spheres, as drawn by Jonathan Koons, inventor of the spirit machine, while in a trance state and under the direction of spirits. According to Koons, G represents "the Star of Light and Beauty beneath the throne of God. It signifies the vast celestial realms of unknown and perhaps illimitable extent, filled with *the subtler fluid,* the impenetrable, the inconceivable, the source, fountain, and centre of all light, heat, life, force, gravitation and attraction…the central sun of being, the profound mystery."

Drawing showing a comparison of the Aero Goeit, designed by Sonora Aero Club member Adolf Goetz, and the Aero Goosey designed by Peter Mennis. These aircrafts were designed in secret in 1858 and drawn from recollection by Dellschau in 1911.

A page of Dellschau's manuscript showing a cut-section of an airship, purportedly built in secret by the Sonora Aero Club.

A page from Dellschau's encoded manuscript, depicting Peter Mennis and his dog next to the airship.

"I saw the machine...It is made of metal...It is equipped with two canvas wings...and a rudder shaped like a birds tail." Illustration from Jules Verne's *Robur le Conquerant*, 1886.

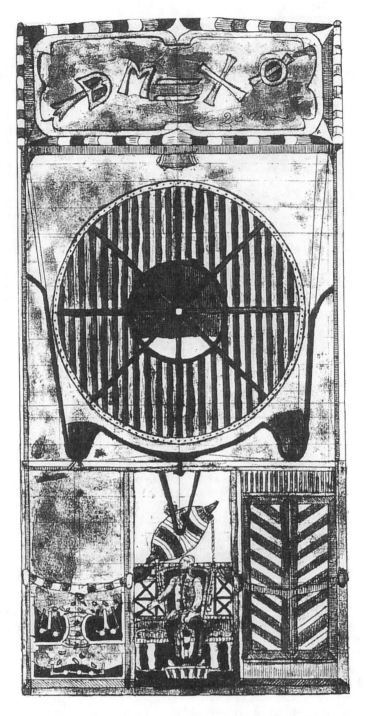

Example of a later design for Dellschau's Airpress Motor (inside view)

Example of press cuttings which form a part of Dellschau's strange encoded manuscripts.

William Colville (1859-1917)
Freemason, medium, occultist and theosophist. Co-editor of *The Gnostic*. He published Keely's remarkable ideas on interplanetary travel and anti-gravity in 1894. A long-time friend of Keely, he wrote the memorial speech at his funeral.

Paschal Beverly Randolph (1825-1875)
Founder of the Brotherhood of Eulis and co-founder of the Hermetic
Brotherhood of Luxor.

Karl Kellner (1850-1905)
Freemason and co-founder of the Ordo Templi Orientis. He constructed
instruments that collected and produced electricity directly from the ether.

Eliphas Levi (1810-1875)
The great French magus and renovator of 19th century continental occultism.

Franz Hartmann (1838-1912)
Occultist, freemason and co-founder of the Ordo Templi Orientis.

Count Louis Hamon, alias Cheiro (1866-1936)
The most famous palmist of his time. He visited Keely in 1890, and
trusted his strange memoirs concerning him to paper.

John Jacob Astor (1864-1912)
Eccentric inventor, and one of the wealthiest men on earth. At one
time he planned to support Keely with several million dollars. He died
on the ill-fated voyage of the Titanic.

Past Lives, Dreams and Soul Travel

"I welcomed this Soul back as I picked up the baby. With intent alertness, his little untrained eyes made an effort to focus on me."

Quote from, *Past Lives, Dreams and Soul Travel*, page 69 by Harold Klemp

Jules Verne (1828-1905)
Author, master of prophecy, alleged psychic and keenly aware of
Keely's discoveries.

"Inside the barn was a strange looking affair, made of aluminum having wings and a rudder." Illustration from Fritz Holten's *Das Aeromobil,* 1912.

12

The Great 19th Century Airship Wave

"I saw...gliding quickly through the air, what appeared a small boat, impelled by sails shaped like wings."
Bulwer-Lytton, *The Coming Race*, 1871

"You can take my word for it, that the airship is no myth."
Rabbi A. Levy, *New Orleans Picayune*, April 24, 1897

"...should we, the aeronauts, forever stay the stepchildren of this Mother Earth?"
Fritz Holten, *Das Aeromobil*, 1912

Two years after Colville's remarkable description of what an airship might look like built according to Keely's antigravity ideas, and two years after Astor published his book in which he treated precisely such a space voyage, something inexplicable happened in the skies over large parts of America. From November 1896 until April 1897, thousands of people saw UFOs, described as brilliant lights and vague aeroforms, which were nicknamed "airships" in the press. During what has been termed "the great 19th century airship wave," all the characteristics of later UFO waves were apparent. There were contacts; eyewitnesses would often meet the alleged builders of these airships, terrestrial and extraterrestrial, with some who claimed to represent the inventors. There would even be an occasional abduction, and tales of crashes and hoaxes.

Thanks to the diligence of several ufologists, the 19th century airship wave is well documented, and we now have hundreds of pages at our disposal, rescued from slowly deteriorating microfilm.[1] Notwithstanding all the reports and all that has been written about it, the wave has never been explained and has remained a ufological enigma. It is not the intention to document this wave, as others have done so.[2]

It is however, significant that the airship wave occurred at a time when Keely was involved in his antigravity research and had actually built several devices for his proposed system of aerial navigation. But before we jump to conclusions, it must be taken into account that from the documentation that is at hand, news about Keely's inventions and news about the airship wave were treated by the contemporary press as two separate events; nowhere in those accounts did I find a reference to each other, either by way of explanation or otherwise.

Moreover, the 19th century airship wave was not the first time that inexplicable sightings occurred. As far back as the 1880s, UFOs were seen that were described as "electric balloons," it is also suggested that in 1892 a similar wave struck Poland.[3] The American airship wave of 1896 had its precedents, but not on such a grand scale.

Various explanations have arisen, such as misidentifications of the planet Venus, hoaxes, the doings of secretive inventors and a massive wave of UFO activity. Undoubtedly, the airship wave consists of all of these components. It has also been pointed out that the 19th century climate helped, too; man dreamt of achieving flight, writers of fiction often used airships in their stories, and there are accounts of dozens of inventors, Keely amongst them, who were working on plans, schemes and designs of airships.

Daniel Caulkins, for instance, published his *Aerial Navigation: The Best Method* at his own expense in Toledo in 1895, a year before the airship wave started. In Caulkins' book is to be found, or so the author states, a "concise description of a new airship which the author believes will be the accepted plan for successful aerial navigation (including) an entirely new motor and new application of wings." Caulkins describes himself on the title page as "the discoverer of the circulation of the nerves and the inventor of the electro-magnetic circular power."

Many eyewitnesses saw strange aeroforms in the years and months leading up to the wave, but the consensus is that the airship wave started on the night of November 17, 1896, with a sighting of an object over Sacramento that was described as having the "shape of a balloon," and having a bright, white electrical arc light. The object was seen by hundreds of people. Human voices were distinctly heard coming from the object.[4]

A few days after the remarkable incident, a letter to a newspaper recounted the astonishment and excitement that was felt: "This is truly an age of invention. ...These navigators of the ethereal regions must have had an experience far more interesting than that of Dr. Nansen in his search for the north pole."[5]

Almost from the beginning, the general opinion was that an unknown inventor had solved the problem of aerial navigation. As a newspaper at that time wrote: "The opinion of the masses is...that some lucky inventor, having solved the mystery of aerial navigation, is, with his companions, testing his invention in secret, with no intent of allowing a curious public to view it until his rights are fully protected by letters patent."[6]

William Jordon surely thought so. In a letter to a newspaper, he recounted an incident that allegedly had taken place in August, while he was out hunting in the Tamalpais mountain range. Chasing a deer, he stumbled upon a "circular brushpile about ten feet in height" in a region of the mountain that "was seldom visited even by hunters." There he discovered to his amazement a "perfect machine shop and an almost completed ship," on which six men worked. "I was sworn to secrecy and have kept it till this moment," Jordon wrote.[7]

Perhaps this general view prompted a San Francisco attorney, George D. Collins—five days after an airship startled the residents of Sacramento—to come forward with a fantastic tale. Collins asserted that yes indeed, the airship existed, and that its inventor was his client. "I have known the affair for some time and I am acting as an attorney for the inventor. He is a very wealthy man, who has been studying the subject of flying machines for fifteen years, and who came here seven years ago from Maine in order to be able to perfect his ideas away from the eyes of other inventors. During the last five years he has spent at least $100,000 on his work. He has not yet secured his patent, but his application is now in Washington. ...I saw the machine one night last week at the inventor's invitation. It is made of metal, is about 150 feet long, and is built to carry fifteen persons. There was no motive power as far as I could see; certainly not steam." According to Collins, the airship was built on "the aeroplane system" and was equipped with two "canvas wings eighteen feet wide and a rudder shaped like a bird's tail."

Collins also witnessed a test run of the device: "The inventor climbed into the machine and after he had been moving some of the mechanism for a moment I saw the thing begin to ascend from the earth, very gently. The wings flapped slowly as it rose and then a little faster as it began to move against the wind. The machine was under perfect control all the time. When it got to a height of about ninety feet the inventor shouted to me that he was going to make a series of circles and then descend. He immediately did so, beginning by making a circle about 100 yards in diameter, and gradually narrowing in till the machine got within thirty feet of the ground. It then fell straight down, very gracefully and touched the earth as lightly as a falling leaf."

This remark is similar to Bloomfield-Moore's descriptions of Keely's model-airship that also was able to rise or descend or remain stationary midway with a motion that was "as gentle as thistledown floating in the air." Collins, however, did not elaborate on the propulsion and from his statements alone it is impossible to ascertain if he was perhaps subconsciously referring to Keely's alleged antigravity experiment. The only thing Collins ever stated about the method of propulsion was that he believed that "as near as I can recollect the propelling power is compressed air,"[8] in connection with which veteran UFO-researcher John Keel remarked that, "a few years previously," Keely built "a strange contraption which could bend bars of steel, and do other things considered impossible for ordinary machines of the period. Detractors claimed that the Keely engine really operated on compressed air."[9]

Collins also confided that the unnamed inventor had "forsaken the ideas of Maxim and Langley entirely in building the machine, and has constructed it on an entirely new theory."[10] A statement that is analog to that of theosophist

Scott-Eliott when he published his opinion of Keely the same year concerning vril that "more nearly approached that which Keely in America is learning to handle than the electrical power used by Maxim."

Collins also stated that the airship was also able to "rise to any altitude desired, describe circles or take an abrupt angle,"[11] remarkable flight-characteristics that once again bring to mind Bloomfield-Moore's description of Keely's experiment with a model airship.

The only problem that was left to the inventor to solve was the wave-like motion of the ship that made him seasick. But, Collins confidently concluded, "In another six days the trouble will be done away with, and it is then his intention to immediately give the people of San Francisco a chance to see his machine."[12]

Collins intimated that none of the larger parts of the airship were made in California; instead they were manufactured in various parts of the East and shipped to Oroville and Stockton, where they were gradually assembled. Naturally I am forced to think of that intriguing mention of Keely having constructed a secret "device for lifting heavy weight's for a person in California."

When a reporter asked Collins how the inventor could hide an object of 150 feet long in a barn in the vicinity of Berkeley or Oroville, where it was guarded by "three men," Collins answered that the barn was "tacked on to an old dismantled two-story dwelling. The partitions have been knocked out, making the place practically one long room."[13]

Finally Collins provided some leads as to the identity of the inventor, and reporters soon thought they had discovered him in the person of one Elmer H. Benjamin. In the end, however, Benjamin turned out to be a dentist and not the airship inventor after all.

In the meantime, residents from Oroville did recall that "parties residing about three miles east of the town and in a thickly wooded section not traversed by roads, have been experimenting with different gases for some time past. One man, who is an expert chemist here, was heard to let slip a word or two about parties who were experimenting with a new and very light gas which would supersede the one now in use for balloons. It is impossible however, to get more than rumors, mere whispers, and it is impossible to say whether they are pure fabrications. It is asserted by more than one person that comparatively unknown parties with abundant means have been experimenting for some weeks with different gases, and that they were sanguine of having solved the problem of aerial navigation. The form of the airship they are supposed to have constructed is a matter of uncertainty. No one can say how the vessel is regulated or what means are used to govern its speed. It is rumored that such a vessel has been constructed near here by Eastern people, who desired to escape prying eyes and sought seclusion. It is evident that more than one man was engaged in the project, but it is impossible to learn any more of the names or give any details about the vessel itself."[14] A rumor this may be, but as we will see, one that has a significant twist to it.

Another story is also significant because of one detail: George Carleton, a city electrician, singer and pioneer of Oakland told a reporter that he heard the name of the inventor through a friend of his. How did this come about? "I heard the story last night of a friend of mine who is also a member of my lodge of masons. It was he who talked with the man who saw the machine, and I was told the name of the inventor by my friend. As I understand it, the trip was made near Oroville. The inventor made an ascension of several hundred feet, flew off four or five miles, circled around a few times and landed as nicely as could be. This my friend's friend saw." But Carleton was not going to give any name. In a curious reference to freemasonic ritual, Carleton said: "I was pledged not to divulge them, and I'd saw my leg off first."[15]

While Collins alleged that the mysterious airship was guarded by three men, which could very well be a reference to the freemasonic societies, we could further speculate that Collins meant that the airship was not so much guarded by three persons, but by an esoteric order, possibly modeled after a Masonic institution. While this is speculation, we have already seen the deep-rooted interest of several orders with avant-garde technology. Also, Carleton's Masonic friend could have had a conversation with Colville, who was also a freemason and who had visited California as early as 1886.

Throughout this part of the airship wave, reporters would unsuccessfully try to hunt down the elusive inventor. All that Collins ever said about his identity was that he was "a resident of Oroville, and a man of wealth, about 47 years of age, and a fine looking fellow."[16] And to add to the high strangeness of it all, Collins would change his story considerably. At first he claimed to have seen the ship itself, then he would suddenly state that, "I have not seen the new wonder."[17] In the end he would even say that he was "sorry to see that the newspapers have been attributing to me such an extensive knowledge relative to this airship. In truth I know very little about it. ...I know nothing of the airship. I do not know what it is made of, what power propels it, nor where its inventor now is. I am just as skeptical and incredulous regarding it as anybody can be."[18]

What then was Frederick Bradley, an old friend of Collins, to make of what Collins had told him? According to Bradley, Collins confided to him that he had seen a "wonderful airship invented by a client of his." Collins also told Bradley that "a man whom he had never seen before had come into his office...and said that he desired to get a patent on an airship." When Collins pressed for details, the mysterious stranger invited him to take a look at his invention: "They went, taking the Oakland ferry and then the train. Collins told me just the place, that I do not like to state. They walked some distance after leaving the train and stopped finally at an old barn. It was locked, but Collins' client had the key." Inside the barn, Collins saw what he described as "a strange-looking affair made of white metal," which, he was told, was aluminum. Aside from the wings and the rudder, Collins also noted a "big light forward, like a searchlight," and some sort of cabin. Collins' reaction was also recorded for posterity: "Bradley must have been drinking to have imagined I told him all such rubbish."[19]

Not that Collins' withdrawal mattered: A new person cropped up that would take his place. His name was William Henry Harrison Hart, former attorney general of the State of California. Hart met with Collins and talked with him about the discovery. At that time Hart emphatically stated that he had no doubt that the affair was "bona fide." "I have seen the thing in the air myself, and believe the idea of this Oroville inventor has proved to be practicable."[20]

Hart, we learn, took Collins' place since the unknown inventor had decided that Collins talked too much. But what a poor choice Hart would prove! While admitting that he had not seen the airship himself, he made the sensational announcement that the airship was to be used in the service of the Cuban insurgents and he intimated that Havana was to be the first point of attack.

Reporters learned through other sources that the name of the alleged inventor was Catlin, and that he was assisted by Benjamin and George Applegate. Hart also confided that there were two airships, "...and they are very much alike. One was perfected in the East and the other in California. I have been concerned in the Eastern invention for some time personally." Hart stated that the aim was now to consolidate both interests. "I have seen the machine invented in the East and I am convinced that it will work all right, and from what I have been told I don't see any reason why the machine invented in California cannot be worked. ...The whole trouble in the problem of aerial navigation has been a question of motive power." But, according to Hart, the California invention had "the right motive power," which was generated through electric storage batteries.

But when Hart's statement is looked upon more closely, it only adds to the mystery, for it appears that he stated that the Eastern invention "will work." Did he mean then that he had seen a device that still "had to be perfected," as Keely had claimed a year earlier concerning his aerial propeller?

Hart stated that the machine, which was tested in California since the climate was of "favorable character," had one drawback: "that is that the inventor cannot cause it to stand still; it must be kept moving like an arrow. Otherwise it is under perfect control. ...It can be made to rise from a dead standstill. I cannot go into details about its construction, but will admit that it is of cylindrical shape, is built of aluminum and has wings."[21]

Hart also explained that it was his interest in storage batteries that brought him and the Eastern inventor together. The only things Hart ever said about this Eastern inventor were that he worked in the vicinity of New Jersey, was a foreigner, "an Italian, I should judge, from his appearance,"[22] a personal friend of the inventor of the Fargo storage battery,[23] and in the end would even identify the mysterious inventor as being "a cousin of John Linn, the electrician of the Cuban patriot general, Antonio Maceo. ...The inventor is not a Californian."[24] and "a man of dark complexion, dark eyed and about 5 feet 7 inches in height and weighed about 140 pounds."[25] Nevertheless, the mysterious inventor has never been identified.

Hart did tell the reporter more about the two airships. "I will say to you frankly that I believe the Californian invention is better than the Eastern invention. The only trouble with the airship in New Jersey is that it will not stay in the air.

It moves through the air, but its course is towards the ground. The Californian inventor has a machine that will stay in the air. The appliance that keeps it in the air is a great parachute that opens automatically when the ship descends and closes automatically when the ship ascends. ...This hat shaped parachute keeps the ship in the air and makes the descent very gentle when the ship comes to the ground. It seems, however, to impart the ship with a wavering motion as it leaves the ground, upward bound. This is a defect to be remedied. This parachute is the principal advantage that the Californian invention has over the New Jersey invention. But the New Jersey man has so constructed his ship that should it drop into the sea it becomes a water boat. The Californian airship has not this advantage. ...The thing necessary to send up an airship is sufficient power. Heretofore this power could not be obtained without overweighing the airship. Storage batteries were too heavy. Compressed air could not be used satisfactorily. Now light and powerful storage batteries can be obtained. That solves the problem. No gas is used. Gas would require a material used in construction that would expand. This is not a balloon. The material of which the airship is made is aluminum. The motive power is electricity in storage batteries."[26]

Hart's allegations of the New Jersey device, meaning the Eastern invention, being able to be used as a seaworthy vessel is remarkably similar to Colville's statements on Keely's airship, which he wrote down two years before: Keely's proposed airship could "travel to other planets in this system as easily as the same ship could navigate the depths of the ocean."

Not only that, Hart's descriptions of the "wavering motion" and his emphasis on "light and powerful storage batteries" also have its parallels. For this we must turn to a passage in Astor's novel. "For flying" Astor fantasized, "we have an aeroplane that came in when we devised a suitable motor power. This is obtained from very light paper-cell batteries that combine some qualities of the primary and secondary type, since they must first be charged from a dynamo, after which they can supply full currents for one hundred hours—enough to take them around the globe—while partly consuming the elements in the cells. The power is applied through turbine screws, half of which are capable of propelling the flat deck in its inclined position at sufficient speed to prevent its falling. The moving parts have ball bearings and friction rollers, lubrication being secured automatically, when required, by a supply of Vaseline that melts if any part becomes hot. All the framing is of thin but very durable galvanized aluminum, which has superseded steel for every purpose in which weight is not an advantage, as in the permanent way on railways. The airships, whose length varies from fifty to five hundred feet, have rudders for giving a vertical or a horizontal motion, and several strengthening keels that prevent leeway when turning. They are entirely on the principle of birds, maintaining themselves mechanically, and differing thus from the unwieldy balloon. Starting as if on a circular railway, against the wind, they rise to a considerable height, and then, shutting off the batteries, coast down the aerial slope at a rate that sometimes touches five hundred miles an hour. When near the ground the helmsman directs the prow upward, and, again turning on full current, rushes up the slope at a speed that far exceeds the eagle's, each drop of two miles serving to take the machine twenty or thirty; though, if the pilot does not wish to soar, or if there is a fair wind at a given height, he can remain in that stratum of the atmosphere by moving horizontally. He can also maintain his elevation when moving very slowly, and though the headway be entirely stopped, the descent is gradual on account of the airplane's great spread, the batteries and motors being secured to the under side of the deck. The motors are so light that they develop two horsepower for every pound of their weight; while, to keep the frames thin, the necessary power is obtained by terrific speed of the moving parts, as though a steam engine, to avoid great pressure in its cylinders, had a long stroke and ran at great piston speed, which, however, is no disadvantage to the rotary motion of the electric motor, there being no recipro-cating cranks, etc., that must be started and stopped at each revolution. To obviate the necessity of gearing to reduce the number of revolutions to those possible for a large screw, this member is made very small, and allowed to revolve three thousand times a minute, so that the requisite power is obtained with great simplicity of mechanism, which further decreases friction. The shafts, and even the wires connecting the batteries with the motors, are made large and hollow. Though the primary battery pure and simple, as the result of great recent advances in chemistry, seems to be again coming up, the best aeroplane batteries are still of the combination-storage type. These have been so perfected that eight ounces of battery yield one horsepower for six hours, so that two pounds of battery will supply a horsepower for twenty-four hours; a small fifty-horsepower aeroplane being therefore able to fly four days with a battery weight of but four hundred pounds."[27]

Hart solemnly stated that it was he who advised the inventor to use his airship as a vessel of war, since he thought that "an airship might not be a very profitable vessel of communication for ordinary commercial purposes." Instead, Hart related to a reporter, the vessel should be used for military purposes. "I advised the inventor and he agreed with my views, that he could employ his invention in war. I assure you that I believe that by means of this airship a great city could be destroyed in forty-eight hours." Apparently Hart's idea was so well taken that Hart even declared: "You know the idea is not to get it patented, but to use it for war purposes."[28]

If Collins talked too much and changed his statement in the end, Hart outdid Collins in both respects. During another interview he stated that the airship was "about 125 feet and of a width in proportion to his length." Also, the inventor was now planning an airship of 50 feet in length, "to be used for war purposes principally," since the 125 feet one was "too large" and "used up too much power in running against the wind—that is, it presents too much surface to the wind." The second airship would be constructed in the locality of Bolinas.[29] The proposed attack would only cost the Cuban junta $10,000,000, and the United States would not become involved in the affair, since "the parties could go outside the jurisdiction on a chartered or purchased steamer and sign the contract and make all of the agreements necessary."[30]

During another interview, Hart changed his description of the airship considerably. Now it did have the characteristics of a balloon: "The second ship will resemble the first. ...The sustaining power is supplied from gas tanks, which are in the hull of the vessel and which are connected with the balloon which flies over the airship by a pipe. When the inventor wants to go up higher he lets more gas into the balloon out of the tanks, which are filled with condensed gas. When the inventor wants to fly lower he simply opens a valve in the balloon and the contrivance naturally descends, just as an ordinary balloon does."[31]

Also, around this time Hart stopped making references to the Eastern invention. Instead he was wholly focused on the Californian airships, of which, we learn to our amazement, not one or two, but now three were in existence: "One...was of large size, capable of carrying three persons, the machinery, the fixtures and 1,000 pounds of additional weight, and another that was much smaller, capable of carrying one man, the machinery, fixtures and 500 or 600 pounds of other matter."

The inventor was also at work on the construction of a third airship "which is to be more commodious and more perfect than the other two." The airship was so constructed that if it fell in the water it could be used as a boat "by detaching a portion of the airship." The inventor had three assistants with him, all of whom were mechanics. Hart also told that the inventor used two kinds of power, gas and electricity. Sometimes gas or sometimes electricity would produce the light.[32]

But, as history has decreed, Havana would not be bombed by an airship, and towards December, 1896 Hart too would sink in a sea of contradictions and ultimately disappear from the pages of history. Several authors have pondered over his strange role in the whole affair. It is suggested that Hart, who owned "the only mine in the world where osmium is found in metal form," was a hustler: the osmium of his mine was used in the manufacture of electric storage batteries, which, as Hart stressed on several occasions, were parts of the construction of the airship. It is suggested therefore that Hart was perhaps planning a scheme to enrich himself, although no evidence has surfaced that he ever did.[33]

Why Collins came forward in the first place is even more uncertain. The opinion of eminent UFO historian Jerome Clark in his recent excellent summary of the 1896 airship wave is that a century later, it has become impossible to judge what, if any, truth lay behind the claims of Collins and Hart. "Conceivably they were truthfully passing on what they had been told by clients who for their part were less than honest. ...All that is clear...is that *someone* was lying." (Clark's italics)[34]

But in order to understand that not all that Collins and Hart had said was nonsense, and to fully appreciate their statements, which quite possibly originated from Astor, we must look at the curious history of Charles August Dellschau and his encoded manuscripts, with which he left a legacy of continuing puzzlement.

Charles August Dellschau (1830-1923) will probably always remain a

.mystery, a shadowy figure for whom history has reserved no place. What we know of his life easily fits in a paragraph or two. Dellschau was born in Germany and emigrated to the United States in 1850. He then went to Galveston, Texas, but his own testimony places him in the mining towns of Sonora and Columbia, California, in the 1850s. It is not known what Dellschau did after these years, or where he went. We do know that he went to Houston, Texas, in 1886. There he married a widow and worked at the company that her family owned. In his latter years he would visit the company's store and would occupy himself by sweeping the floors. In 1908 he began with his manuscripts, homemade scrapbooks, eventually devoting all his time and becoming a furtive recluse. He spent his last years in a room, still working laboriously on his scrapbooks until he died in 1923.

In the end he would leave a dozen or more scrapbooks, each filled with numerous designs of airships, each drawing carefully laid out, and some drawn on a grid and watercolored. It is suggested that at one time Dellschau had help with these drawings, since some of them are drawn in a different hand than his. As aviation history expanded, he adorned every page of his scrapbooks with newsclippings, eventually leaving no room for his drawings with the exception of fanciful borders and patterns with which he decorated all the pages. During World War I, he filled some of his scrapbooks with war pictures from the newspapers.

Interspersed among all of the clippings and his drawings were notations and cryptic notes, which obviously carried a hidden message. It was from these notes that researcher P.G. Navarro was able to reconstruct the story that Dellschau apparently wanted to make known, but only after someone had deciphered the code. This code was written in symbols which were unique, although Navarro found superficial resemblances with the secret writings of several known secret societies. It seems that Dellschau was only reluctantly writing down his fantastic story, written down in a manner that would discourage all but the most determined researcher. Dellschau wrote: "You, Wonder Weaver, will unriddle these writings...which are my stock of open knowledge." What his writings yielded after their decipherment was the amazing story of the enigmatic Dellschau and his activities with the equally mysterious Sonora Aero Club, and those who were members of this most secret society with him. And while all this was fantastic, the riddle posed another mystery; for behind the secret Sonora Aero Club loomed another, even more secret powerful organization, known only as NYZMA.

From the interpretation of Dellschau's writings, he and a small group of associates gathered in Sonora, California, where they formed an "Aeroy Club" which they called "the Sonora Aero Club" in 1858 after becoming associated with the even more mysterious organization NYZMA. The Sonora Aero Club eventually grew to a membership consisting of some 62 members or associates, mostly German immigrants, Englishmen, some Spanish or Mexican members and one Frenchman. There are indications that they built their first airship in 1857 and that the motors were invented or made in 1856.

The Sonora Aero Club was dedicated to the designing and construction of navigable aircraft. The group worked in secrecy, and the rules of secrecy had to be strictly observed. "They were obviously members of a larger organization of which the Sonora Aero Club was but a small branch," Navarro writes. Members were not permitted to talk openly about their work or their airships, called "Aeros," nor were they permitted to use the aeros for profit making. Anyone who went against the dictates or rules of the society was summarily dealt with.

About the identity of NYZMA, Dellschau said little, if anything. The only allusions in many of his notations is that the Sonora Aero Club was under the direction of an organization called NYZMA, which was located "somewhere back East." Its unnamed superiors were overseeing and financing the activities of the Sonora Aero Club. The superiors were definitely not government authorities, for Dellschau, in his ambiguous way, states that a government official who somehow learned of their work once approached club members with the suggestion that they design and sell their aeros as weapons of war. The suggestion was turned down. The Sonora Aero Club was against war and dreaded the eventual conversion of their aircraft designs for war purposes, or, as Astor writes about his airships, "Having as their halo the enforcement of peace, they have in truth taken us a long step towards heaven."[35]

The Sonora Aero Club designed and proposed a large number of aeros, including many ideas for motor designs, stabilizers and other gadgetry. But judging from Dellschau's drawings, it is difficult to believe that these machines were ever capable of flight. For instance, the bodies of the airships are radically out of proportion to the gas bag or balloon, which was supposed to lift the airship in the air. The gas that was used to lift the airships was no ordinary gas. According to Dellschau, the substance that they used was a fuel which, when injected into a chamber containing a drum-like device which soaked up the liquid fuel, produced a gas which Dellschau designates as "NB" gas. This gas not only provided the motive power for the airship, but also had the capacity to negate weight. "Incredible as it may seem," writes Navarro, "Dellschau was talking here about antigravity. ...It does not say that it lifted the weight, but that it 'negated' or 'eliminated' weight." The NB gas was produced from a substance called "supe," and apparently only one man in the organization knew the complete formula for the manufacture of this important ingredient.

This man was Peter Mennis and he not only designed and built the first motor, but he probably also discovered the substance known as NB. The group seems not to have been a harmonious one; Mennis, for instance, did not share the formula of the gas with other members, and Dellschau's tale is riddled with strange and violent deaths of group members and sudden accidents. In fact, in some cases, the group members were even working against each other. Nevertheless, if we are to believe Dellschau's incredible tale, it was Peter Mennis who was the first person to successfully fly a navigable aircraft called the Aero Goosy.

This airship was a small craft with a basket-like affair in which the pilot was seated. At both ends of the basket were rigid chambers in which the power units were located. There the gas was produced and then directed into the two balloons on either side of the basket. In the center and attached to a pole was the air-pressure motor and the fuel container. Above this was an umbrella-like device, and once again we see a curious relation with Hart's descriptions, and at the bottom of the airship was a shock absorber which was called a "Falleasy." The airship had wheels to enable it to drive on the ground. Other drawings from Dellschau's scrapbooks suggest that the airship was also capable of sea travel.

Peter Mennis died during a fire accident. It is not clear if his death was foul play, but after his demise, the Sonora Aero Club was unable to produce the so-important NB gas. After his death the picture emerges that the group fell in disarray, although Dellschau is not too clear on this point. What happened to the mysterious NYZMA is also unknown. Dellschau's scrapbooks are not light-hearted reading, instead he paints a somber picture. The symbol of the crossbones and skull, which is also used by the freemasonic fraternity, is predominant in many of Dellschau's plates. There is also a vertical line of ciphers which contain the word "todt," which is the German word for death. Throughout the scrapbooks, he often uses the symbol of the skull and crossbones and the reference to death.[36]

Navarro's thorough research failed to uncover any evidence that there had ever existed a Sonora Aero Club in or around Sonora that was involved in the construction and flying of airships. He did find, though, that certain events which Dellschau described as having taken place in Sonora did actually happen, and although a local towns historian suggested to Navarro that if something might indeed have taken place in total secrecy, definite evidence until now has not been found.

But it is important at this point to remember rumors that are now a century old. At the time that Collins was making wild claims in the San Francisco press and Keely was busy trying to perfect his system for aerial navigation in Philadelphia, certain residents from Oroville alleged that somewhere in the vicinity of their town in a thickly wooded area where no one ever ventured, unnamed secretive parties were experimenting with a new kind of gas. While this rumor was started at least 40 years after the period in which the Sonora Aero Club allegedly operated, the similarities are striking and there is no exact date as to when the Sonora Aero Club—if it existed—fell in disarray.

Yet somebody in America carefully monitored the developments in California, this much we can distill from the few statements that Dellschau made concerning NYZMA. Perhaps that somebody was also instrumental in forging a cloak of secrecy that surrounded the device for lifting heavy weights that Keely allegedly made in 1881 for an unidentified gentleman in California.

Perhaps somebody was also carefully following Keely's progress in Philadelphia, for if there ever was a strictly secret organization involved in the overseeing of aeronautical endeavors, Keely's antigravity research, regularly published in newspapers across America, would certainly have drawn the attention of such an alleged group. This in turn might explain the sudden withdrawal of Keely's antigravity instruments in 1898, coincidentally the year after the great airship wave stopped.

Also, Delisle Hay's cryptic 1881 novel *Three Hundred Years Hence* suddenly obtains a different dimension, especially the parts on the zodiacal force and of lucegen, the astonishingly lighter-than-air gas which alters its volume and thus its buoyancy when an electrical current is passed through it. And Hart's statements about "an invention in the East," or of the employment of gas and electricity, suddenly make more sense when seen against the backdrop of Keely's researches in Philadelphia, Dellschau's strange tale, and when compared to Delisle Hay's cryptic novel.

Dellschau never disclosed the identity of NYZMA, except that he vaguely stated that this parent organization was located in the East. In the latter part of 1848, R. Porter & Co., a firm which listed its address as Room 40 of the Sun Building in New York, distributed an advertising flyer in the Eastern parts of America. The flyer read in part that the company was making active progress in the construction of an "Aerial Transport" or an "aerial locomotive" that would take passengers between New York and California. The Aerial Transport would be put into operation on April 1, 1849, but as far as we know, no such thing happened in 1849.[37]

Also living in New York was millionaire John Jacob Astor, inventor and acquaintance of both Keely and Tesla, who almost half a century later chose to use the term apergy which he borrowed from Greg's *Across the Zodiac*. It is inviting to speculate that Astor's literary borrowing was, on another level, meant as a subtle hint; that NYZMA perhaps means the New York Zodiacal Motor Association. We might envision behind that name an extremely wealthy and ultra-secret group that had mastered atmospheric and perhaps even space flight with exotic propulsion systems partially based on Keely's concepts, but this we do not know for sure. What we have seen, however, is that there is a large part of history that offers enough clues inviting us to theorize that in fact there always has been an exclusive underground that did possess certain advanced alternative technical means. Those means often surpassed conventional well-known technical achievements of the time-period or took a radical departure from known scientific doctrine.

What else are we, for instance, to make of an incident that occurred on Friday in April 1897, during the great 19th century airship wave? A Mr. Hopkins, described as an "elderly Christian Gentleman," encountered the airship when it landed in a valley in the vicinity of Springfield. "As the sun shone upon it the rays were reflected as from burnished aluminum. It rested upon four legs or supports, which raised it from the ground sufficiently to give room for two wheels like the propeller of a ship lying horizontally; one at the bow and one at the stern. Another at the stern lying perpendicularly was evidently for the purpose of propelling the vessel ahead, while the other two raised the vessel. The vessel itself was about twenty feet long and eight feet in diameter and the propellers about six feet in diameter."

Hopkins also had a chance to study the interior of the strange craft: "In the side was a small door. I looked in. ...From the ceiling was suspended a curious ball, from which extended a strip of metal, which he struck to make it vibrate. Instantly the ball was illuminated with a soft, white light, which lit up the whole interior. ...At the stern was another large ball of metal, supported in a strong frame-work and connected to the shaft of the propeller. At the stern was a similar mechanism attached to each propeller and smaller balls attached to a point of metal that extended from each side of the vessel and from the prow. And connected to each ball was a thin strip of metal similar to the one attached to the lamp. He struck each one and when they vibrated the balls commenced to revolve with intense rapidity, and did not cease till he stopped them with a kind of brake. As they revolved intense lights, stronger than any arclight I ever saw, shone out from the points at the sides and at the prow, but they were of different colors. The one at the prow was an intense white light. On one side was green and the other red. ...I pointed to the balls attached to the propellers. He gave each of the strips of metal a rap, those attached to the propeller under the vessel first. The balls began to revolve rapidly, and I felt the vessel begin to rise."[38]

The remarkable description immediately conjures up images of Keely's vibratory system. Was this then a description of Keely's antigravity propulsion in all its grandiose completeness? This seemingly easy-to-find explanation only loses its easiness when we learn that the two human-looking occupants, a man and a woman, whom poor Hopkins encountered, were stark naked. They were able to communicate a bit through sign language though, and the male occupant pointed to the sky and made a noise that sounded to Hopkins like "Mars."

Was Hopkins referring to Greg's, Delisle Hay's, Cromie's or Astor's novels in a veiled sense? And how was it then that Hopkins, a devote Christian, would possibly describe two totally naked, golden-haired persons in connection with a highly advanced and sophisticated aerial craft?

A possible solution might be that Hopkins was hinting at something deeper while using a metaphor; for in Hopkins' letter he speaks of the two occupants as "Adam and Eve." We could dismiss this phrase as a typical 19th century expression of nakedness. We could interpret Hopkins words as a hint at Keely's inventions, which were animated by "the breath of God," the same breath that also gave life to Adam. It remains open to conjecture if Hopkins indeed was alluding to the beautiful innocence of Keely's vibratory science. While the naked man and woman are age-old alchemical symbols for the opposing forces, thirty-four years later and buried deep in the German occult substrata, the German Reichsarbeitsgemeinschaft which would take both Keely's and Bulwer-Lytton's concepts into the twentieth century, devoted a whole chapter in their second booklet with the title, "The World Apple. Why Were Adam and Eve Not Allowed to Break the Apple of the Tree of Knowledge?" Significantly, in this chapter the Berlin group made certain technical and physical conclusions concerning vril-energy.

And what are we to make of an incident that occurred in Mitchell, South

Dakota, years after the airship wave? Herbert V. DeMott claimed that in 1906, when he was 10 years old, he saw a "craft" land near a well. A door opened and the young DeMott was invited inside the object. There he was welcomed by two ordinary-looking men sitting on what looked like camp stools. During their conversation, the men refused to tell him where they came from, but they did tell him something about the propulsion system of their "airship." "The outer shell of the craft was filled with helium gas, and when the lever was moved the magnetism from the earth was cut off, allowing the craft to rise." As a strange allusion to Keely's early experiments involving water to be used to obtain his vaporic force, the men took water through a hose "to be used in making electricity."[39] What is also disturbing, since this account stands not on its own, is that these more or less "normal" accounts are interspersed with even stranger encounters, of which plenty are found in UFO lore.

What we do know is that Astor was intimately connected to Keely, who used the same term to describe his force. Hart's puzzling statements of the airship going to Cuba obtains another dimension when we learn that Astor, who wrote about spaceships traveling through the solar system on apergy, was at one time a staff officer in Cuba to General Shafter in the Spanish-American War. Interestingly, Kinraide's researches in x-rays and his invention of the Kinraide coil also have a minor connection to the Spanish-American War: four ships involved in the conflict were outfitted with x-ray equipment, and a detailed summary of the U.S. Army's experience with x-rays in this war was eventually compiled, showing the importance of their use for medical ends. While Hart alleged that the airship inventor was of Cuban descent, we must not forget that New Yorker Louis Senarens too, who also profusely wrote of air and spaceships, was of Cuban descent. Also, Hart's stressing of the usage of electric storage batteries has been interpreted as a scheme of promoting the use of these batteries. However, it was Senarens who was the first person to propose that an air vessel be driven by electric engines powered by storage batteries. Under closer examination, Senarens' literary work suddenly yields a link to Keely.

It would also be easy to dismiss Dellschau's incredible tale as the ranting of an unstable mind, the idea of an exclusive underground with advanced technology as pure speculation, and Keely's inventions as a mere delusion, even when we take the strange whispered rumors of the Oroville community and the accomplishments of other free-energy inventors into account.

Since no positive evidence has surfaced that the Sonora Aero Club and the even more mysterious NYZMA did exist outside Dellschau's coded plates, others have suggested that Dellschau was indeed drawing solely from his imagination. And since only fragments of all those other incredible inventions are all that is left to us now, we are not in the position to draw any definite conclusion. But when we take all those similarities and possible metaphors into account, it very well could be that, through widening the scope of our research, we have by chance stumbled upon a very secretive underground of which we have for the first time uncovered its dim outlines.

It is alleged that Dellschau was a furtive man, as if in fear for his life. He also took take great pains to hide his messages in code. We therefore might as well scan all data for codes, metaphors, and hidden messages of which there are enough.

It is quite possible that Greg, Astor, Delisle Hay and Senares employed codings in their fanciful tales. This notion is not easily discarded; their writings have never before been the subject of such a search. In Astor's *A Journey In Other Worlds*, for instance, we find a company of the novel's heroes more than once seated at Delmonico's. Delmonico's was a restaurant that very much existed in the real world; Tesla used to dine there quite often. The names of the four protagonists each begin with a successive letter of the alphabet - a, b, c, and d.

In the titles of two of Senarens' tales that were influenced by Keely's inventive mind, we find at least a symbolic summary, a metaphor so to speak if we allow some poetic insight, and perhaps here we intuitively feel an expression of the growing hope and bitter disillusionment of the whole episode without precisely knowing what its sad contents are. We do feel, however, that in this highly complex episode there is enough symbology to somehow suggest that various and as yet unidentified opposing historical forces or fractions were at work. In one tale, published in 1897, the year that the airship wave struck large parts of the United States and reached its peak, Senarens named his spaceship "The Shooting Star." A year later, when the great wave stopped, Keely died and his antigravity devices disappeared from the pages of history. Senarens partially titled his tale involving yet another spaceship "The Sinking Star."

But there are other, spidery connections of a more tangible nature in this strange episode. Keely's alleged exposure was published for the first time in a New York newspaper that was owned by wealthy multimillionaire William Randolph Hearst. Hearst also played a significant role in the great airship wave, a role that has never been properly explained but that has been labeled by one author as "deliberate misdirection."

One of the newspapers that most heavily covered news about the airship was Hearst's *San Francisco Examiner*. Its tone was one of complete skepticism and even scathing sarcasm. But meanwhile across the continent, the *New York Morning Journal* took a completely opposite viewpoint and even went so far as to positively assert that the airship was a reality. Strangely, this newspaper was also owned by Hearst.

Who then was Hearst? Not only puissantly rich, he also desperately wanted to be the president of the United States, and he was the prime catalyst which sent America into the Spanish-American War. The prelude to that conflict was fanatically developed by Hearst in his newspapers side by side with news about the great airship. In this light, Hart's statements of the airship as a weapon for stamping out Spanish rule in Cuba, juxtaposed with Dellschau's assertion that the Sonora Aero Club dreaded the use of their airship designs as a possible weapon, obtains an unnerving meaning.

The question, asked by author Paris Flammonde, looms large: why was

Hearst, an extremely wealthy publisher and financier with political ambitions and deeply involved in America's international and military affairs, directing his editors on each coast to present different and even contradictory news on the airship?

Adding to the strangeness is that while the scathing cynicism came from a newspaper that published news on the airship-sightings in the same locale as the events, assurances of authenticity came from a newspaper that was published on the other side of the country. "Hearst was a man of singular ability, a gambler for the highest stakes and a man of sweeping imagination. He did nothing without a very good reason. What did America's most aggressive newspaperman know about the great airship that he wished to have obscured?" muses Flammonde.[40] And we might also add, what, if anything, could Hearst possibly have known about the nature of Keely's inventions that he so desperately wished to have obliterated from the pages of history?

13

Into the Realms of Speculation
Anomalous Documentation and Mythological Tales

"It is in the arcana of dreams that existences merge and renew themselves, change and yet keep the same—like the soul of a musician in a fugue."

Bram Stoker, *The Jewel of Seven Stars,* 1903

"In every name there is a hidden force, and when we repeat that name over and over again...we draw into our blood that spiritual force which...in time, finally transforms our whole body."

Gustav Meyrink, *Das Grüne Gesicht,* 1916

Recently, Joscelyn Godwin raised the important question of interpretation. He asks if one should draw the line between what one can read into an author's work, given a certain key or method, and what the author actually wrote.[1] The same may be said for any amount of data or information. Should one arrive at its conclusions in a linear and, by the absence of other data, limited sense, or should one try to create other possibilities by leaving the rigid, linear trail by application of different methods?

The Keely history, which at times has been a journey into little-expected areas, has delivered an amount of data that can be divided into two classes: the non-anomalous data and the anomalous data. With the non-anomalous data concerning Keely and his discoveries at hand, I have reconstructed development of the Keely mystery in a historical sense.

The amount of anomalous data asks for a different approach. To approach these in a confined and orthodox sense would mean to discard the strange details, analogies and correlations as if they were irrelevant detritus that all history has clogging at its periphery. But one might choose to ignore; in an unguarded moment, one is uncomfortably tapped on the shoulder and again reminded of these incongruencies. With the anomalous data, we are able to create a parallel history, a speculative framework of alternatives and possibilities. Those possibilities based on speculations and suppositions may be eliminated or substantiated in due time through careful and thorough research. What remains after such an examination is valid and may be added to the set of non-anomalous data that we have now.

At the end of the previous chapter I left the trail of that which is verified by hard documentation, and instead concentrated on applying a different method of interpretation by pointing out similarities in text and content. This might be a point of criticism, but the very nature of the events demand that we take that course. The historian may be reasonable and orderly; history is not. History is far more than documents in archives; history is also largely composed of that for which there is no means of documentation. History is not a set of well-outlined and sharply defined incidents, arranged in a logical fashion. Some of its most far-reaching causes—such as the hint, the metaphor, the influence of ideas, the sudden impulse or the interchange of concepts—often come and go unacknowledged, especially in the fields of the occult, the irrational and the alternative sciences.

When we further overstep the boundaries of our conventionalism and our consensus of what we think that reality consists of, we may also glimpse that other reality, which is partly the world of the highest occultist. We may perceive another surrealist dimension: not so much a tangible one that is backed by historical documentation, but one that stretches out across the ages, across little-known events, in the air that is still stale of incense after a ceremony or a ritual; in a room that is still resonating after a heated discussion, in the unspoken philosophies of forgotten dreamers and in the curious encrypted analogy of legend, symbol, name, locality, and language.

Verne was keenly aware of Keely's discoveries, and some of it was printed in *The Hunt for the Golden Meteor* that was published posthumously in 1908. How could he not have been aware as he wrote the foreword to Cromie's novel. He admitted that he could only read "those works which have been translated into French"[2] but we have seen elsewhere that information on Keely's researches also appeared in the French language. "I esteem myself fortunate as having been born in an age of remarkable discoveries, and perhaps still more wonderful inventions," Verne told an interviewer in the autumn of 1894, "I always took numerous notes out of every book, newspaper, magazine or scientific report that I came across. ...I subscribe to over twenty newspapers...and I am an assiduous reader of every scientific publication. ...I keenly enjoy reading or hearing about any new discovery or experiment..."[3]

There are circumstances, however, that suggest that Keely could have been affiliated with Verne in a distinctly other way, by sharing a membership in a little-known society called the Angelic Society. The Angelic Society truly bore the mark of a secret society as there exists no membership roll accessible in the archives of the Western world. The name of this order lacks in even the most learned studies of secret societies. Verne's membership and the existence of the

Angelic Society itself were established relatively recently by French author Michel Lamy, through an original method of interpretation, the careful dissection of their writings, and the unriddling of codings therein.

Lamy argues convincingly that Verne's writings were "entirely dedicated to the transmission of a message," and have "reflected the thought not of one man, but of a community."[4] After all, of Verne's life Lamy remarked that, "The end of his life has been marked by a profound loneliness, a curious melancholy of being," and he further wonders: "But the whole of his existence was inscribed by the sign of the unknown. His wife, Honorine, felt haunted by some incomprehensible mystery that he would not share with anybody. ...Why did Jules Verne, before he died, burn hundreds of letters, personal papers, his unedited manuscripts and his account books? ...What has become of the 3,000 or 4,000 square words that he wrote and left his son Michel? Who has destroyed them? Are they really lost?"[5]

And indeed, Lamy makes us question what we really know, or thought we always knew about the great French author. When interviewed in the autumn of 1894, Verne showed the greatest reserve as to biographical details and it was with reluctance that he discussed his life or his books.[6] Naturally we find no mention of Verne's membership in the Angelic Society in the interviews, and for this we must again turn to Lamy's researches. Set on the trail of this mysterious society by the writings of little-known late 19th century writer on cryptography Grasset d'Orcet, Lamy compiles the membership of this society as consisting of Andre Dumas, Gerard de Nerval, George Sand, Jules Verne, numerous painters and artists of all nationalities and others.[7]

The Angelic Society was founded in Lyon in 1562 by a German named Sebastian Greif. After a while the society, closely allied to freemasonry and Rosicrucianism, named itself simply "Le Brouillard" meaning "mist" or "fog." Lamy points towards "Phileas Fogg" in Verne's *A Journey Around the World in Eighty Days* and his membership of the Reform Club, of which its initials "R" and "C" stand for Rosy Cross. We might add that the legend of the Rosicrucians first appeared during the reformation. Phile-as, according to Lamy, has the same meaning as poli-philo, "lover of all," and he asserts that the Angelic Society possessed a curious book titled *Songe de Poliphile*.[8]

Lamy points out a multitude of similarities between Verne's initiatory novel *Journey to the Center of the Earth* and Sand's *Laura or the Journey in the Crystal*. Both are full of polar symbology. Verne's protagonist is called "Axel," Sands heroine "Alexis." Professor Hartz in Sand's novel has his Vernean counterpart in Professor Liddenbrock. Both are described as German scholars.[9] Curiously the most famous of the Hartz mountains in Germany is the sacred Brocken peak. The Brocken is known for an optical phenomenon known as "the Brocken spectre." In 1938, a strange device called a "transmitter" was erected on the top of the peak, a tower surrounded by an array of posts with pear-shaped knobs on top. At the same time similar constructions were erected in other places in Germany.

When the device on Brocken peak began operating, there were reports of engine failure by cars that traveled in the vicinity of the Brocken transmitter.[10]

Returning to the Keely history, we also came across the name of Jules Verne in additional instances. In connection with Astor's *A Journey In Other Worlds*, a slight reference to Verne appeared in the periodical *The Unknown World,* issued by circles surrounding the Golden Dawn. Later, occult grail seeker Rensburg referred to Verne in his writings and neo-Rosicrucian Surya dropped what might be seen as a slight hint. Verne corresponded with his visionary literary counterpart Louis Senarens, and Bloomfield-Moore writes how a scientist while witnessing Keely's demonstrations, is given to exclaim: "What would Jules Verne say if he were here?"[11] In his 1886 tale involving an airship, "Robur le Conquerant," of which Lamy pointed out reference to the Rosicrucians—the initials of the title "R" and "C" again meaning Rosy Cross—Verne casually remarks that the most beautiful Masonic temple is erected in Philadelphia.[12] Four years before Verne published his tale involving an airship, Keely had "given no attention whatever the occult bearing of his discovery." But in 1887, a year after Verne's singular novel, and at a time that Keely was working on his airship, "a bridge of mist" formed itself before him, "connecting the laws which govern physical science with the laws which govern spiritual science..."[13] This "bridge of mist" is synonymous with the Angelic Society, also known as Mist, that initiated Keely, thus connecting his researches with a spiritual foundation.

There are also circumstances which indicate that the Angelic Society was not an exclusive French affair, but also had its foothold in England and the United States and that a substantial part of its message involved an intricately hidden symbology of the poles and the zodiac. Blavatsky, while investigating spiritists, was living at Girard Street, and Keely's memorial address, written by spiritist and freemason Colville, was delivered at Girard Avenue, both in Philadelphia. In that city, a Stephen Gerard had been the principal financial backer of the new Masonic temple built in 1819 on the premises of a much older Masonic temple that was destroyed by fire,[14] where in 1731 Benjamin Franklin was initiated.[15]

We will leave the application of name-analogy there, and concentrate on the Masonic Temple instead. We find that the temple is the residence of the Grand Lodge of Pennsylvania and the same one that Verne referred to. The temple was located at Broad Street,[16] the same street where the warehouse was located containing Keely's devices that were not taken away to Boston. Other orders, such as the Knights of the Temple and The Knights of Malta also held their meetings in the Masonic Temple at Broad Street.[17]

The temples of the Freemasons are often adorned with depictions of the zodiac. In the 18th century a Masonic grade called Knight of the Zodiac was in existence,[18] and towards the beginning of the 19th century, the existence of a secret society called Les Illumines Du Zodiaque became known.[19] The oldest depiction of the zodiac is found in the ancient Egyptian temple at Denderah that was consecrated to Hathor, the Mother of Light. The temple at Denderah has other, very strange features, including several illustrations on the walls of subterranean chambers. These illustrations are the only ones found in the all of Egypt depicting what modern writers have variously interpreted as being a symbol of "that which is to be expected and is preordained from times immemorial."[20] It has also been theorized that the illustrations were of electrical devices or possibly electric light sources, or at least held as a clue that points toward the use or knowledge of an unknown power source known by the ancients.[21]

In a number of these wall designs, the figure of a snake is depicted inside oblong tubes. Thierens, who pondered on Keely's supposed psychic abilities, departs from theosophical ground while explaining that, "Astral primal atoms appear as snakes. This is the reason that in the occult sciences left behind by Egyptian and Indian symbolism, the astral beings are depicted as snakes and that one calls the Higher Beings, that lead the evolution of the soul, the Masters of Wisdom, Nagas or snakes."[22] Thierens' "Astral primal atoms," conceived in 1913, are clearly the next phase in the evolution of Levi's concept of "the astral light," which he also called "the great serpent," an idea which is said to have also influenced Bulwer-Lytton's concept of vril. Bulwer-Lytton in turn influenced Greg, who in 1880 introduced the term apergy for the first time in his book *Across the Zodiac*. As a zodiacal echo, in his novel, Greg writes how the Martians use a system of the multiplication of twelve.[23] But we are also reminded of that highly curious Martian initiatory ritual that the protagonist in *Across the Zodiac* has to undergo to become a member of the "Children of the Star": "A bright mist of various colors intermixed in inextricable confusion, an image of chaos but for the dim light reflected from all the particles, filled a great part of the space before us. ...Presently, a bright rose-colored point of light, taking gradually the form of an Eye, appeared...beyond the mist; and emanating from it, a ray of similar light entered the motionless vapor. Then a movement...commenced in the mist. Within a few moments the latter had dissolved, leaving in its place the semblance of stars, star-clusters, and golden nebulae, as dim and confused as that in the sword-belt of Orion, or as well defined as any of those called by astronomers planetary.

"'What seest thou?'" said a voice whose very direction I could not recognize. 'Cosmos evolved out of confusion by Law; Law emanating from Supreme Wisdom and irresistible Will.'

"'And in the triple band?'

"'The continuity of Time and Space preserved by the continuity of Law, and controlled by the Will that gave Law.'

"When I spoke a single nebula grew larger, brighter, and filled the entire space...stars and star-clusters gradually fading away into remote distance. This nebula, of spherical shape—formed of coarser particles than the previous mist, and reflecting or radiating a more brilliant effulgence—was in rapid whirling motion. It flattened into the form of a disc, apparently almost circular, of considerable depth or thickness, visibly denser in the center and thinner towards the rounded edge. Presently it condensed and retracted, leaving at each of the several intervals a severed ring. Most of these rings broke up, their fragments conglomerated and forming a sphere; one in particular separating into a multitude of minute spheres, others assuming a highly elliptical form, condensing here and thinning out there; while the central mass grew brighter and denser as it contracted; till there lay before me a perfect miniature of the solar system, with planets, satellites, asteroids and meteoric rings."

There the vision does not stop, for after answering the ritual questions, Greg's protagonist is given to witness the beginning of life in the solar system itself, all inside a "small transparent sphere within the watery globe, containing itself a spherical nucleus. From this were evolved gradually two distinct forms, one resembling very much some of the simplest of those transparent creatures which the microscope exhibits to us in the water drop...The other was a tiny fragment of tissue, gradually shaping itself into the simplest and smallest specimens of vegetable life. The watery globe disappeared, and these two were left alone. From each gradually emerged, growing in size, complexity, and distinctness, one form after another of higher organism."[24]

Initiated, Greg's hero is now given to behold the emblem of the "Children of the Star": "Towards the roof, exactly in the center, was a large silver star. ...Around this was a broad golden circle or band; and beneath, the silver image of a serpent—perfectly reproducing a typical terrestrial snake, but coiled, as no snake ever coils itself, in a double circle or figure of eight, with the tail wound around the neck. On the left was a crimson shield or what seemed to be such, small, round, and swelling in the center into a sharp point; on the right three crossed spears of silver with crimson blades pointed upwards."[25]

We have encountered that suspicious phrasing "mist" several times in Greg's visionary description. His Martian secret society "The Children of the Star" is also called "The Children of the Light." The Angelic Society also called itself mist or "le Brouillard," alluding to its status as a "Church of Light."[26]

William Delisle Hay described zodiacal force and zodiamotors in his 1881 *Three Hundred Years Hence*. When the numbers 1881 and 300 are added, the number 2181 and then 12 is obtained. Twelve is of course the number of the houses of the zodiac; when the numbers 1 and 2 are added, the number 3 is obtained. Greg painted a vivid picture of a three-fold symbol, part of it consisting of three crossed spears. While Keely asserted that everything was composed of triune streams, the number 3 also has a special significance in freemasonry. Freemasons often refer to themselves as "brothers of the three points," a custom that was introduced for unknown reasons in the 18th century,[27] the period when the Masonic grade Knight of the Zodiac simultaneously sprang into existence. Delisle Hay was not the first author to use the title *Three Hundred Years Hence*. He was preceded in 1836 by Mary Griffith who published a futuristic utopian novel with the same title, which involves time travel through suspended animation.[28]

Interestingly, Griffith's story is based in Philadelphia, in the years 1835 and 2135. Amongst others, Griffith described a new power source "not steam, not muscle," developed by a woman, that is widely used in ships, government-operated trains, aircraft and ground vehicles equivalent to automobiles. The force however is not further explained.[29] The fact that the power was developed by a woman may serve us to see a certain influence of ideas on Bulwer-Lytton's *The Coming Race*, in which he writes that women are superior to men in handling vril. As with Delisle Hay's novel, the year of publication, 1836, and the title, 300, deliver the values 12 and 3.

According to Collins, the airship of the never-identified inventor was guarded by "three men" and Hart stated that the mysterious inventor had "three assistants." Hart also alleged that there were "three airships" in existence. A freemason claimed to have learned the true identity of the inventor in a Masonic lodge. In the second part of the airship wave the inventor identified himself as Hiram Wilson. In freemasonic tradition, Hiram, also known as Hiram Abiff or the widow's son, is an important figure that was the builder of the temple of Salomo. As freemasonic legend relates, Hiram was murdered by "three unworthy men." In Dellschau's manuscripts, the mysterious NYZMA is frequently connected by the symbol of the skull and crossbones. Not only is this symbol used by the Chapter of the Skull and Bones, but it also features prominent in Masonic symbolism. The zodiac was not only depicted on the ceilings of the Masonic temples and the temple at Denderah; it was also neatly engraved on Wronski's globular Prognometer and depicted on d'Alveydre's Archeometer.

In 1912, Count Von Rosen purportedly sent Keely's secrets via Scotland to Stockholm, Sweden. It is said that the original Knights Templar fled to Scotland. It is in Stockholm that the grand lodge of the Swedish Freemasons is settled.[30] The Swedish Rite affirms in its traditional history that Jacques de Molay, last Grand Master of the Knights Templar, committed the order into the hands of his nephew, the Comte Beaujeu, who carried it to Sweden, together with the ashes of his uncle. Another branch of Swedish freemasonry was derived from the "system of the strict observance," an 18th century Masonic system also based on the legend of the perpetuation of the Knights Templar.[31] The word "Rose" is the Masonic symbol for the search for a higher life. In ancient Greece and Egypt the rose was the symbol for secrecy. Something experienced or learned by an adept adorned with a rose, or Sub Rosa, had to stay secret.[32]

Freemason and Keely's friend Wiliam J. Colville published a book called *Our Places in the Universal Zodiac* in 1895.[33] His book was followed a year later by a book called *Across the Zodiac*, written by an Edwin Pallander about whose life nothing is known. In the book a spaceship, the Astrolabe, is made operative by a giant gyroscope that nullifies gravity. Flights to the moon are made, where remnants of a dead civilization are found, and to Saturn, where life is discovered.[34] Pallander used not only the same title as Greg's book, but also elements of Greg's plot, which he fused with certain elements from Jules Verne's *Twenty Thousand Leagues Under the Sea*.[35] Greg's *Across the Zodiac* was published exactly thirteen years before Bloomfield-Moore's book on Keely was issued by the same publishing house. 1880 plus 13 delivers again the numbers 12 and 3.

Cheiro visited Keely in 1890. In one of his remembrances of Keely that he hid in an otherwise fictional tale, he writes: "In the inscrutable wisdom of the Thought Force of Purpose, and the Creator of all Design, the Zodiacal system that controls this earth compels it to alter its axis once in every 25,000 years. At the end of each of these periods of Time called by men "the Precession of the equinoxes" the tilt or inclination of the Poles causes oceans to alter, continents to be swept away, civilizations to be destroyed, and new ones to appear."[36]

Cheiro, strapped in the strange device which we have seen described earlier, travels to a remote time when the temple of Atlantis still proudly stood: "I saw stretching into illimitable distance a wide avenue of giant figures of stone leading to a vast temple, of which every part had an astrological meaning. Formed like a circle, this temple appeared divided into twelve parts symbolizing the Twelve Signs of the Zodiac and their influence on human life. In the middle of this majestic temple appeared a throne on which the Sun as the Giver of Life reigned in the form of a mangod. Rays of light charged with ions of magnetism radiated from this center to each of the twelve signs and from them again flooded the Earth-Planet as it swept through each Sign in its annual pathway through the heavens. Stars sang to stars and suns to suns in one universal vibration of harmony. Designs, with threads of gold, linked planet to planet. Purpose radiated from the Sun-God bearing life and Death within its hands. ...The wide space before the throne was filled with myriads of people...all were drawn by some mysterious magnetic force to the Sign of the Zodiac under which they were born; every man, every woman and every child was robed in the same color as their Sign and on each forehead was their own distinctive jewel."[37]

When we leave Cheiro's visionary description of Atlantis and its polar and zodiacal symbology, we find that Greg's *Across the Zodiac*, published long before Cheiro's account, exerted its influence on other writers. It is asserted that Robert Cromie's novel about a trip to Mars, *A Plunge into Space*, published in 1890, owed an obvious debt to Greg's work, from which Cromie borrowed his antigravity mechanism.[38] The novel was endorsed by Jules Verne, who wrote his only foreword to the second edition of the book published a year later.[39]

Lord E.'s remark in Surya's account of having had as his task to fathom the "double sphinx of force and matter" reminds us of Verne's 1895 tale *le Sphinx des Glaces* or *The Sphinx of the Ice Fields*, which Verne not only dedicated to the literary genius Edgar Allan Poe, but also to "his American Friends."[40] In the story, Verne's protagonists encounter, behind a curtain of mist, a huge magnetic mountain in the form of a sphinx on the south pole. Verne also wrote a little-known story in 1889 about the tilting of the axis, *The Purchase of the North-Pole.*[41] It is also argued on several occasions that Verne's stories are in fact a huge and elaborate code,[42] and that one of Verne's characters in *Journey to the Center of the Earth,* called Axel, is merely a disguised form for the word AXIS.[43] The same may be said for Sand's heroine Alexis.

The depiction of the zodiac at Denderah has a striking feature; it shows a deviation of the axis in relation to its current position, a detail that evokes Cheiro's visions. Greg's term "apergy" was again used by Astor in *A Journey In Other*

Worlds. In his novel, Astor named his spaceship "Callisto," after the moon that circles Jupiter which is the subject of a visit in his novel. However, he chose Jupiter with a very good reason, since Jupiter and Callisto are both a reference to the axes or the poles, which in his novel are such an important theme. In Greek mythology Callisto was loved by Zeus, by the Romans named Jupiter, and she bore him Arcas. For this Callisto was turned into a she-bear; in the end Zeus turned her into the constellation of seven stars known as the Great Bear. But Zeus also placed their son Arkas or Arcas in the heavens as Arktouros, the Polestar.

A year before Astor published his striking novel, John O'Neill published the first volume of *The Night of the Gods,* in which we find the layer that Astor carefully covered in name-analogy, explained: "...the Most High, the deity symbolically worshipped on High Places, was the God of the Polestar, who was seated at the Highest celestial spot of the Cosmos, the North Pole of the Heavens."[44]

In the year that O'Neill published the first volume of his study on polar lore, Bloomfield-Moore published her book in which Keely stated the polar forces consisting of magnetism, electricity and gravital sympathy, "each stream composed of three currents, or triune streams, which make up the governing conditions of the controlling medium of the universe."[45] Magnetism Keely described as "polar attraction," gravity as "polar propulsion."[46]

Where O'Neill saw the Most High as the God of the Polestar, the son of Callisto and also the name of Astor's spaceship, and Cheiro called this the Sun-God, Keely envisioned the streams of polar force originating from God; "So God created man in his own image, in the image of God created he him; male and female created He them. ...All sympathetic conditions, or streams of force, are derived (if we dare to make use of such a term in speaking of Deity) from the cerebral convolutions of the Infinite; from the center of the vast realm of the compound luminous. From the celestial intermediate, the brain of Deity, proceed the sympathetic flows that vitalize the polar terrestrial forces,"[47] an imagery similar to Hopkins' mystical allusion to "Adam and Eve." Bloomfield-Moore proclaimed that "The great polar stream, with its exhaustless supply of energy, places at our disposal a force. ...We have but to hook our machinery on to the machinery of nature, and we have a...force, the conditions of which when once set up remain for ever, perpetual molecular action the result."[48]

While O'Neill saw the pole as the supreme Arcanum or secret, Arx being the celestial pole and Arcadia the secret of the polar sanctum, he wrote "Arkas...was the father of the Arkades or Arcadians, who claimed to be the first men. Hermes...was *the* (his italics) Arcadian...and the caduceus of Mercury was therefore called the Arcadian rod."[49] The last image is synonymous with the earliest perpetuum mobile designs—material representations of an eternal cosmology—with their hollow rods filled with mercury. The first men are synonymous with Adam and Eve.

Analogous to Keely, O'Neill developed a polar interpretation for all triplets, trinities and triple figures. All fours and their multitudes, including 12's—and we have seen how from several novels the numbers 12 and 3 arise—are the symbols of the directions of space around the pole, the zodiac. In O'Neill's cosmoconception, all sevens—including 21's—refer to the twice-seven stars of the Great and Little Bears.

Between the Great and the Little Bear is Draco, which may be imagined as the dragon serpent guarding the apples of the Hesperides which grow on the axial tree. The serpent evokes images of Levi's Great Serpent or Great Dragon that he envisioned to be the carrier of cosmic life-force. The seasonal positions of the Little Bear around Draco formed a swastika around 4,000 BCE. The swastika is held as the prime symbol of the Pole in its aspect as center of the celestial or terrestrial circle; in the West, the caduceus is the prime symbol of the World axis that joins the two.[50] While it is suggested that the Brotherhood of the Swastika was a possible influence on Bulwer-Lytton, it has been asserted that from this order the Brotherhood of Luxor originated.[51]

The highest grade of the Chevaliers bienfaisants de la Cite Sainte that was founded around 1770 on Knights Templar tradition, and to which possibly Bulwer-Lytton belonged, was named after the patron of the Merovingian Empire. The symbol of the Merovingian empire is the bear and also the bee, which in time would become the symbol of the fleur-de-lis. In the fleur-de-lis, O'Neill saw a polar interpretation as it is habitually drawn at the north point of the compass.[52]

The planet Mars in astrology corresponds with the human head, and it is asserted that Keely's liberator corresponded in its parts to the human head. In astrology Mars is also synonymous with the reproductive organs, which play an important part in Isis symbology. Isis symbology, the painting "Et in Arcadia Ego" by Nicolas Poussin with its reference to Arcadia or Arcas, the supreme secret, and the symbology of the bear—which is held as the symbol of the Merovingian Empire—are considered of special importance in the riddle of Rennes-le-Château.[53] "Et in Arcadia Ego, you know," writes Sand in a letter to Gustave Flaubert,[54] and "...more like Bedlam than Arcadia."[55] a protagonist in Cromie's novel is given to utter. Cheiro visited Keely in 1890, but aside from having met with Bloomfield-Moore at her London home, he also numbered amongst his clientele Emma Calvé, the famous French singer. "Cheiro told me of things terribly true in the lines of my hand, through his advise he saved me from big misfortunes,"[56] she commented. Calvé had a very intimate relationship with Parisian occultist Jules Bois, who at one time lectured in Paris on the hidden Isis symbolism found in the geometry of that city. But it is also asserted that she had a relationship with abbe Saunière.

In the history of Rennes-le-Château in which Saunière was one of the pivotal characters and that started during Keely's lifetime, we find numerous connections of the strangest nature. There are those with the ancient Merovingian Empire and the Royal house of the Habsburgs, with secret services and societies, with the Vatican and with other highly placed clerical circles; there are rumors of occult orders such as the Prieure du Sion, of alchemists and Rosicrucians and it is asserted that the Knights Templar knew what the nature of the secret was. There are coded documents of uncertain origin and references to the bloodline of Christ, and the whole landscape surrounding Rennes-le-Château seems to have been carefully arranged in an immense sacred geometrical riddle forming a complex pentagram. At the heart of the riddle lies what possibly is an energy phenomenon of unknown nature.

Adding to the mystery, above the entrance of the church of Rennes-le-Château the following description is placed: "terribilis est locus iste," meaning, "this place is terrible." In the introduction to Cromie's book, Verne wrote: "Certainly, it is a terrible venture, but they need not fear; their guide is skillful and bold."[57] Verne is considered an important factor in the riddle of Rennes-le-Château.[58]

In a coded manuscript of uncertain origin that is said to have been found inside the church of Rennes-le-Château but which may also be a modern forgery, mention is made of "blue apples," which may be synonymous to the apples guarded by the dragon seated between the Great and the Little Bears.

Verne also writes in the foreword about the Steel Globe, the spaceship that Cromie fantasized and that was the first of its kind to be globular in shape,[59] and "almost a perfect sphere, with only a certain flattening at the top and the bottom—like the polar depressions of the Earth."[60] which echoes Wronski's Prognometer, Keely's Globe Motor, Schappeller's primal force machine constructed as a miniature earth and the strange ball-shaped devices of the Reichsarbeitsgemeinschaft which they called "World Globes" and their depiction of the world as an apple, "vertically sliced in halves" to demonstrate the magnetic currents and the magnetic axis. This in turn echoes the major theme in Astor's *A Journey In Other Worlds*.

There is an abundance of references in the writings of Albert Ross Parsons to the world-axis, the zodiac, the Pleiades, and the two lost planets called Quan and Habel from whose collision and disruption the asteroid belt was formed. We find Parsons' lost worlds-idea coupled with Bulwer-Lytton's *The Coming Race*, in Engel's strange novel *Mallona*.

According to Parsons, the center of the universe is a star in the Pleiades, and one of the galleries in the great pyramid is aligned to that star. Parsons then travels to "a remote and unknown period of prehistoric time," where he finds a humanoid race, superior to us, as in the writings of Blavatsky, ariosophist Lanz von Liebenfels and forgotten Grail seeker Rensburg. But in the fashion of Cheiro's later tale, a giant cataclysm has swept that ancient civilization away; the fall of Lucifer or Satan is to Parsons the symbol of the collision and destruction of two planets, forming the asteroid belt, inflicting the earth and tilting the axis.[61] And after having pondered over an experiment of wireless transmission of sound, he writes: "there is a possibility of inter-planetary communication."[62] In all this, Parsons delivers a cryptic message: the zodiac is our only means of salvation.[63]

While it is asserted that Verne was a member of a secret society, an intellectual underground of avant-garde occultists, scientists and writers, called The Brouillards, or the Angelic Society, it is also claimed that he was closely associated with such occult orders as the O.T.O., The Theosophical Society and The Golden

Dawn.[64] The Golden Dawn founded the Ahatoor Temple in Paris in 1894[65] in the year that Astor's and Colville's books were printed.

Verne had also read Luis P. Senarens' stories. The two corresponded irregularly and Senares and Verne both used each other's ideas in their stories.[66] Verne wrote about airships and his imagery is often held responsible for being highly influential in respect to the 19th century airship wave.

Verne's message appeared only in the second edition of Cromie's book that was published the same year as the book *The Vril Staff* was published. The main theme of this book is the strange magic wand-like device operating on vril-force that Bulwer-Lytton so aptly described in *The Coming Race,* the book that was one of the inspirations of the Reichsarbeitsgemienschaft. More codings: *The Vril Staff* was written by somebody who chose to make himself known only as XYZ,[67] which is reminiscent of the legendary and secret NYZMA group that oversaw the equally legendary Sonora Aero Club.

It has been suggested that the ultimate secret of Rennes-le-Château may be a technical device with awesome abilities, fanatically guarded by secret societies and orders as the Knights Templar, the Rosicrucians and the Priory Of Sion. While we have seen how some of the most prominent 19th century occultists clustered around Keely, it is also alleged that Keely himself was a member of a secret society—if so, possibly the Brouillards—and that his devices are hidden in a church in France.[68]

Levi tells of a secret object that the Knights Templar possessed which is met here and there in the world of dreams, but after the manner of bare allusions only.[69] There is an interesting fictional tale about the alliance of Knights Templar with a crashed extraterrestrial named Baphomet, who supplies them with fantastic technology.[70]

While it is said that the Knights Templar were deeply involved in the riddle of Rennes-le-Château, an early 19th century Templar order was involved in the protection of the secret of Wronski's Prognometer. It is also claimed that in 1984 a group of commandos, all members of a secret society, excavated a grave from the cemetery at Millau in France and retrieved a device called a Planetary Talisman or the Talisman of Set, whose immense power is said to be solar.[71]

This reminds us of Keely's planetary system engine, the whereabouts of which are not known, Philipp's alleged solar-power-motor and the primal machine of the Reichsarbeitsgemeinschaft, being a configuration of seven ball-shaped devices. Seven is also the number of the stars of the Pleaides, and AXIS can be recognized as a sigil of Set,[72] Set being Satan or Lucifer, according to Parsons the symbol of the destruction of two worlds and the tilting of the earth's axis. The Talisman of Set or the Planetary Talisman was then brought to Bear Island off the coast of Norway. Mention is made of Psychic Capacitators in the Rennes area, and the members of a secret society move around by levitation, UFOs and the manipulation of time.[73] While it is claimed that members of the secret society involved in retrieving the planetary talisman travel around in UFOs, it is well to remember that Sweden, the country to which Keely's secrets were sent, set up the world's first UFO bureau in 1910, following a UFO wave in 1909.[74]

The title of the book of alleged Golden Dawn initiate Bram Stoker that was published in 1903, *The Jewel of the Seven Stars,* is analogous to Hockley's *Crowned Angel of the Seventh Sphere,* a work that unfortunately was never printed, and to the seven stars that form the Pleaides or the Great or the Little Bear, another seven-star group that once formed a swastika. As in Griffiths' book *Three Hundred Years Hence* that involved Philadelphia, Stoker's novel, which is burdened with occult doctrine, involves suspended animation in which an Egyptian princess, adept in an ancient science, rests.[75] We have already seen that the oldest depiction of the zodiac was in Egypt. Stoker visited Philadelphia in 1896[76] and Lamy asserts that Stoker's *Jewel of the Seven Stars*—according to Lamy its title refers to the seven stars in the Great Bear—is analogous to Verne's 1882 novel *The Green Ray.* Not only that: Bulwer-Lytton's vril is nothing else than the Green Ray, Lamy concludes.[77]

At Verne's funeral in 1905, the year that Steiner started his lectures about Keely, an incident occurred during the ceremony that has never been properly explained. It may have a special significance, considering Verne's alleged involvement with the earlier mentioned secret societies. Among the mourners was a strange, immaculately dressed Englishman who silently approached each member of Verne's family, shook their hands, and solemnly declared in French: "Be brave, be brave in your heavy hour of trial." As soon as he passed all those along the line and gave his symbolic allusion to a terrible venture, the mysterious figure disappeared into the crowds.[78]

There is a curious analogy with the mysterious incident at Verne's funeral. Levi too was visited by an unknown young man, somewhere in 1865. An account of this strange visit, written by Levi in a letter to the cabalist baron Spedalieri is as follows: "Between three and four in the afternoon, I heard seven short knocks on my door, in this fashion: oo-o-oo-oo. I opened the door, and a young man, very distinguished and immaculately dressed with a sarcastic air, entered." Levi told how the young man said that he knew all of Levi's past, present and future life, that it was ruled by the number five, and went on to relate several little-known facts of Levi's private life, including details of his visit to Bulwer-Lytton. The mysterious man even predicted the year of Levi's death, 1875, accurately, something that was unknown at that time.

The stranger identified himself as Juliano Capella, of Italian parentage, and went on to discuss the laws of nature, and how these could be influenced by the powers of the mind. Not a very original concept, as we have already tasted something of a deep stratum of occultists, obsessed with mind-driven or telepathic technology.

Capella offered to visit Levi another time, but Levi refused, having obtained a strange disliking for the person. Capella then left, while stating that he would "go on a perilous journey," to never return. Chacornac complained that, although he encountered numerous persons who visited Levi, he never was able to trace the mysterious Capella: "I even knew that one of the last original Rosicrucians still living in Paris, sometimes visited Levi, but of Capella, nothing."[79] Capella's perilous journey is synonymous to a terrible venture or a heavy hour of trial, the number five is associated with the pentagram, the sign of Baphomet and the giant configuration of the Rennes area,[80] the number seven, the times that Capella knocked on Levi's door, is also the number of the stars in the Pleaides and the Great and the Little Bears. Capella is the name of a particular star, also briefly mentioned in Cromie's novel, and stars and constellations are strewn all over: Arcturus, Aldebaran, Orion, the Pleaides, The Great and Little Bears, and Sirius. Interestingly, the ether-driven ship—the "Sirius" of Surya's account—holds a smaller vessel, thus symbolically mimicking the twin-star system of Sirius.

The year 1912 was strange and significant in the parallel history of Keely's inventions. The never-explained UFO wave, named "scareships," that would haunt England, France, Germany, and Holland, was but a year away. The terrible First World War was but two years away, and one cannot help thinking of Steiner's exclamation: "What would have become of this war when this Keely-ideal had become a reality in those days!" 1912 was a year of revelation; Steiner made his first statement meant for the general public concerning a future technology in his *Hüter an der Schwelle,* performed in Munich. It was a year of loss; John Jacob Astor and William Thomas Stead would both perish during the terrible Titanic disaster,[81] and Franz Hartmann and Bram Stoker died.

It was a year of hiding as well. The encoded name "NYZMA" was also featured prominently on every page of Dellschau's strange manuscripts, until 1912. After that year, Dellschau's coded references to NYZMA suddenly ceased. In 1912, Count Von Rosen sent Keely's secrets via Scotland to Stockholm, Sweden. Several members of Von Rosen's family were employed by the Swedish navy. It is alleged that when the Titanic sank, a ship was in the direct vicinity, but dimmed its lights and silently sailed away. The ship has never been identified.

In that strange year of 1912, inventor Otto Witt (1875-1923) chose to go to Sweden. Witt is considered the Swedish Hugo Gernsback, but with 10 times the ego. Witt wrote dozens of novels, all bursting with new and unusual technical ideas.[82]

Witt was originally a mining engineer and through his profession would almost certainly have met that other Austrian mining engineer, Hans Hörbiger, who was called the Prophet of the Welteis Lehre or, Sacred Ice. Hörbigers cosmogony became the official pseudoreligion in Nazi Germany. His book was published in 1913, a year after Witt's travel to Sweden. Hörbiger supposedly was once involved in a project of the tilting of the axis; allegedly there was a secret Nazi project to tilt the axis with a device not unlike Verne's "neutral helicoidal ray," called a "reflector of telluric waves," controllable at will and the invention of Dutch scientist Willibrod. Willibrod collaborated with Hörbiger. Hitler even asked Hörbiger at one time if it would be possible to displace the north magnetic pole, to which Hörbiger replied positively.[83] Adherents to Hörbiger's philosophy are still to be found today, and in one of their books Keely is mentioned alongside vril.[84]

Witt studied at the Technicum in Bingen in Germany. At the same time Karl Hans Strobl (1877-1946), the Austrian writer of unusual, dark tales, who launched *Der Orchideengarten,* the first magazine devoted to the fantastic in Austria, as well as Hugo Gernsback (1884-1967) studied there.[85]

Gernsback, who is called "a would-be inventor" and "a prophet of the new technology," was a close friend of Tesla. He would often publish articles by or about Tesla in his magazines. When Tesla died, it was Gernsback who organized a death mask to be made. It is suggested that in 1924 Gernsback probably met with Harry Grindell-Mathews, inventor of a death ray who, through Gernsback, may have also met with Tesla.[86] Grindell-Mathews had at one time been in South Africa, as had Robert Pape, Dutch inventor of the Life Wave generator.

Gernsback emigrated to the United States in 1904. Intensely interested in electricity and radio, he designed batteries and by 1906 was marketing a home radio set. He would coin the phrase "science fiction" and launched his first magazine in 1908,[87] the same year that Verne's story *The Hunt for the Golden Meteor* was published posthumously. Gernsback's magazine was called *Modern Electric,* subsequently called *Electrical Experimenteer* and *Science and Invention.*[88] Among others, Gernsback proposed the idea in 1917 to equip Mars with an artificial atmosphere in order to make it habitable.[89] He too employed codings of some sort; his most important character, which he featured in several of his tales published in his magazine between 1911-1912 to promote unusual and avant-garde scientific ideas, was called Ralph 124C41, which is pronounced as "one to foresee for one."[90] When the numbers are added, we once again obtain the number 12, the number of the houses of the zodiac, and again 3, which amongst others is of special significance in freemasonry.

In 1913, a booklet was published holding a large fold-out reproduction of the zodiac at the temple of Denderah, which somehow reminds us of Parson's claim. On its very last page under the heading, "The promise of the zodiac," it is written that, "It may be that a new dispensation is at hand, and that the promise of the zodiac, that has never failed us, yet, will not fail us now. But so long as the old dispensation is with us we may remember that "out of Egypt I have called my son,"[91] a cryptic foreshadowing of Gernsback's "one to foresee for one" and the modern interpretation of certain illustrations at the temple of Denderah as being symbolic for that which has been preordained since times immemorial.

Gernsback also knew Howard Philips Lovecraft, and he would publish several of Lovecraft's stories. Lovecraft (1890-1937) had read Arthur Edward Waite, and refers to him in veiled sense in three of his tales.[92] Lovecraft also read the works of Algernon Blackwood, and Arthur Machen, whose writings he admired. Waite, Machen and Blackwood were all members of the Golden Dawn. A review of Astor's *A Journey In Other Worlds* was published in the occult periodical of which Waite was its editor. Machen had been involved in the compilation of the catalogue of the library of Hockley,[93] who at one time worked for Denley, who in turn owned a bookstore that Bulwer-Lytton often visited.

In the tale, "The Festival," written in 1923, Lovecraft describes a church in which a tomb is found beneath an altar. Lovecraft modeled his church after a real church. It was not until 1976 that during restoration of that church, Lovecraft's seemingly fanciful tale of a crypt just before the pulpit was proven to be accurate. The crypt was not located under the present altar, but under an older, hidden altar.[94] Lovecraft's fictional tale, based on fact that was later known, echoes Saunière's discovery of a crypt in his church at Rennes-le-Château.

Lovecraft lived most of his life in Providence, Rhode Island, the town of Mrs. Staunton, the Paris Golden Dawn member. Lovecraft knew about Keely and his discoveries through the writings of Charles Fort and partly modeled his NYARLATHOTEP on Keely; the word holding shades of NYZMA, THOTH and SET.

The last that is known about Keely's inventions is that they were shipped to Boston. Lovecraft is known to have made several trips to Boston. In his 1924 tale, "The Shunned House," he refers to Dr. Chase.[95] In his 1937 story, "The Evil Clergyman," Lovecraft describes a person who finds himself in a house in London. Through the use of a strange little device that he finds in his pocket, he summons the forms of a group of clergymen, dressed as Anglicans, but clearly the members of a secret magical cult. It is alleged that this is a parallel with the Men In Black.

In 1924, Algernon Blackwood wrote a short story, "The Pikestaffe Case," in which he described the non-Euclidian geometry of a dimensional trap lurking within a mirror, reminiscent of Lovecraft's tale, "The Dreams in the Witch House." Lovecraft travelled to Philadelphia once, and Blackwood had long held an interest in the occult. He not only became a member of the Golden Dawn, but also helped to establish the Canadian Theosophical Society. In 1892, Blackwood became a reporter for the *New York Evening Sun* and later worked for *New York Times* newspapers, both of which regularly published articles about Keely.

The year before Lovecraft penned the tale, "The Rats in the Walls," in which he describes a large cavern beneath an old manor house where time is warped into reverse. In connection with Rennes-le-Château, it is alleged that hidden in the region a device might be located that can only be described as a "Time Portal."[96] In 1923, the same year that Lovecraft trusted to paper the tale of a cave where time is warped into a reverse, Moholy-Nagy began the construction of his "Light-Space Modulator," and forgotten Dutch clairvoyant, Grail-seeker, theosophist and author J.K. Rensburg wrote a remarkable preface in his book *Wereldbouw* or, *The Building of a World*.

In it, Rensburg states that the entire solar system is an organism with a central consciousness, that metals have a consciousness, and that higher, material beings live on the sun and on Mars. Rensburg praises Verne as a forerunner of "the inter-astral direction." Rensburg then writes: "there are material, superhuman people; who, like us and the animals, feed, procreate and die in higher developed worlds than our own...material gods and goddesses who may divulge their decisions and knowledge to clairvoyant persons by means of inter-astral telepathy, meaning, marconigraphy of organical nature, directly from their to our nerve-system,"[97] which echoes Parsons' concept of interplanetary communication and Cromie's fictional Inter-Planetary Communication Company Limited.

In Lovecraft's 1920 tale, "From Beyond," an inventor who blended science with metaphysics builds a device that emanates waves which enable the person to see beyond that what our sense organs usually perceive. "The waves from that thing are waking a thousand sleeping senses in us; senses which we inherit from eons of evolution from the state of detached electrons to the state of organic humanity," the inventor claims. But that is not all, for the inventor states "You have heard of the pineal gland? ...That gland is the great sense organ of organs - I have found out. It is like sight in the end, and transmits visual pictures to the brain."[98] Through the rays that this device emits, people are able to see beings that live in another dimension or plane of existence.

But years before Lovecraft painted his haunting picture, in 1882 a woman said to Keely: "You have opened the door into the spirit-world." He answered, "Do you think so? I have sometimes thought I might be able to discover the origin of life." At this time, Keely gave no attention whatsoever to the occult bearing of his discovery; and it was only after he had pursued his research, under the advantages which his small Liberator afforded him for such experiments, that he realized the truth of this woman's assertion.

It was then, in 1887, that Keely was enabled to walk into the light, to cross the last barriers and jump the glittering chasm, to attend and have his membership received. There he would learn the great metaphysical truths that underpin his wonderful discoveries. He would learn of the shining, spinning zodiac and of the space-time axis. He would learn of the secrets of Atlantis, thought lost for such a long time, and of the sacred geometry that is so essential. He would come to know the mysteries of the Rosy Cross and the riddle of Rennes-le-Château. He would study the principles of the magnificent giant world machine and the living, breathing universe that creation is.

But we must now part company with John Worrell Keely. It has been such a pleasant company, with many a strange tale. It has been an incredible journey indeed, but we must leave the brilliant inventor and discoverer for now, turn away and silently close the door of his workshop in the Philadelphia of a century ago, and we must not disturb him any longer.

Notes

Chapter 1. Discoverer of the Ether: The Early Life of John Keely

1. William Mill Butler, 'Keely and the Keely Motor,' *The Home Magazine*, 1898, page 104. Only reference to Chester found in: 'Keely, The Inventor, Dead,' unspecified clipping, November 18, 1898, Sympathetic Vibratory Physics Homepage, Internet.

2. Various sources give a different birth year and when not specifically mentioning this, give his age at the time of his death in 1898 as 72: *Public Ledger Almanac*, 1900, or sometimes as 71 years: *Public Ledger and Daily Transcript*, November 19, 1898, *American Machinist*, vol.21, no.47, November 24, 1898, and *Locomotive Engineering*, December 1898. In 1895 it was claimed that Keely was 68 years old. In: 'Two Hours With Keely,' *Public Ledger and Daily Transcript*, November 11, 1895. It is also claimed that Keely was 63 years at the time of his death! While describing the casket in which Keely's body lay, it was noted that "A silver plate on the lid bore the simple inscription, 'John Worrall Keely, in his sixty-third year.'" (note the different spelling of the middle name). In: 'John W. Keely Laid to Rest,' *The Times*, November 24, 1898. *The Times* November 19, 1898 edition, gives as his birth date September 3, 1837, as do later Theosophical sources including *H.P. Blavatsky Collected Writings*, vol. VIII, 1960, page 267, and *H.P. Blavatsky Collected Writings*, vol. XIII, 1982, page 384. 1827 is claimed to have been Keely's birth year by *The New Encyclopedia Britannica*, 'Micropedia Ready Reference' 15th edition and Schribner's *1961 Dictionary of American Biography*, Frank Edwards, *Strangest of All*, Citadel 1956, Carol Paperbacks, 1991, page 160, Gaston Burridge, 'The Baffling Keely 'Free Energy Machines,' *Fate*, vol.10, no.7, 1957, page 47, Carl Sifakis, *Hoaxes and Scams*, Michael O'Mara Books,' 1994, page 140.

3. William Mill Butler, 'Keely and the Keely Motor,' *The Home Magazine*, 1898, page 114. See also: 'Keely Motor Man Dead,' *Public Ledger and Daily Transcript*, November 19, 1898.

4. 'The Keely Motor,' *International Cyclopedia*, vol. VIII, 1899, page 458.

5. William Mill Butler, 'Keely and the Keely Motor,' *The Home Magazine*, 1898, page 104. See also: 'No Other Has Ever Been So Shrouded In The Mist Of Publicity,' *New York Herald*, November 27, 1898.

6. 'Public Opinion,' vol. XXV, 1898. See also: 'No Other Has Ever Been So Shrouded in the Mist Of Publicity,' *New York Herald*, November 27, 1898.

7. Clara Bloomfield-Moore, *Keely and His Discoveries, Aerial Navigation*, Kegan Paul, Trench, Trübner & Co., 1893, page 290. Also in: William Mill Butler, 'Keely and the Keely Motor,' *The Home Magazine*, 1898, page 104.

8. 'Keely's Secret Known,' *The Times*, January 26, 1899, See also: 'Another Theory About The Motor,' *The Evening Bulletin*, January 26, 1899.

9. A. Wilford Hall, 'John Keely - A personal interview,' *Scientific Arena*, January, 1887.

10. William Mill Butler, 'Keely and the Keely Motor,' *The Home Magazine*, 1898, page 104.

11. 'Keely of Motor Fame is Dead,' *The Times*, November 19, 1898.

12. Clara Bloomfield-Moore, *Keely and His Discoveries, Aerial Navigation*, Kegan Paul, Trench Trübner & Co., 1893, page 333.

13. 'Keely's Secret,' *The World*, May 11, 1890.

14. Clara Bloomfield-Moore, *Keely's Secrets*, T.P.S. 1888, page 17. in *Keely and His Discoveries, Aerial Navigation*, on page 84. Also in: Afra, 'John Worrell Keely,' *Theosophia*, no.13, May, 1893.

15. Charles Morris, 'Apergy: Power without Cost,' no.10, 1895.

16. A. Wilford Hall, 'John Keely - A personal interview,' *Scientific Arena,* January, 1887. Elsewhere it was written that 'Keely's explanation of how he came to make his reputed discovery has a certain interest. Once, when a boy, he saw the windows of a house vibrate long after the wagon that had caused the vibration was out of sight. Also that once some drummers were driven into a hall by a storm. The storm made all the drums beat for an instant into a concerted roll, and the windows were broken by the vibration. Thus he held to have discovered that there was such a thing as sympathetic vibration, by which, under certain conditions, a force could be communicated from one thing to another. Later on he concluded that this vibratory motion was present in nearly everything, and announced that if a substance be vibrated by a musical note in harmony with it, not only will the distance between the molecules composing it be greatly augmented, but be dissipated into their component atoms, and thus exert a dynamic force double that of steam. So the Keely Motor, or harmonic engine, came into existence...' In: 'What Is The Force, Hidden and Unseen, of Keely's Motor?,' *New York Herald,* August 22, 1897.

17. 'Keely and his Motor,' *The Evening Bulletin,* August 22, 1887.

18. Clara Bloomfield-Moore, *Keely and His Discoveries, Aerial Navigation,* Kegan Paul, Trench, Trübner & Co., 1893, page 290.

19. 'Keely of Motor Fame is Dead,' *The Times,* November 19, 1898.

20. 'Everybody's Column,' *Philadelphia Enquirer,* October 16, 1931, H.P. Blavatsky, *H.P. Blavatsky Collected Writings, Vol. VIII,* Theosophical Publishing Society, 1960, page 267.

21. Megargee, 'Seen and Heard in Many Places,' *The Times,* March 11, 1898.

22. Charles Morris, 'Apergy: Power without Cost,' *New Scientific Review,* no.10, 1895.

23. 'Men and Things,' *The Evening Bulletin,* November 21, 1898. Also in: William Mill Butler, 'Keely and the Keely Motor,' *The Home Magazine,* 1898, page 104.

24. 'Hidden Tubes Show How The Keely Motor Worked,' *New York Herald,* January 20, 1899.

25. 'Keely, Motor Man, Dead,' *Public Ledger and Daily Transcript,* November 19, 1898. About Keely's mechanical turn of mind, see also short reference in: William Mill Butler, 'Keely and the Keely Motor,' *The Home Magazine,* 1898, page 104.

26. 'What Will become of Keely's Motor,' *The Evening Bulletin,* November 19, 1898.

27. William Mill Butler, 'Keely and the Keely Motor,' *The Home Magazine,* 1898, page 104. See also: 'No Other Has Ever Been So Shrouded in the Mist of Publicity,' *New York Herald,* November 27, 1898. Also in: Frank Edwards, 'John Keely's Mystery Motor,' *Strangest of All,* Citadel, 1956, Carol Paperbacks, 1991, page 160.

28. Letter by H.R. Borle to Megargee, 'Seen and Heard in Many Places,' *The Times,* November 26, 1898.

29. Letter in *The Evening Bulletin,* November 26, 1898. Collier said that Keely was 'a skilled musician on so small an instrument as the flute...' In: 'There Was No Meeting,' *The Times,* November 27, 1898.

30. ibid.

31. *Public Opinion,* vol. XXV, 1898.

32. Megargee, 'Seen and Heard in Many Places,' *The Times,* March 11, 1898.

33. Megargee, 'Seen and Heard in Many Places,' *The Times,* November 26, 1898. The story of Keely being a cannon ball-tosser began to lead a life of its own; a year after Keely died it was written that 'There is a story also that he used to be the 'cannon ball man' in a circus...' In: 'Keely's Sphere Not His Secret,' *The Evening Bulletin,* January 25, 1899.

34. *The Dictionary of American Biography, Vol.V,* Scribner's, 1961, remarks that, 'He had been for a time leader of a small orchestra and in certain more or less apocryphal stories he figured as a circus performer.' *The New Encyclopedia Brittanica,* Vol. 6, 15th edition of the *Micropedia Ready Reference* writes that Keely, 'is said to have been an orchestra leader, a circus performer and a carpenter.' Equally unsubstantiated is the following yarn: 'Obviously the ex-carnival man had run fun and mystery house contraptions in his youth...' In: Carl Sifakis, *Hoaxes and Scams,* Michael O'Mara Books, 1994, page 141.

35. This confusion was already noted in 1901 by a certain George Canby who had met Keely one time in his workshop; he collected various contemporary newspaper clippings concerning Keely and arranged these chronologically in three scrapbooks, now preserved in the Franklin Institute. In his 'memorandum' Canby wrote that, 'The various confused newspaper accounts, with their many discrepancies, very fairly demonstrate the manner in which the public mind was puzzled, for over twenty-five years...' In: George Canby, 'Keely Motor Scraps,' no.1, not published. For his recollection of his meeting with Keely, see chapter 2, also chapter 2, note 104.

36. Frank Edwards, 'John Keely's Mystery Motor,' *Strangest of All*, Citadel, 1956, Carol Paperbacks, 1991, page 160. Anecdote also found in 'No Other Has Ever Been So Shrouded in the Mist Of Publicity,' *New York Herald*, November 27, 1898.

37. 'Keely's Secret,' *The World*, May 11, 1890. Keely never explained how he exposed the mediums. It is possible that here we have the nucleus of the tales that Keely was connected to a circus, a sleight-of-hand performer or showed amazing dexterity with card tricks, since mediums often were—and still are—exposed by magicians or stage conjurors. See: Carl Sifakis, *Hoaxes and Scams*, Michael O'Mara Books, 1994, page 241.

38. R. Harte, introduction to *Keely's Secrets*, Clara Bloomfield-Moore, Theosophical Publishing Society, July, 1888.

39. A. Wilford Hall, 'John Keely - A personal interview,' *Scientific Arena*, January 1887. See also: William Mill Butler, 'Keely and the Keely Motor,' *The Home Magazine*, 1898, page 105.

40. Letter by H.R. Borle to Megargee, 'Seen and Heard in Many Places,' *The Times*, November 26, 1898.

41. 'The Motor Gets Into Court,' *New York Times*, January 3, 1888. 'Keely Motor Suit Ended,' *Philadelphia Press*, February 26, 1890. A year before Keely had admitted that he knew Wilson; see note 45.

42. 'The Keely Motor,' *International Cyclopedia*, vol.VIII, 1899, page 458.

43. 'The Keely Motor Criticized, a republication in pamphlet form of a series of editorials which appeared in the *Public Record* of Philadelphia, August 3rd, 4th, 5th and 6th, 1875,' no place, no date, but in all probability published in 1875 by the *Public Record*, pages 2-5.

44. 'Says He Knows Keely's Secret,' *The Times*, January 1, 1899. Also: 'Says he Knows Keely's Secret,' *New York Daily Tribune*, January 2, 1899, a shorter article the gist of which is that the motor secret did not die with Keely, and that he (Repetti) is the only living man in possession of it.

45. Keely himself stated that, 'In 1866, I was residing and in business in this city at No. 817 Market Street, and was associated with Bennett C. Wilson, as the testimony in the suit of Wilson against me will evidence...' In: 'It Was Not Keely,' *The Times*, January 22, 1889.

46. A. Wilford Hall, 'John Keely - A personal interview,' *Scientific Arena*, January 1887.

47. Clara Bloomfield-Moore, *Keely and His Discoveries, Aerial Navigation*, Kegan Paul, Trench, Trübner & Co., 1893, pages 10, 11, 320, 150.

48. ibid. page 336. According to Bloomfield-Moore, this happened in 1884: 'It was not until Macvicar's *Sketch of a Philosophy* fell into Mr. Keely's hands that he realized he had imprisoned the ether. This was in 1884.'

49. 'The Keely Motor Criticized,' a republication in pamphlet form of a series of editorials which appeared in the *Public Record* of Philadelphia, August 3rd, 4th, 5th and 6th, 1875, no place, no date, but in all probability published in 1875 by the *Public Record*, pages 2-5.

50. A. Wilford Hall, 'John Keely - A personal interview,' *Scientific Arena*, January 1887.

51. 'The Keely Motor Criticized,' a republication in pamphlet form of a series of editorials which appeared in the *Public Record* of Philadelphia, August 3rd, 4th, 5th and 6th, 1875, no place, no date, but in all probability published in 1875 by the *Public Record*, pages 2-5.

52. Undated (but probably around 1892) letter by C.G. Till of Brooklyn, New York. In: Clara Bloomfield-Moore, *Keely and His Discoveries, Aerial Navigation,* Kegan Paul, Trench, Trübner & Co., 1893, page 320.

53. Megargee, 'Seen and Heard in Many Places,' *The Times,* March 11, 1898. A large part of the text was exactly repeated several months later in 'Seen and Heard in Many Places,' *The Times,* November 21, 1898.

54. 'The Keely Motor Criticized,' a republication in pamphlet form of a series of editorials which appeared in the *Public Record* of Philadelphia, August 3rd, 4th, 5th and 6th, 1875,' no place, no date, but in all probability published in 1875 by the *Public Record,* page 5.

55. William Mill Butler, 'Keely and the Keely Motor,' *The Home Magazine,* 1898, page 106.

56. 'The Keely Motor Criticized,' a republication in pamphlet form of a series of editorials which appeared in the *Public Record* of Philadelphia, August 3rd, 4th, 5th and 6th, 1875,' no place, no date, but in all probability published in 1875 by the *Public Record,* page 5.

57. 'The Keely Motor,' *New York Times,* November 6, 1875.

58. William Mill Butler, 'Keely and the Keely Motor,' *The Home Magazine,* 1898, page 105.

59. 'Keely, Motor Man Dead,' *Public Ledger and Daily Transcript,* November 19, 1898.

60. 'Patent Application of John Ernest Worrell Keely,' *Sympathetic Vibratory Physics,* vol.4, issue 12, September 1989, pages 7-9.

61. 'The Keely Motor Criticized,' a republication in pamphlet form of a series of editorials which appeared in the *Public Record* of Philadelphia, August 3rd, 4th, 5th and 6th, 1875,' no place, no date, but in all probability published in 1875 by the *Public Record,* pages 2-5.

62. 'The Motor Gets Into Court,' *New York Times,* January 3, 1888.

63. According to Sykes, Keely also requested a U.S. patent in 1876, but it was denied. The reason was that officials of the patent bureau had asked Keely to build a working device, but Keely refused to do so. 'Up to recently the specifications and drawings were on file with the U.S. patent office.' In: Egerton Sykes, *The Keely Mystery,* 2nd revised edition, Markham House, 1972, pages 3-4. However, Keely never applied for a patent in 1876. The source for Sykes' yarn probably was a letter of a certain Edward N. Dickerson dated November 30, 1888, which he wrote to *The Tribune,* and which involved Keely's patent application in 1872: '..Keely applied for a patent before 1876, but did not assign to the purchasers their shares; whereupon some of them protested against the issue of the patent unless their shares were recognized in the grant. The Patent Office replied to these protests that it could not recognize the rights claimed unless there was a written assignment filed in the office, which the claimants did not have. The Commissioner, however, called upon Keely to furnish a 'working model' of his invention, which, of course, he could not do, and his application was rejected. The specification and drawings of this apparatus show a very silly form of the common perpetual motion machine, of which there are thousands. It was open to the public for some years, when, under a new rule of the office, it, along with all other rejected applications, was withdrawn from inspection; but it is in the office, together with the protests of those who had paid Keely for a share in it. I examined it years ago.' In: Clara Bloomfield-Moore, *Keely and His Discoveries, Aerial Navigation,* Kegan Paul, Trench, Trübner & Co., 1893, pages 108-109. Another variant has it that the rights to his invention, the Hydro-Pneumatic-Pulsating-Vacuo-Engine, were assigned to five individuals on February 24, 1872. This partnership then evolved into the Keely Motor Company. In: Robert Schadewald, 'The Perpetual Quest,' *The Fringes of Reason, a Whole Earth Catalogue,* 1989, page 123.

64. Egerton Sykes, *The Keely Mystery,* 2nd revised ed., Markham House, 1972, page 2. Very likely Sykes confused the date of the demand by the patent office for a working model, made November 26 the year before.

65. Correspondence with Dale Pond, dated August 29, 1995.

66. 'Keely Claims at Last to Have Harnessed a New Force,' *The Times,* March 6, 1898. Keely's workshop was also described as a 'stable or stable-like structure' in 'Men and Things,' *The Evening Bulletin,* November 21, 1898. Also as 'a queer-looking workshop.' In: Megargee,

'Seen and Heard in Many Places,' *The Times,* November 22, 1898. The workshop apparently was built originally for the Keely Motor Company, was publicly sold around 1886 and bought by a certain Daniel Dorey, who then rented it to Keely; the sale was made 'about the time that Keely and the motor people were not getting along harmoniously, and when Mrs. Moore commenced supplying Keely with funds for his experiments...' In: *Public Ledger and Daily Transcript,* January 9, 1899.

67. Although there is some confusion as to the year in which Collier met Keely, in *The Times,* November 19, 1898 edition, Collier is quoted as saying about Keely: 'I have known him since 1872.,' and in *The Times,* November 27, 1898, it is written that Collier was 'the firm friend and believer in Keely, the inventor, from 1870 until the latter's death...' Yet in the sources of note 53, it is written on both occasions that before the exhibition on November 10, 1874, Collier had never met Keely.

68. 'The Keely Motor,' *Scientific American,* July 17, 1875, pages 2-3. Facsimile reprinted as *Collier's Letter to Scientific American,* Delta Spectrum Research, no date.

69. ibid. page 3, 'The Keely Motor Criticized,' a republication in pamphlet form of a series of editorials which appeared in the *Public Record* of Philadelphia, August 3rd, 4th, 5th and 6th, 1875,' no place, no date, but in all probability published in 1875 by the *Public Record,* page 13.

70. ibid. page 4, ibid. page 13.

71. Megargee, 'Seen an Heard in Many Places,' *The Times,* March 11, 1898.

72. 'The Keely Motor Criticized,' a republication in pamphlet form of a series of editorials which appeared in the *Public Record* of Philadelphia, August 3rd, 4th, 5th and 6th, 1875,' no place, no date, but in all probability published in 1875 by the *Public Record,* page 6.

73. ibid, page 9: Collier's report stated 'say 2,000 pounds per square inch,' the report that was endorsed by J. Snowdon Bell, the mechanical assistant of Collier, stated 1,430 pounds per square inch.

74. ibid. pages 8-9.

75. ibid. page 9.

76. ibid. pages 6-11.

77. Megargee, 'Seen and Heard in Many Places,' *The Times,* November 21, 1898.

78. 'Men and Things,' *The Evening Bulletin,* November 21, 1898.

79. Megargee, 'Seen and Heard in all Places,' *The Times,* March 11, 1898.

80. William Mill Butler, 'Keely and the Keely Motor,' *The Home Magazine,* 1898, page 106.

81. Bloomfield-Moore writes that the company's origins were in 1872. In: *Keely and His Discoveries, Aerial Navigation,* Kegan Paul, Trench, Trübner & Co., 1893, page 2. Also in her 'Aerial Navigation,' 'The Arena,' 1894, page 387, reprinted by Delta Spectrum Research, no date. Her view is shared by *The Times,* November 19, 1898, Julius Moritzen, 'The Extraordinary Life Story Of John Worrell Keely,' *The Cosmopolitan,* vol.XXVI, no.6, April 1899, page 635, and Alexander Klein, 'Atomic Energy, 1872 - 1899: R.I.P.,' in: *The Grand Deception,* Lippincott, 1955, page 74, a highly inaccurate article concerning historical data. Besides the main company there also existed a Keely Company of Mexico, 'to control patents in that country,' and a 'New England Keely motor corporation' which 'controlled all the New England States.' In: 'Keely's Motor Not Dead Yet,' *New York Herald,* January 1, 1899.

82. 'The Keely Motor. Feats Of Which It Is Capable,' *New York Times,* July 3, 1875.

83. ibid.

84. 'The Keely Motor,' *Public Ledger Almanac,* 1900, page 101. Sometimes court proceedings would take place between parties of stockholders, fighting for stock. See: 'That Celebrated Motor,' *Philadelphia Evening Bulletin,* August 5, 1876.

85. Clara Bloomfield-Moore, *Keely and His Discoveries, Aerial Navigation,* Kegan Paul, Trench, Trübner & Co., 1893, page 3.

86. See: 'Keely Motor Co. of Philadelphia. Incorporated under the laws of Pennsylvania.,' G.V. Town & Son, printers, 529 Chestnut Street, 1875. Facsimile reprint by Delta Spectrum Research, no date.

87. 'The Keely Motor,' *Public Ledger Almanac,* 1900, page 101.

Chapter 2. Where the Molecules Dance: The First Decade

1. William Mill Butler, 'Keely and the Keely Motor,' *The Home Magazine,* 1898, page 107.

2. 'The Keely Motor Deception,' *Scientific American,* June 26, 1875. Keely's reply appeared in *Scientific American* and from there as a short notice titled: 'What Is Claimed For The Keely Motor,' *New York Daily Tribune,* July 9, 1875.

3. Charles Collier, 'Collier's Letter to Scientific American,' July 17, 1875. Facsimile reprint by Delta Spectrum Research, no date.

4. 'The Keely Motor. What is Claimed for it.,' *New York Times,* June 11, 1875. Also: 'A New Motor,' *New York Daily Tribune,* June 10, 1875.

5. 'The Keely Motor,' *The Bulletin,* June 29, 1875.

6. 'The Keely Motor. Feats of Which It is Capable.,' *New York Times,* July 8, 1875.

7. 'Possibilities of the Keely Motor,' *The Evening Bulletin,* July 8, 1875.

8. Edward N. Dickerson, 'The Keely Motor Craze,' letter to the *Tribune,* November 30, 1888. Also in: Clara Bloomfield-Moore, *Keely and His Discoveries, Aerial Navigation,* Kegan Paul, Trench, Trübner & Co., 1893, pages 108-109.

9. William Mill Butler, 'Keely and the Keely Motor,' *The Home Magazine,* 1898, page 107.

10. 'Keely, Motor Man Dead,' *Public Ledger and Daily Transcript,* November 19, 1898.

11. *Frank Leslie's Illustrated,* November 3, 1877.

12. ibid.

13. Letter dated February 25, 1878, in the Edison Archives, Madison, New Jersey.

14. See: V.K. Chew, M.A., *Talking Machines 1877 - 1914,* A Science Museum Book, 1967.

15. Gerry Vasilatos, 'Vocal Motors, Sound Mills and Phonomotors,' *Borderlands,* no.4, 1995, pages 19-22.

16. Letter in the Edison Archives, Madison New Jersey. Copy obtained through Dale Pond.

17. See: 'Keely's Secret Was Preserved,' *The Evening Bulletin,* November 21, 1898, and Gaston Burridge, 'The Baffling Keely Free Energy Machines,' *Fate,* vol.10, no.7, 1957, page 47. Alexander Klein, 'Atomic Energy, 1872 - 1899: R.I.P.,' in: 'The Grand Deception,' Lippincott, 1955, page 77.

18. 'Mr. Keely and his Motor Exploded,' *The Engineering and Mining Journal,* March 30, 1878, page 221.

19. ibid. Also: 'Keely and his Motor,' *New York Daily Tribune,* March 29, 1878.

20. ibid.

21. O.M. Babcock, 'Keely Motor on the Defence,' *Public Ledger and Daily Transcript,* April 13, 1878. For the publication of the two professors in question, see *Public Ledger and Daily Transcript,* April 6, 1878.

22. Clara Bloomfield-Moore, *Keely and His Discoveries, Aerial Navigation,* Kegan Paul, Trench, Trübner & Co., 1893, page 130.

23. William Mill Butler, 'Keely and the Keely Motor,' *The Home Magazine,* 1898, page 108.

24. ibid.

25. O.M. Babcock, *The Keely Motor, Financial, Mechanical, Philosophical, Historical, Actual, Prospective,* privately printed, Philadelphia, June, 1881, pages 15-16. Facsimile reprint by Delta Spectrum Research. Also in: *H.P. Blavatsky's Collected Writings,* vol. VIII, 1960, page 385.

26. 'The Keely Motor in China,' *The Practical American,* February, 1880. Although on March 20, twenty 'New York and Boston capitalists' came to Keely's workshop where an exhibition was held 'with closed doors.' In: 'Keeley (sic) Motor Experiments,' *New Haven Journal and Courier,* March 22, 1880.

27. 'Keely confident of success. His Motor to be Ready for Use in about Six Weeks. What He Will do With Five Drops of Water.,' *New York Times*, March 25, 1880.
28. 'Expressing Confidence in Keely's Motor. Stockholders declaring by Resolution, that the Machine is nearly perfected,' *New York Times*, December 9, 1880.
29. 'The Secret Out,' *New York Times*, June 3, 1881.
30. According to the source of note 29. I have not been able to locate this publication. While the source of note 29 does not mention Babcock, he did 'a series of lectures' on Keely and his discovery. In: Clara Bloomfield-Moore, 'Aerial Navigation,' *The Arena*, 1894, page 388, reprinted by Delta Spectrum Research, no date. The lecture in New York was probably done by Babcock, who started to lecture on Keely that year. Some fragments of Babcock's lectures have survived; Bloomfield-Moore published several parts of his lectures in her 1894 article 'Aerial Navigation' in *The Arena*.
31. 'The Secret Out,' *New York Times*, June3, 1881.
32. Clara Bloomfield-Moore, *Keely and His Discoveries, Aerial Navigation*, Kegan Paul, Trench, Trübner & Co., 1893, page 33.
33. ibid. page 2, page 11, page 10.
34. 'The Secret Revealed,' New York Times, April 25, 1881. 'The Keely Motor Deception,' *Scientific American*, May 14, 1881. See also: William Mill Butler, 'Keely and the Keely Motor,' *The Home Magazine*, 1898, page 108.
35. ibid.
36. ibid.
37. ibid.
38. ibid. See also: William Mill Butler, 'Keely and the Keely Motor,' *The Home Magazine*, 1898, page 108.
39. Clara Bloomfield-Moore, *Keely and His Discoveries, Aerial Navigation*, Kegan Paul, Trench, Trübner & Co., 1893, page 241. See also: 'Keely, Motor Man, Dead,' *Public Ledger and Daily Transcript*, November 19, 1898, William Mill Butler, 'Keely and the Keely Motor,' *The Home Magazine*, 1898, page 109. On October 18, what apparently was 'the first public exhibition of his vibratory engine' was held in the company of several 'prominent members' of the Keely Motor Company and the press. In: 'Public Test of the Keely Motor,' *New York Daily Tribune*, October 19, 1881, 'Keely Shows His Motor,' *New York Times*, October 19, 1881.
40. John H. Lorimer, 'Keely Motor Company. Minority Report to the Stockholders from the Board of Directors. Dec. 8, 1880, to Dec. 14, 1881.' Grant, Faires & Rogers, 1881, reprinted by Delta Spectrum Research, no date.
41. Clara Bloomfield-Moore, 'Aerial Navigation,' *The Arena*, 1894, page 388, reprinted by Delta Spectrum Research, no date.
42. William Mill Butler, 'Keely and the Keely Motor,' *The Home Magazine*, 1898, page 109.
43. Clara Bloomfield-Moore, 'Aerial Navigation,' *The Arena*, 1894, pages 388, 391.
44. 'The Keely Motor,' *Public Ledger Almanac*, 1900, page 101. See also: 'Noted Woman Dead. Mrs. Bloomfield-Moore Passes Away In London,' *Public Ledger and Daily Transcript*, January 6, 1899. There was some uncertainty in the contemporary newspapers concerning the year in which she came to Keely's aide; *The Times* wrote that 'In the late eighties, 1887 or 1888, Mrs. Moore became interested in the Keely inventions...' In: 'Keely of Motor Fame is Dead,' *The Times*, November 19, 1898. About Clara Bloomfield-Moore, see: Dale Pond, *Universal Laws Never Before Revealed: Keely's Secrets*, The Message Company, 1995, pages 228-230.
45. Clara Bloomfield-Moore, 'Aerial Navigation,' *The Arena*, 1894, page 388, reprinted by Delta Spectrum Research, no date.
46. ibid.

47. Clara Bloomfield-Moore, *Keely and His Discoveries, Aerial Navigation,* Kegan Paul, Trench, Trübner & Co., 1893, pages 241-242. See also: 'Men and Things,' *The Evening Bulletin,* January 6, 1899.

48. Concerning Keely, she published, apart from the sources already referred to, 'Keely's Progress,' Theosophical Publishing Society, Vol.V, no.1 of 'Theosophical Siftings,' 1892, and of course her 1893 apologia, in which 'Keely's Secrets' was incorporated. She also submitted articles about Keely to magazines as *New York Home Journal, New Science Review, Scientific Arena* and *Lippincott's Magazine.*

49. 'The Keely Motor,' *Public Ledger Almanac,* 1900, page 101.

50. Clara Bloomfield-Moore, *Keely and His Discoveries, Aerial Navigation,* Kegan Paul, Trench, Trübner & Co., 1893, page 11.

51. ibid. page 2.

52. 'Keely, Motor Man, Dead,' *Public Ledger and Daily Transcript,* November 19, 1898.

53. Clara Bloomfield-Moore, *Keely's Secrets,* T.P.S., 1888, page 19. Also in Clara Bloomfield-Moore, *Keely and His Discoveries, Aerial Navigation,* Kegan Paul, Trench, Trübner & Co., 1893, page 88.

54. Keely Confident of Success. His Motor to be Ready for Use in about Six Weeks. What He will do with Five Drops of Water.' *New York Times,* March 25, 1880.

55. Clara Bloomfield-Moore, *Keely and His Discoveries, Aerial Navigation,* Kegan Paul, Trench, Trübner & Co., 1893, page 78.

56. Gaston Burridge, 'The Baffling Keely Free Energy Machines,' *Fate,* vol.10, no.7, 1957, page 49.

57. Charles Fort, *Wild Talents,* Claude Kendall, 1932, Pages 339-340.

58. 'A Panick Stricken Company,' *The Railway Age,* January 19, 1882.

59. 'Keely Wishes to keep His Secret,' *New York Times,* January 21, 1882.

60. 'Seeking Keely's Secret,' *New York Times,* March 28, 1882.

61. 'Keely to Divulge his Secret,' *New York Times,* April 2, 1882.

62. 'Keely's Alleged Motor,' *New York Times,* May 25, 1882.

63. 'A Trustful Stockholder,' *New York Times,* June 11, 1882. This was already proposed in the end of 1881. See: 'Keely's Secret Demanded,' *New York Times,* December 15, 1881.

64. 'One Man to Know Keely's Secret,' *New York Times,* June 8, 1882.

65. Wm. Boekel, 'Communication of Wm. Boekel,' dated June 25, 1875, in: Collier's Letter to *Scientific American,* July 17, 1875, page 12. Reprinted by Delta Spectrum Research, no date.

66. 'Mr. Boekel's Report,' *New York Times,* August 12, 1882. About these mysterious explosions, see Bloomfield-Moore, *Keely and His Discoveries, Aerial Navigation,* Kegan Paul, Trench, Trübner & Co., 1893, page 11.

67. 'The Keely Motor. An Expert Reports That Keely Has Discovered All That He Has Claimed.,' *Engineering News & American Contract Journal,* December 16, 1882. Also 'The Keely Motor Revived,' *New York Times,* December 14, 1882. Apparently Keely was still in communication with Boekel in April, 1883: 'Mr. Keely is now engaged in telling the secret of his motor to William Boekel, said Mr. Frank G. Green of the Keely Motor Company.' In: 'Keely's Motor,' *New York Herald,* April 8, 1883.

68. 'Will Keely Succeed. The Much Harrassed Inventor's Achievments.,' *The News,* July 2, 1883.

69. 'The Keely Motor Completed,' *New York Times,* August 29, 1883. 'The Keely Motor Completed,' *Engineering News & American Contract Journal,* September 15, 1883.

70. 'Mr. Keely's Performances. Another Postponement of the Motor Test Announced,' *New York Times,* August 30, 1883.

71. 'Disgusted Keely Motor Men,' *New York Times,* October 30, 1883.

72. 'Keely Explains Again,' *New York Times,* October 31, 1883. 'The Motor Man Granted Two Months,' *New York Times,* December 13, 1883. I have not been able to uncover more specifics about the death of Keely's first wife, her burial, or his marriage to his second wife.

73. 'The Keely Motor,' *New York Times,* March 17, 1884, 'Keely Nearing the End,' *Scientific American,* March 29, 1884, page 196.

74. 'Keely Not Yet Ready,' *New York Times,* March 26, 1884, 'The Keely Motor Stuck Again,' *Scientific American,* April 5, 1884, page 213.

75. Clara Bloomfield-Moore, *Keely and His Discoveries, Aerial Navigation,* Kegan Paul, Trench, Trübner & Co., 1893, page 12.

76. Fort Lafayette mentioned in 'Lieut. Zalinski and Mr. Keely,' *New York Times,* November 18, 1884. Also slight reference in: Megargee, 'Seen and Heard in Many Places,' *The Times,* March 21, 1898. *American Machinist,* vol. 21, no. 47, November 24, 1898, wrongly dates the experiments at Fort Lafayette in 1887.

77. 'Keely's Vaporic Force. Experiments With a Mysterious Gun at Sandy Hook,' *New York Times,* September 21, 1884. About the gun, see also: Clara Bloomfield-Moore, *Keely and His Discoveries, Aerial Navigation,* Kegan Paul, Trench, Trübner & Co., 1893, page 77.

78. 'The Keely Motor Deception,' *Scientific American,* October 11, 1884, page 230. See also 'The Keely Motor Fraud,' *Scientific American,* January 28, 1899. Interestingly, while the 1884 issue dated the Sandy Hook experiment in 1884, the issue of the *Scientific American,* December 3, 1889, mistakenly dated the experiment in 1888.

79. 'Keely's Vaporic Force. Experiments With a Mysterious Gun at Sandy Hook,' *New York Times,* September 21, 1884. See also: William Mill Butler, 'Keely and the Keely Motor,' *The Home Magazine,* 1898, page 109.

80. A. Wilford Hall, 'A Visit to Mr. Keely,' *Scientific Arena,* July, 1886.

81. 'Keely's Vaporic Force. Experiments With a Mysterious Gun at Sandy Hook,' *New York Times,* September 21, 1884.

82. 'The Keely Motor Deception,' *Scientific American,* October 11, 1884, page 230.

83. ibid.

84. Clara Bloomfield-Moore, *Keely and His Discoveries, Aerial Navigation,* Kegan Paul, trench, Trübner & Co., 1893, page 32.

85. 'The Keely Motor Deception,' *Scientific American,* October 11, 1884, page 230.

86. 'Keely Was Cornered,' *The Evening Bulletin,* January 28, 1899.

87. A. Wilford Hall, 'A Visit to Mr. Keely,' *Scientific Arena,* July, 1886.

88. 'Keely's Etheric Vapor,' *New York Times,* September 21, 1884.

89. 'The Keely Motor Deception,' *Scientific American,* October 11, 1884, page 230.

90. ibid. The view of *Scientific American* was shared by a G.W. Browne who compared the vaporic gun with 'the pneumatic pop gun of childhood.' In: 'Artillery Branch of Keely's Service,' *New York Herald,* November 1, 1891. Also in: 'The Wizards Latest Highest Mystery Is Explored,' *New York Herald,* November 8, 1891.

91. 'The new Dynamite Gun,' *Scientific American,* April 5, 1884, page 214.

92. 'Lieut. Zalinski and Mr. Keely,' September 24, 1884, *New York Times,* November 18, 1884. 'The Times Retort,' *New York Times,* November 18, 1884.

93. 'Keely Bearded in his Den,' *New York Times,* November 16, 1884. Also present was Colonel Hamilton who had initiated the experiment at Sandy Hook. See: 'No Other Has Ever Been So Shrouded In The Mist Of Publicity,' *New York Herald,* November 27, 1898.

94. 'Keely's Secret Hidden,' *New York Times,* November 20, 1898.

95. 'Zalinski Denounces Keely. An Expert Severely Criticizes The Famous Motor. Offers Which Were Refused,' *The Times,* November 29, 1888.

96. A. Wilford Hall, 'A Visit to Mr. Keely,' *Scientific Arena,* July, 1886.

97. 'Exposing Keely's Secret. A Connecticut Machinist Who Says the Inventor is a Fraud,' *New York Times,* December 29, 1884.

98. 'Where Keely Got His Idea,' *New York Times,* December 30, 1884.

99. 'Thinks Keely Was Huss?,' *The Times,* January 19, 1889. 'Was Keely Ever Huss?,' *The Times,* January 19, 1889.

100. 'It Was Not Keely,' *The Times,* January 22, 1889.

101. Megargee, 'Seen and Heard in Many Places, *The Times,* 'March 11, 1898. Or see: 'Keely, The Motor Man,' *Chicago Tribune,* April 22, 1888: 'He (Keely) has lived comfortably enough all these years, with a handsome horse most of the time to draw him about town and big diamonds to decorate his ample chest. For a little time, when luck was against him, his diamonds disappeared and he was content to walk, but with a return of better times the horse and jewels came back. His fondness for diamonds is a striking trait. He used to wear three big yellow ones in his shirt bossom, but nowadays he is content with one or two, smaller in size, but of finer quality.' See also: 'Seen and Heard in Many Places,' *The Times,* November 21, 1898. Of this enlargement of Keely's knuckles, which also led to the presumption that Keely was a cannon-ball tosser in a circus (same article), see also: 'There Was No Meeting,' *The Times,* November 27, 1898, and A. Wilford Hall, 'A Visit To Mr. Keely,' *Scientific Arena,* July, 1886. Allegedly Keely once said that the enlarged finger-joints 'were caused by varnish.' In: Letter by H.R. Borle, Megargee, 'Seen and Heard in Many Places,' *The Times,* November 26, 1898. About the diamonds, see also: 'The Keely Motor,' 'International Cyclopedia,' vol.VIII, 1899, page 458: 'His weakness was diamonds of which he had a handful, and always sported a large cluster scarf pin...'

102. 'Men and Things,' *The Evening Bulletin,* November 21, 1898.

103. Megargee, 'Seen and Heard in Many Places,' *The Times,* November 22, 1898.

104. Letter by George Canby, in Megargee, 'Seen and Heard in Many Places,' November 26, 1898.

105. Letter by George Mays, dated November 21, Megargee, 'Seen and Heard in Many Places,' *The Times,* November 23, 1898.

106. 'The Keely Motor,' *Public Ledger Almanac,* 1900, page 101.

107. 'Two Hours With Keely,' *Public Ledger and Daily Transcript,* November 11, 1895.

108. 'Keely's Power,' *Public Ledger and Daily Transcript,* January 26, 1899.

Chapter 3. Prophet of the New Force: The Third Decade

1. Letter dated June 1st, in: Clara Bloomfield-Moore, *Keely and His Discoveries, Aerial Navigation,* Kegan Paul, Trench, Trübner & Co., 1893, page 34-35. Also: 'Keely Still Promising Wonders,' *New York Times,* March 27,1885. Elsewhere another reason for Keely's refusal to give more exhibitions was proposed: 'It is stated that a Connecticut man has discovered the alleged secret of the Keely Motor. The story goes that he obtained work in Keely's shops, and, as a spy, discovered that the alleged power was...compressed air. This of course, the inventor denies. Keely is close and suspicious about everything pertaining to his motor. He guards his shop carefully.' In: 'A Visit To The Keely Motor,' *New York Daily Tribune,* February 16, 1885. The Connecticut man was Baker. See chapter 2 and chapter 2, note 97.

2. 'Keely's Red Letter Day,' *New York Times,* June 7, 1885. Also: 'A Motor In B Flat,' *New York Daily Tribune,* June 7, 1885, 'The Keely Motor Humbug,' *New York Times,* June 8, 1885.

3. Letter dated July 15. In: Clara Bloomfield-Moore, *Keely and His Discoveries, Aerial Navigation,* Kegan Paul, Trench, Trübner & Co., 1893, page 35.

4. Letter dated August 5. ibid. page 36.

5. ibid. page 87.

6. ibid. pages 10-11.

7. Letter dated December 17, 1885. ibid. page 39.

8. ibid. page 39.

9. The newspaper in question was the *New York Home Journal,* August 5, 1885. In: Clara Bloomfield-Moore, *Keely and His Discoveries, Aerial Navigation,* Kegan Paul, Trench, Trübner & Co., 1893, page 12.

10. ibid. page 106.

11. ibid. pages 39-40. Bloomfield-Moore writes that Ricarde-Seaver had transported 'the first box of stored electricity' from Paris to Lord Kelvin in Edinburgh.

12. ibid. page 40.
13. ibid. page 87.
14. Letter dated August 17, 1885. ibid. page 39
15. A. Wilford Hall, 'A Visit To Mr. Keely,' *Scientific Arena*, July, 1886. Reprinted in *Sympathetic Vibratory Physics*, vol.III, no.4, January 1988, pages 4-12.
16. ibid.
17. 'Motor Keely Gets Angry,' *New York Times*, May 22, 1886.
18. A. Wilford Hall, 'A Visit To Mr. Keely,' *Scientific Arena*, July, 1886. Reprinted in *Sympathetic Vibratory Physics*, vol.III, no.4, January 1988, pages 4-12.
19. ibid.
20. 'Motor Keely Gets Angry,' *New York Times*, May 22, 1886.
21. A. Wilford Hall, 'A Visit To Mr. Keely,' *Scientific Arena*, July, 1886. Reprinted in *Sympathetic Vibratory Physics*, vol. III, no.4, January 1988, pages 4-12.
22. ibid. See also: Clara Bloomfield-Moore, *Keely and His Discoveries, Aerial Navigation*, Kegan Paul, Trench, Trübner & Co., 1893, page 87.
23. 'Motor Keely Gets Angry,' *New York Times*, May 22, 1886.
24. A. Wilford Hall, 'A Visit To Mr. Keely,' *Scientific Arena*, July, 1886. Reprinted in *Sympathetic Vibratory Physics*, vol.III, issue 4, January 1988, pages 4-12.
25. 'Motor Keely Gets Angry,' *New York Times*, May 22, 1886. 'Mr. Keely Exhibits His Motor,' *New York Daily Tribune*, May 23, 1886. For an account of a demonstration, made in July that year, see: 'Keely's Spinning Motor,' *New York Times*, July 25, 1886.
26. Henry B. Hudson, 'Mr. Keely's Researches - Sound To be Shown A Substantial Role,' *Scientific Arena*, December 6, 1886. Reprinted in: *Sympathetic Vibratory Physics*, vol.III, issue 5, February 1988, pages 2-5.
27. ibid.
28. ibid.
29. ibid.
30. ibid.
31. ibid.
32. ibid.
33. ibid.
34. Clara Bloomfield-Moore, *Keely and His Discoveries, Aerial Navigation*, Kegan Paul, Trench, Trübner & Co., 1893, page 40.
35. William Mill Butler, 'Keely and the Keely Motor,' *The Home Magazine*, 1898, page 110.
36. Henry B. Hudson, 'The Keely Motor Illustrated,' *Scientific Arena*, January, 1887. Reprinted in *Sympathetic Vibratory Physics*, Vol.III issue 6, March 1988, pages 2-6.
37. 'Keely And His Motor,' *The Evening Bulletin*, August 22, 1887.
38. Clara Bloomfield-Moore, *Keely and His Discoveries, Aerial Navigation*, Kegan Paul, Trench, Trübner & Co., 1893, page 41.
39. William Mill Butler, 'Keely and the Keely Motor,' *The Home Magazine*, 1898, page 110.
40. 'Keely's Aerial Navigation,' unspecified clipping, December 12, 1887, Sympathetic Vibratory Physics Homepage, Internet.
41. 'Keely's Change of Base,' *New York Times*, December 14, 1887. Also, 'Gaseous Keely,' *San Francisco Daily Examiner*, May 14, 1888.
42. Clara Bloomfield-Moore, *Keely and His Discoveries, Aerial Navigation*, Kegan Paul, Trench, Trübner & Co., 1893, page 75. Her 1888 pamphlet 'Keely's Secrets' was translated into French and was serialized as 'Les Secrets de Keely' in the esoteric magazine *L'Aurore* subtitled 'organe du Christianisme esoterique,' in No.9, September 1888, pages 478 - 488, no.10, October 1888, pages 530-537, and no.11, November 1888, pages 593-596. A certain 'Lee' published his or her article on Keely entitled 'Quelques experiences de John Worrell Keely' in *L'Aurore*, tome IV, 1889, pages 199-206. Bloomfield-Moore would publish another article on Keely entitled 'Le progress de Keely, in no.4, April 1892, pages 151-155. In

L'Aurore, no.2, February 1892, Colville would also publish an article entitled 'Le mystre du Corps du Christ,' pages 55-61.

43. *Le Lotus,* September 1888. A portrait of Keely accompanied the article. Blavatsky's written statements concerning Keely were printed in an article entitled: 'Occult Light On Keely,' *New York Daily Tribune,* January 15, 1888, some months before they were published in *The Secret Doctrine.*

44. Clara Bloomfield-Moore, *Keely and His Discoveries, Aerial Navigation,* Kegan Paul, Trench, Trübner & Co., 1893, page 102.

45. ibid. pages 100-102.

46. See chapter 1, and chapter 1, notes 41 and 45. 'The Keely Motor,' *Philadelphia Evening Bulletin,* March 13, 1888. 'Keely and his Motor,' *New York Times,* November 4, 1888.

47. 'Keely Motor Suit Ended,' *Philadelphia Press,* February 26, 1890.

48. 'Everybody For Keely,' *New York Times,* September 8, 1888.

49. 'Motor Men Disgusted,' *New York Times,* September 11, 1888.

50. 'To See Keely's Machine,' *New York Times,* September 26, 1888. Also: 'Keely Not Yet in Jail,' *New York Times,* September 19, 1888.

51. See chapter 2 and chapter 2, note 21. Also: 'The Keely Motor Experts,' *New York Times,* April 8, 1888.

52. See chapter 2 and chapter 2, note 19.

53. 'Keely Motor. What Dr. Cresson said to the Reporter of *The Philadelphia Press,* November 12th, 1888,' undated pamphlet, George Canby, 'Keely Motor Scraps,' vol.1, not published.

54. 'Going For Keely,' *The Times,* November 12, 1888.

55. 'Inventor Keely In Jail,' *New York Times,* November 17, 1888. Also: 'Mr. Keely May Go To Jail,' *Williamsport Sun Gazette,* November 12, 1888, 'Motor Keely In Contempt,' *Hartford Courant,* November 12, 1888, 'Keely committed for contempt,' *New York Daily Tribune,* November 18, 1888, 'Mr. Keely's Contempt,' *New York Times,* November 16, 1888, 'Inventor Keely In Prison,' *Hartfort Courant,* November 19, 1888.

56. 'Mr. Keely's Sunday In Jail,' 'The Press, November 19, 1888. See also: Clara Bloomfield-Moore, *Keely and His Discoveries, Aerial Navigation,* Kegan Paul, Trench Trübner & Co., 1893, pages 100-101.

57. ibid.

58. 'Keely's Sunday In Jail,' *New York Times,* November 19, 1888.

59. 'Mr. Keely's Sunday In Jail,' *The Press,* November 19, 1888.

60. 'Keely's Sunday In Jail,' *New York Times,* November 19, 1888. There is some confusion as to where the Supreme Court resided; while the *New York Times* wrote Harrisburg, 'The Press' wrote Pittsburgh.

61. 'Mr. Keely's Sunday In Jail,' *The Press,* November 19, 1888.

62. 'Keely Motor Suit Ended,' *Philadelphia Press,* February 26, 1890. Keely's release was also noted in *Scientific American*: 'The Philadelphia court which thought it could keep Mr. Keely in confinement has seen its error. As the *Tribune* already remarked, Mr. Keely is out of jail and has returned to his motor.' In: 'Mr. Keely's Motor,' *Scientific American,* December 22, 1888, page 388.

63. 'Keely's Discharge,' *The Times,* January 29, 1889. The date February 1 was also given. In: 'Keely Motor Suit Ended,' *Philadelphia Press,* February 26, 1890.

64. 'Keely Not In Contempt,' *New York Times,* January 29, 1889.

65. ibid.

66. ibid. See also: Clara Bloomfield-Moore, *Keely and His Discoveries, Aerial Navigation,* Kegan Paul, trench, Trübner & Co., 1893, pages 111-112.

67. 'The Keely Decision,' *The Times,* January 29, 1890.

68. 'Keely Not In Contempt,' *New York Times,* January 29, 1889.

69. 'Keely Motor Suit Ended,' *Philadelphia Press,* February 26, 1890.

70. 'Fair Play This Time,' *The Times,* March 29, 1889.

71. Clara Bloomfield-Moore, *Keely and His Discoveries, Aerial Navigation*, Kegan Paul, Trench, Trübner & Co., 1893, page 122.

72. ibid. page 162. On his provisional engine and its graduation, see also: 'Keeley (sic) Still Promising,' *New York Times*, December 18, 1889.

73. ibid. page 114.

74. 'Keely's New Force. A Motor That Surprises Scientists,' *San Francisco Chronicle*, April 7, 1890.

75. ibid. See also: Clara Bloomfield-Moore, *Keely and His Discoveries, Aerial Navigation*, Kegan Paul, Trench, Trübner & Co., 1893, page 114. Of course, Leidy by now was a supporter of Keely's claims. See: 'The Keely Motor Again,' *Invention*, October 19, 1889. See also: Clara Bloomfield-Moore, *Keely and His Discoveries, Aerial Navigation*, Kegan Paul, Trench, Trübner & Co., 1893, pages 117-119, and, 'Professor Leidy's Adherence to the New Force,' *Evening Telegraph*, April 13, 1890. Apparently Leidy withdrew his statement, although there is some confusion as to that; Bloomfield-Moore claimed that 'on the contrary he maintained it until his death.' In: 'Keely Motor Once More,' *Public Ledger and Daily Transcript*, November 6, 1895.

76. 'Keely's Secret,' *The World*, May 11, 1890. Also: 'Keely's Secret,' *Los Angeles Sunday Times*, June 15, 1890.

77. ibid.

78. ibid.

79. ibid.

80. 'A Modern Wizard: The Keely Motor and its Inventor,' in: Cheiro, 'Mysteries and Romances of the World's Greatest Occultists,' Herbert Jenkins, 1935, 237-251. Short reference in Marc J. Seiffer, Wizard *The Life and Times of Nikola Tesla*, Birch Lane Press, 1996, page 64.

81. ibid. page 241. According to Cheiro, this happened in the presence of Ricarde-Seaver, who was hearing about the motor for the first time. Together they went to visit Keely in Philadelphia. But in fact Ricarde-Seaver had already done so in 1884. See this chapter, and this chapter, notes 10 and 11.

82. ibid. page 248.

83. Clara Bloomfield-Moore, *Keely and His Discoveries, Aerial Navigation*, Kegan Paul, Trench, Trübner & Co., 1893, pages 114-121. The *Philadelphia Inquirer* of March 30, 1890 copied this article headed: 'The Keely Motor: some observations on the invention from a foreign publication.'

84. Clara Bloomfield-Moore, *Keely and His Discoveries, Aerial Navigation*, Kegan Paul, Trench, Trübner & Co., 1893, page 128. The pamphlet in question was 'The Keely Motor Bubble,' which contained Lorrimer's 'Minority Report to the Stockholders,' published in 1881.

85. Bloomfield-Moore mentions the newspapers *New York Home Journal, Truth, Detroit Tribune, Chicago Herald, Toledo Blade, Atlanta Constitution, The Statesman*, and the Austrian *Vienna News*. I have discovered an article written by one Afra and entitled, 'John Worrell Keely' in the Dutch theosophical magazine *Theosophia*, 1893, no.13. More on this article in chapter 8 and chapter 8, note 63. In 1892 the book *Vera Vita, The Philosophy of Sympathy. Discovery of a New Element and its Connection with Real Life, Practically Demonstrated in Keely's Experiments*, written by David Sinclair also appeared in London.

86. 'Saw The Keely Motor,' *The Times*, January 31, 1893.

87. ibid.

88. ibid.

89. 'Keely's Great Mystery,' *San Fransisco Examiner*, October 22, 1893.

90. 'Keely Motor Progress,' *Public Ledger and Daily Transcript*, December 14, 1893.

91. Letter by Alfred H. Plum to *Philadelphia Enquirer*, January 14, 1894.

92. untitled article, *The Times*, October 27, 1893.

93. Letter by Alfred H. Plum to *Philadelphia Enquirer*, January 14, 1894.

94. ibid.
95. Introduction in William Colville, *Dashed Against The Rock,* Banner Of Light Publishing Company, 1894.
96. The book contains 'The Principles of Matter and Energy,' chapter 4, Keely's 'Forty Laws,' chapter 6, 'The Scale Of Forces,' chapter 8, a short glossary of Keely's terms on pages 57-58 and six drawings of Keely's molecule, inserted between pages 55-56. The book is currently in print by Delta Spectrum Research.
97. On *Dashed Against The Rock,* see chapter 8 on Colville, see chapter 9.
98. About John Jacob Astor, see chapter 4, about *A Journal in Other Worlds,* see chapter 8.
99. 'Two Hours With Keely. Power without Cost - Engines Which Do Not Move,' *Public Ledger and Daily Transcript,* November 11, 1895.
100. Letter by Clara Bloomfield-Moore, dated May 30, 1894. She copied for her correspondent the excerpt of Keely's letter which she received on May 28 with the remark: 'for you alone' and marked 'confidential.'

Chapter 4. The Power Millennium: Keely's Last Years

1. Marc J. Seiffer, *Wizard, the Life and Times of Nikola Tesla,* Birch Lane Press, 1996, page 161.
2. 'Advertisement' dated February 18, 1895, *The New Scientific Review,* undated clipping.
3. Circular dated February 18, 1895, *The New Scientific Review,* undated clipping.
4. ibid.
5. For a full text of the agreement, see: 'Keely's Discoveries. Financial Support Given For Research Only,' *Public Ledger and Daily Transcript,* November 14, 1895.
6. 'Keely Motor Once More,' *Public Ledger and Daily Transcript,* November 6, 1895.
7. ibid.
8. 'Keely-Astor Deal is Off,' *Public Ledger and Daily Transcript,* November 9, 1895. It appears that Astor met with Keely and Bloomfield-Moore in Philadelphia at least once. Reference to Astor's one time visit in: 'Keely Motor To Be Reorganised,' *New York Herald,* April 28, 1896. It also appears that Astor became interested in Keely's inventions due to Bloomfield-Moore. Short reference in: 'Sudden Death of Keely's Friend,' *New York Herald,* January 6, 1899.
9. 'Keely's Motor,' unspecified clipping, August 11, 1895, Sympathetic Vibratory Homepage, Internet. Very likely this 'enthusiastic advocate' was Clara Bloomfield-Moore. 'Keely Motor To Be Reorganised,' *New York Herald,* April 28, 1896. Although Astor's purchase of stock is contradicted: 'Neither the Astor syndicate, of which John Jacob Astor, Dr. Seward Webb and Mr. William C. Brewster were members, nor the Vanderbilts - and the fact that the Vanderbilts examined the Keely motor has been hitherto unknown - made purchase of stock.'
10. Marc J. Seiffer, *Wizard, the Life and Times of Nikola Tesla,* Birch Lane Press, 1996, page 162.
11. 'Keely-Astor Deal is Off,' *Public Ledger and Daily Transcript,* November 9, 1895. There seem to have been more reasons under the surface of the press coverage; a glimpse is perhaps offered in Bloomfield-Moore's statement that, 'When, on Nov. 5, the unauthorized and untrue statements of Col. Astor's transactions were made public I said that plans formed to have the scientific value of Mr. Keely's discoveries acknowledged publicly before commercial success is attained had been refuted for the third time within five years. I then despaired of accomplishing these aims, and Nov. 7, all negotiations with the financiers were abruptly brought to a close.' In: 'Will Retain Her Privileges,' *Public Ledger and Daily Transcript,* November 9, 1895, 'Mrs Moore On The Keely Motor,' *New York Times,* December 9, 1895. Some weeks later an electrician, J.E. Wright, who was invited to test Keely's claims, said that Bloomfield-Moore 'spoke of Mr. Astor's visit to the laboratory rather bitterly. His interest in the first place, followed by a sudden failure, half puzzled and displeased her. ...The capitalists, she said, did not care for anything except the money which was in the enterprise.

Her desire now is to stop speculation, if possible, (and) push the discovery to a final and useful conclusion for the sake of humanity.' In: 'An Expert On Keely's Motor,' *New York Herald*, January 19, 1896. Some months later, an 'Official denial was given to the report that Astor or Vanderbilt money will be invested in the (Keely Motor) company.' In: 'Affairs of the Keely Company,' *New York Times*, April 29, 1896.

12. Marc J. Seiffer, *Wizard, the Life and Times of Nikola Tesla*, Birch Lane Press, 1996, page 146.

13. Possibly Theodore Puskas was the same as Tivadar Puskas, whose brother Ferenc was a friend of Tesla. Both ran the American Telephone Exchange. Short references in Marc J. Seiffer, Wizard, the Life and Times of Nikola Tesla, Birch Lane Press, pages 20, 27.

14. Letter from Nikola Tesla to Bloomfield-Moore, dated December 11, 1894. Copy obtained through Dale Pond.

15. Letter from Nikola Tesla to Clara Bloomfield-Moore, dated April 18, 1895. Copy obtained through Dale Pond.

16. Letter from Nikola Tesla to Clara Bloomfield-Moore, dated April 24, 1895. Copy obtained through Dale Pond. This was generally seen as the reason that Tesla refused to investigate Keely. When Tesla received a telegram from Prof. Dewar of the Royal Institution of Great Britain to represent the institution at Keely's workshop, 'Tesla declined' noted a newspaper, 'His reasons for declining are given in a characteristic letter, addressed to Mrs. Bloomfield-Moore, in which he says: It is my honest conviction that I am far in advance in certain lines of scientific investigation, which I consider of greater importance for mankind. The problems are difficult, even now, after I have investigated them for many years, and my powers are taxed to the utmost, so that I cannot spare the even the smallest effort.' In: 'An Expert On Keely's Motor,' *New York Herald*, January 19, 1896.

17. Letter from Nikola Tesla to Clara Bloomfield-Moore, dated May 14, 1895. Copy obtained through Dale Pond.

18. Marc J. Seiffer, *Wizard, the Life and Times of Nikola Tesla*, Birch Lane Press, 1996, pages 101-102.

19. ibid. page 149.

20. Letter from Nikola Tesla to Bloomfield-Moore, dated October 19, 1895. Copy obtained through Dale Pond.

21. A fact that is missed. See: Marc J. Seiffer, 'Nikola Tesla & John Hays Hammond Jr., A History of Remote Control Robotics,' in: Steven R. Elswick, editor, 'Proceedings of the 1990 International Tesla Symposium,' International Tesla Society, 1991, pages 1-32, 1-33, Marc J. Seiffer, *Wizard, the Life and Times of Nikola Tesla*, Birch Lane Press, 1996, pages 63-65. While adorning Tesla with unanimous praise, which of course he rightfully deserves since he was a staggering genius, at the same time the need is felt to refer to Keely as a 'mountebank inventor,' thus missing the real cause; John Jacob Astor.

22. Marc J. Seiffer, *Wizard, the Life and Times of Nikola Tesla*, Birch Lane Press, 1996, page 152.

23. Derek Wilson, *The Astors*, Weidenfeld and Nicholson, 1993, page 200.

24. Marc J. Seiffer, *Wizard, the Life and Times of Nikola Tesla*, Birch Lane Press, 1996, page 152.

25. ibid. page 162.

26. Letter from Clara Bloomfield-Moore, dated June 28, 1893. The mother of Bloomfield-Moore's grandfather was a daughter of General Collins, whose sister married Oliver Wolcott, one of the signers of the Declaration of Independence. In: Baron Harold De Bildt, 'Ancestry of Clarence Bloomfield Moore of Philadelphia, by the late Clara Jessup Moore,' *National Genealogical Society Quarterly*, March 1940, page 4.

27. Letter from Clara Bloomfield-Moore, Atlantic City, dated September 14, 1895. Copy obtained through Dale Pond.

28. Letter from Clara Bloomfield-Moore, probably written circa August or September, 1895, Sympathetic Vibratory Physics Homepage, Internet.
29. Marc J. Seiffer, *Wizard, the Life and Times of Nikola Tesla,* Birch Lane Press, 1996, pages 162-190.
30. ibid. pages 178-179.
31. ibid. pages 209-210.
32. ibid. pages 210-211.
33. ibid. page 244, also: Marc J. Seiffer, 'Nikola Tesla & John Hays Hammond Jr., A History of Remote Control Robotics,' writes: 'Astor was dismayed...' In: Steven R. Elswick, editor, 'Proceedings of the 1990 International Tesla Symposium,' International Tesla Society, 1991, page 1-34.
34. ibid. page 361.
35. 'Two Hours With Keely. Power Without Cost - Engines Which Do Not Move,' *Public Ledger and Daily Transcript,* November 11, 1895.
36. 'Keely Engine Not To Be Exhibited,' *Public Ledger and Daily Transcript,* November 16, 1895.
37. 'Keely Motor Company. An Enterprise Practically Without Assets,' *Public Ledger and Daily Transcript,* December 10, 1895.
38. ibid.
39. 'Will Retain Her Privileges,' *Public Ledger and Daily Transcript,* November 9, 1895. Her nom-de-plume, when conducting business with Keely, such as the signing of contracts or the issuing of public statements, was H.O. Ward. Also: 'Mrs. Moore on the Keely Motor,' *New York Times,* December 9, 1895.
40. 'Keely To Explain. He Agreed To Teach Astor's Expert How To Sensitize Metal,' *Public Ledger and Daily Transcript,* December 14, 1895.
41. 'Clarence B. Moore's Answer,' *The Evening Star,* December 14, 1895.
42. 'The Keely Motor,' *Evening Bulletin,* December 17, 1895.
43. 'Keely Motor Company,' *Public Ledger and Daily Transcript,* December 12, 1895.
44. 'The Keely Motor,' *Public Ledger and Daily Transcript,* December 25, 1895.
45. 'The Keely Motor. Patents On The Vibratory Machinery To Be Applied For,' *Public Ledger and Daily Transcript,* March 3, 1896. Reference to runaway horse in: William Mill Butler, 'Keely and the Keely Motor,' *The Home Magazine,* 1898, page 114.
46. ibid.
47. Letter by John Keely dated March 31, 1896. In: 'Keely Wins. Spirited Meeting Of Stockholders In The Mysterious Motor,' *Public Ledger and Daily Transcript,* April 3, 1896.
48. 'The Keely Motor. Patents On The Vibratory Machinery To Be Applied For,' *Public Ledger and Daily Transcript,* March 3, 1896.
49. 'Keely Wins. Spirited Meeting Of Stockholders In The Mysterious Motor,' *Public Ledger and Daily Transcript,* April 3, 1896.
50. ibid.
51. ibid.
52. 'Keely Motor May Soon Mote,' *The Times,* April 3, 1896.
53. ibid.
54. 'The Keely Motor Again,' *Public Ledger and Daily Transcript,* April 17, 1896.
55. 'New Life In Keely Patents,' *The Evening Telegraph,* April 26, 1896. Also: 'Keely Motor To Be Reorganized,' *New York Herald,* April 28, 1896, 'Affairs of the Keely Motor Company,' *New York Times,* April 29, 1896.
56. Gaston Burridge, 'The Baffling Keely Free Energy Machines,' *Fate,* vol.10, no.7, 1957, page 43, Carl Sifakis, *Hoaxes and Scams,* Michael O'Mara Books, 1994, page 141.
57. 'Support Withdrawn,' *Public Ledger and Daily Transcript,* May 9, 1896.
58. See chapter 2 and chapter 2, note 44.
59. 'Keely's Vision,' *The Evening Telegraph,* May 9, 1896.

60. Letter by Clara Bloomfield-Moore to her son-in-law, Count C.G. von Rosen, September 25, 1891. Although this letter was printed in a pamphlet, I have not been able to discover its bibliographical details. Currently the letter resides in the Franklin Institute.

61. Letter of Bloomfield-Moore, dated September 9, 1894. Copy obtained through Dale Pond.

62. On Hartmann's occult connections and his visit to Keely, see chapter 9.

63. Francis King, *The Secret Rituals of the O.T.O.*, Samuel Weiser, 1972, page 24. More biographical details on Hartmann found in Nicholas Goodrick-Clarke, *The Occult Roots of Nazism*, The Aquarian Press, 1985, pages 24-27.

64. 'Keely's Vision,' *The Evening Telegraph*, May 9, 1896.

65. Letter by Clara Bloomfield-Moore to her son-in-law, Count C.G. von Rosen, September 25, 1891. In the Franklin Institute.

66. *H.P. Blavatsky's Collected Writings*, vol. XIII, 1982, page 385.

67. 'Keely's Vision,' *The Evening Telegraph*, May 9, 1896.

68. Letter by Clara Bloomfield-Moore, dated September 14, 1895. Copy obtained through Dale Pond.

69. 'Keely's Vision,' *The Evening Telegraph*, May 9, 1896.

70. ibid.

71. 'She Lays Down A Great Burden,' *The Evening Telegraph*, May 11, 1896.

72. 'The Keely Motor,' *The Evening Bulletin*, May 11, 1896.

73. ibid.

74. 'A Keely Motor Meeting,' *Public Ledger and Daily Transcript*, June 5, 1896.

75. 'Keely Motor Meeting,' *Public Ledger and Daily Transcript*, December 10, 1896.

76. *Harper's Weekly*, January 2, 1897.

77. 'Saw Keely Motor Work,' *Public Ledger and Daily Transcript*, May 6, 1897. Also: 'Not Yet Doth The Motor Mote,' *The Evening Bulletin*, May 7, 1897. Also in: 'Keely of Motor Fame is Dead,' *The Times*, November 19, 1898. In this account, Fransioli is called Francioli, and apart from McNally, the article alleges that he was accompanied by Chief Engineer Pierson of the Metropolitan Traction Company, Chief Electrical Engineer Brown of the Western Union Telegraph Company 'and others of equal note.' Also: 'Not To Use The Keely Motor,' *New York Daily Tribune*, May 7, 1897, 'Keely Motor Gets A Move On,' 'Wilkes Barre record,' May 11, 1897, 'New Yorkers Visit Keely,' *New York Herald*, June 20, 1897, 'A Keely Motor Tested,' *New York Times*, June 20, 1897, 'A Sure Success,' *Wilkes Barre Weekly News Dealer*, June 20, 1897. Later, Fransioli 'displayed great interest in Mr. Browne's explanation of the mystery behind Keely's long heralded motor.' In: 'What Is The Force Of Keely's Motor?,' *New York Herald*, August 22, 1897. On G.W. Browne, see chapter 7 and chapter 7 note 118.

78. Letter by Clara Bloomfield-Moore, dated May 16, 1897.

79. 'Keely's Etheric Force,' *The Evening Bulletin*, June 30, 1897. Also: 'A Puzzle For Scientists,' *Wilkes Barre Record*, June 21, 1897.

80. The paper was published in the *Proceedings of the Engineers' Club of Philadelphia*, vol. XIV, no.4, January 1898, and reprinted as a separate pamphlet titled 'The Keely Motor,' by E.A. Scott, active member, read January 8, 1898,' no place, no date. Scott's opinion was preceeded by G.W. Browne, who had 'studied the subject' and published his negative views on the subject in large, illustrated articles. On Browne and his articles, see chapter 7 and chapter 7, note 118.

81. 'Keely Motor Discussed,' *Public Ledger and Daily Transcript*, January 10, 1898.

82. ibid.

83. 'The Keely Motor,' *The Times*, March 26, 1898. Also: 'Personal,' *New York Times*, March 26, 1898.

84. 'Keely Claims At Last To Have Harnessed A New Force,' *The Times*, March 6, 1898.

85. ibid.

86. See: 'An Interesting Crank,' *New York Tribune*, November 11, 1898.

87. 'John W. Keely Ill,' *The Evening Bulletin,* November 18, 1898. Also: 'John W. Keely Dead,' *New York Daily Tribune,* November 19, 1898.

88. In an earlier account the accident happened at Chestnut street. See also note 45.

89. 'Keely, Motor Man, Dead. Succumbed After A Week's Illness Of Pneumonia,' *Public Ledger and Daily Transcript,* November 19, 1898.

90. 'Pall Bearers for Keely's Funeral,' *Public Ledger and Daily Transcript,* November 22, 1898, 'Inventor Keely's Funeral,' *The Times,* November 23, 1898.

91. 'Keely's Funeral,' *The Evening Bulletin,* November 23, 1898, 'Mr. Keely's Funeral,' *Public Ledger and Daily Transcript,* November 24, 1898, 'John W. Keely Laid To Rest,' *The Times,* November 24, 1898.

92. 'Keely Of Motor Fame Is Dead,' *The Evening Bulletin,* November 19, 1898.

93. 'No Word Of Keely's Will,' *The Times,* November 20, 1898.

94. See: 'Mrs. Moore Dead Keely's Patron,' *The Evening Bulletin,* January 5, 1899. Also: 'Mrs. Moore Dies Broken-Hearted,' *The Times,* January 6, 1899, 'Her Anxiety Brought Death,' January 6, 1899. 'Mrs. Bloomfield-Moore Dead,' *New York Times,* January 6, 1899. 'Sudden Death of Keely's Friend,' *New York Herald,* January 6, 1899.

95. 'Noted Woman Dead,' *Public Ledger and Daily Transcript,* January 6, 1899.

Chapter 5. Into the Void: The Final Stage of the Keely Mystery

1. Apart from the quotation at the top of this chapter, in Clara Bloomfield-Moore, *Keely and His Discoveries, Aerial Navigation,* Kegan Paul, Trench, Trübner & Co., 1893, page 74, see also page 29: '...when the writings of John Worrell Keely...are about to be given to the world,' page 39 where she quotes Keely: 'I have been writing out some of my theories as to sound and odour...I see the time approaching when I will be able to write up my system of the true philosophy of nature's grandest force...,' page 80: 'Keely's Theoretical Exposé is in preparation for the press, and, when these volumes are published...,' page 97 where she repeats this statement, and page 151: 'He has determined and written out a system of the vibratory conditions...' See also: Clara Bloomfield-Moore, *Keely's Secrets,* Theosophical Publishing Society, July, 1888, page 13: 'Mr. Keely's 'Theoretical Exposé is nearly ready for the press, and when these volumes are issued...' (she changed this passage in her 1893 book, in which her 1888 pamphlet was incorporated in: 'Keely's Theoretical Exposé is in preparation for the press...,' on page 80, so that at that time Keely's volumes weren't published), and page 36; 'Keely's Secret,' *The World,* May 11, 1890. The reporter also quotes from these proof sheets, and this is one of the rare instances that a confirmation of the existence of a manuscript from a newspaper source exists (see also note 11). There is a possibility however that these were proof sheets of Bloomfield-Moore's 1893 book, or her 1888 pamphlet. Other hints are found in an editorial in the *Philadelphia Enquirer,* January 14, 1894, consisting of a letter written by one Alfred H. Plumb, a scientist from Boston, dated 1894: '...though his writings show a familiarity with scholarly works...,' and in Megargee, 'Seen and Heard in Many Places,' *The Times,* March 21, 1898: '...For instance, listen to this from one of Keely's manuscripts...'

3. William Mill Butler, 'Keely and the Keely Motor,' 1898, page 113.

4. Clara Bloomfield-Moore, *Keely and His Discoveries, Aerial Navigation,* Kegan Paul, Trench, Trübner & Co., 1893, amongst others on pages 32-36, 59, 70, 86, 163, 177-179, 201, 226, 280. Of this, Burridge writes 'It is recorded, Keely wrote extensively about his inventions and theories.' If these writings were ever published, I have found no trace of them. It is quite evident that Mrs. Moore had full access to their manuscript form, for she quotes *Pendulum, a Monthly Digest of Radiesthesia,* vol.5, no.3, December 1954, page 539. Article also published in *Round Robin,* September-October, 1954. Of this, Pond writes: '...a large segment of Bloomfield-Moore's 1893 book is composed of quotes from letters of Keely. She made an attempt to place his name on some of these quotes...' Correspondence with Dale Pond, dated January 9, 1997.

5. 'Confession by Mrs. Keely's Attorney. Statement from Charles S. Hill,' in 'Keely The Monumental Fraud of the Century,' *New York Times,* January 29, 1899.
6. 'Fortune Was Squandered In Keely Motor,' *Philadelphia Press,* January 22, 1899.
7. 'Clara Bloomfield-Moore, *Keely and His Discoveries, Aerial Navigation,* Kegan Paul, Trench, Trübner & Co., 1893, pages 156-157. See also: Clara Bloomfield-Moore, *Keely's Secrets,* Theosophical Publishing Society, 1888, page 36, where she also writes that 'These volumes are to be published by the Lippincott Publishing Company, of Philadelphia, as soon as Keely has completed his mechanical work.' Of these treatises Sykes writes that 'none of these have been traced; it is possible that the originals are lying in some private library.' In: Egerton Sykes, *The Keely Mystery,* Markham House, 2nd revised edition, 1972, page 9. Pond writes that Lippincott denies they ever saw or published these books. See: Dale Pond, 'Keely Bibliography,' Delta Spectrum research, 1995, pages 3-4.
8. C.W. Snell, *The Snell Manuscript,* Delta Spectrum Research, 1995, page 1. A large part of the book consists of Bloomfield-Moore's 1888 pamphlet and her 1893 book. On this, Pond writes 'While the Snell manuscript contains many of these same quotes there are a large number of paragraphs that I have not seen elsewhere. This is why I maintain a certain coincidence with Snell's allegation of the source being a book by Keely. However...there is a great possibility these quotes came from elsewhere.' Correspondence with Dale Pond, dated January 9, 1997.
9. Dale Pond, *Universal Laws Never Before Revealed: Keely's Secrets,* The Message Company, 1995, page 260.
10. See: 'National Union Catalog, pre-1956 imprints,' vol. 61, page 563, 598, and 'British Museum General Catalog of Printed Books to 1955, compact ed.,' vol.13, page 1067.
11. 'Two Hours With Keely,' *Public Ledger and Daily Transcript,* November 11, 1895.
12. Alexander Klein, 'Atomic Energy, 1872-1899: R.I.P.,' 'Grand Deception,' Lippincott, 1955, page 77. Klein also writes that 'an attempt, at the writers request, to play the notes resulted, disappointedly, neither in some haunting melody nor in a jarring bedlam, but rather a weak and vapid disconnectedness...,' which is of course a matter of taste.
13. William Mill Butler, 'Keely and the Keely Motor,' *The Home Magazine,* 1898, page 115. See: Julius Moritzen, 'The Extraordinary Story of John Worrell Keely,' *The Cosmopolitan,* April, 1899, vol.XXVI, no.6, page 638, and 'Keely Monumental Fraud of the Century,' *New York Journal,* January 29, 1899.
14. For illustrations see: Julius Moritzen, 'The Extraordinary Story of John Worrell Keely,' *The Cosmopolitan,* April, 1899, vol.XXVI, no.6, page 638, and 'Keely Monumental Fraud of the Century,' *New York Journal,* January 29, 1899.
15. 'They are diligently searching for Keely's papers. His complete diary is in existence,' *New York Herald,* January 1, 1899. From this article, it appears that Anna Keely turned these over to Kinraide: 'It will take Mr. Kinraide about two weeks to go over his papers and effects in Philadelphia before he can begin working on the motors at his laboratory here...' See also this chapter, for conflicting statements. On Kinraide, see this chapter and chapter 6.
16. ibid.
17. 'Keely's Secret was Preserved. President of the Company Says Inventor's Widow Has The Manuscript,' *The Evening Bulletin,* November 21, 1898.
18. 'The Motor's Future. After the Inventor's Funeral All Doubts Will Be Dispelled.,' *Philadelphia Inquirer,* November 23, 1898.
19. 'Mr. Keely's Funeral. Sympathetic Tributes Delivered By Several Ministers,' *Public Ledger,* November 24, 1898.
20. 'John W. Keely Laid to Rest. Many Friends attended the Dead Inventor's Funeral. Words of Eulogy Spoken,' *The Times,* November 24, 1898.
21. 'May Fight Over The Keely Motor. Wife of Dead Inventor Wants to Keep His Secret. A Protest From the Company,' *The Evening Bulletin,* November 25, 1898.

22. 'Keely's Papers. Uncertain as to What Disposition Will Be Made of Them,' *Philadelphia Inquirer*, November 26, 1898.

23. 'Keely Stockholders Still in the Dark,' *The Times*, November 29, 1898.

24. 'No Keely Developments,' *Public Ledger and Daily Transcript*, November 29, 1898.

25. 'No Keely Motor Meeting Today,' *The Evening Bulletin*, November 29, 1898.

26. 'No Keely Developments,' *Public Ledger and Daily Transcript*, November 29, 1898.

27. 'No Keely Motor Meeting,' *The Times*, November 30, 1898.

28. 'Death of John W. Keely,' *Scientific American*, December 3, 1898.

29. 'Keely's Will Reveals No Secret. No Mention Of His Motor,' *The Evening Bulletin*, December 1, 1898. See Also: 'Keely's Will Probated,' *Public Ledger and Daily Transcript*, December 2, 1898, 'Inventor Keely's Will Filed,' *New York Times*, December 21, 1898. Also: 'Motor Not Mentioned in Keely's Will,' *New York Daily Tribune*, December 2, 1898.

30. 'Hope For The Motor's Future,' *The Evening Telegraph*, December 5, 1898. The article erroneously names Anna Keely's lawyer as 'Charles S. Smith.'

31. 'Keely's Friends Confer,' *The Times*, December 6, 1898. The article erroneously names Anna Keely's lawyer as 'Charles S. Smith.'

32. 'Keely Motor Plans,' *The Evening Bulletin*, December 7, 1898.

33. 'Keely Motor Meeting To-Day. It is Not Expected That a Quorum Will be Present to Transact Business,' *The Times*, December 14, 1898.

34. 'Mrs. Keely And The Motor Co. Annual Meeting To-Day Postponed on Account of Small Attendance,' *The Evening Bulletin*, December 14, 1898. See also: 'Keely Meeting Had No Quorum,' *Evening Telegraph*, December 14, 1898.

35. 'Fighting Over Keely's Secret. Annual Meeting of the Company's Stockholders Ends in a Row. Strong Language Used,' *The Evening Bulletin*, December 20, 1898.

36. 'To Finish The Motor. Turbulent Meeting Of The Keely Motor Company Stockholders,' *Public Ledger and Daily Transcript*, December 21, 1898.

37. 'Keely's Friends Confer,' *The Times*, December 6, 1898.

38. 'Boston Man in Keely's Shoes. J.B. Kinraide Will Try To Make Motor Mote,' *The Times*, December 21, 1898. Also: 'Will Work on Keely's Motor,' *New York Times*, December 25, 1898.

39. 'T.B. Kinraide Will Endeavor to Complete the Machine,' unspecified Boston newspaper clipping, December 24, 1898, Sympathetic Vibratory Physics Homepage, Internet.

40. 'Keely Directors Skeptical,' *The Times*, December 22, 1898.

41. 'Confession by Mrs. Keely's Attorney. Statement from Charles S. Hill,' in 'Keely, The Monumental Fraud of the Century,' *New York Times*, January 29, 1899.

42. 'Noted Woman Dead. Mrs. Bloomfield-Moore Passes Away In London,' *Public Ledger and Daily Transcript*, January 6, 1899. See also: 'Her Anxiety Brought Her Death,' *Philadelphia Inquirer*, January 6, 1899.

43. Gaston Burridge, 'The Baffling Keely Free Energy Machines,' *Fate*, vol.10, no.7, 1957, page 46. Of the Keely files allegedly in Sweden, Sykes suggested that 'Keely's papers are locked up in the cellars of a castle in Sweden just waiting for somebody to go and look at them.' In: Egerton Sykes, 'Keely Once Again,' *Atlantis*, vol.21, no.3, May/June 1968, page 68. Burridge's tale resurfaced in 1972; that year Swedish publisher Wendelholm issued the ring-stapled volume 'Keely - Pictures of his Discoveries' in a limited edition of 500 copies, now out of print. According to a certain A.L. who wrote the forward, '...Keely had a young relative who is apparently still alive. This relative had no great faith in Keely's work, but he related that Mrs. Moore passed on many of Keely's secrets, however good or bad, to Count von Rosen, at that time a resident of Scotland. The last information we have of this material is that it was brought to Stockholm in 1912. But it has recently come to light again.' The book consists of 54 photographs of some of Keely's devices and charts, some culled from contemporary articles on Keely. Although the writer suggests that the book is a selection of, or all of the Swedish Keely materials, it is by no means certain that these photo's were part of what

was allegedly sent to Sweden. Pond communicated with the publisher in 1985; the written answer was: 'The original (sic) for reprinting was remitted us from a secret society in Sweden (sic). As we remember the material was transmitted to the society by Count von Rosen there was in family contact with Mrs. Bloomfield-Moore (sic). The material was in old blueprints and brown-colored pictures of bad quality, - particularly the diagrams (sic)...We mean, that the only reason we was contacted for this mission was that we in long time have been produced books of alkemistic caractere, and the society was solicitous to bring the Keely-tradition to the public...(sic)' Copy of letter of publisher dated November 6, 1985, in my possession. My attempts to contact the publisher to clarify this matter were to no avail. Letters were returned, stating that the P.O. Box address was no longer in use. The circumstantial evidence is that some of the photographs sport Bloomfield-Moore's handwriting, identifying several parts on Keely's engines and requesting the return of the photo's to her London home address.

44. 'Count Von Rosen Talks Of Keely,' *The Evening Bulletin,* March 15, 1899.
45. 'Thinks Keely Was A Fraud,' *The Times,* March 10, 1899. The name of the grandson appears in: 'Count Von Rosen Talks Of Keely,' *The Evening Bulletin,* March 15, 1899.
46. 'Count Von Rosen Talks Of Keely,' *The Evening Bulletin,* March 15, 1899. See also: 'Bloomfield-Moore Estate,' *Public Ledger and Daily Transcript,* March 9, 1899.
47. 'Mrs. Moore's Will. Arrival Of Count Von Rosen, Executor Of Her Estate,' *Public Ledger and Daily Transcript,* March 16, 1899.
48. Gaston Burridge, 'The Baffling Keely 'Free Energy Machines,' *Fate,* vol.10, no.7, 1957, page 49. An acquaintance told to Pond that nothing was found in Meyers' legacy concerning Keely.
49. The library of Bloomfield-Moore, 'numbering some 3,000 volumes' and 'rich in works of science and medieval history,' with 'some rare books in the latter class,' was transferred to the University of Pennsylvania; 'This acquisition is being catalogued according to the Dewey system...' In: 'Mrs. Bloomfield-Moore's Library,' *The Evening Bulletin,* February 10, 1899. The University of Pennsylvania informed me that they do not have Bloomfield-Moore's correspondence.
50. 'Fortune Was Squandered In Keely Motor,' *Press,* January 22, 1899. See also: 'Keely Motor,' 'Everybody's Column,' in *Philadelphia Enquirer,* October 16, 1931. The numbers vary; elsewhere it was stated that Keely had invented 150 different engines. In: 'Keely Of Motor Fame Is Dead,' *The Times,* November 19, 1898. Edwards writes that 'Between 1885 and 1888 he made more than seventy devices, testing some before witnesses, destroying others after a few runs seen only by himself and his machinists.' In: Frank Edwards, *Strangest of All,* Citadel, 1956, Carol Paperbacks, 1991, page 164.
51. 'Keely's Secret,' *The World,* May 11, 1890.
52. O.M. Babcock, *The Keely Motor, Financial, Mechanical, Philosophical, Historical, Actual, Prospective,* privately printed, Philadelphia, June, 1881, page 20. Facsimile reprint by Delta Spectrum Research, no date. Babcock's statement also in Clara Bloomfield-Moore, 'Aerial Navigation,' *The Arena,* 1894, pages 388, 391, reprinted by Delta Spectrum Research, no date.
53. Keely's Vaporic Force. Experiments With A Mysterious Gun At Sandy Hook,' unspecified newspaper clipping, September 21, 1884, Sympathtic Vibratory Physics Homepage, Internet.
54. 'Keely Sphere Was No Secret,' *Philadelphia Press,* January 8, 1899. About the destruction of his engines, see also: Clara Bloomfield-Moore, *Keely and His Discoveries, Aerial Navigation,* Kegan Paul, Trench, Trübner & Co., 1893, page 11, 130, 165, 263. See also: 'Fortune Was Squandered In Keely Motor,' *The Press,* January 22, 1899, and the sources of note 52.
55. O.M. Babcock, *The Keely Motor, Financial, Mechanical, Philosophical, Historical, Actual, Prospective,* privately printed, Philadelphia, June, 1881, page 20. Facsimile reprint by Delta Spectrum Research, no date.

56. *Madame Blavatsky's Writings,* vol. XIII, Theosophical Publishing Society, 1982, page 385. See also William Mill Butler, 'Keely and the Keely Motor,' *The Home Magazine,* 1898, pages 109, 111.

57. Afra, 'John Worrell Keely,' *Theosophia,* no.13, May, 1893.

58. Clara Bloomfield-Moore, *Keely and His Discoveries, Aerial Navigation,* Kegan Paul, Trench, Trübner & Co., 1893, page 130. See also William Mill Butler, 'Keely and the Keely Motor,' *The Home Magazine,* 1898, page 109.

59. ibid. pages 11, 165.

60. Gaston Burridge, 'The Baffling Keely Free Energy Machines,' *Fate,* vol.10, no.7, 1957, page 46.

61. Egerton Sykes, *The Keely Mystery,* 2nd revised edition, Markham House, 1972, page 14.

62. 'Keely, Motor Man Dead. Succumbed After A Week's Illness Of Pneumonia,' *Public Ledger and Daily Transcript,* November 19, 1898.

63. Kinraide was misspelled as 'Kincaid,' 'T.B. Kinrade,' 'Kinraid,' 'J.B. Kinraide,' and even 'Y.B. Kineraide.' That his name was misspelled in print on several occasions, is not so mysterious, when one considers that in the contemporary newspaper articles, Charles W. Schuellermann's name was sometimes misspelled as Schnellermann, Schullerman or Scheullermann, and other names would be sometimes misspelled as well.

64. Documents of the West Laurel Hill Cemetery Company, dated December 20, 1898, copies received through Dale Pond.

65. 'Did Another Share Keely's Secret? A Mysterious Stranger Whose Presence is a Puzzle to the Directors,' *The Evening Bulletin,* November 26, 1898.

66. 'Official Statement To The Journal By A Director,' in: 'Keely Monumental Fraud of the Century, *New York Journal,* January 29, 1899.

67. 'Did Another Share Keely's Secret? A Mysterious Stranger Whose Presence is a Puzzle to the Directors,' *The Evening Bulletin,* November 26, 1898.

68. 'Keely Stockholders Still in the Dark,' *The Times,* November 29, 1898.

69. 'Fighting Over Keely's Secret,' *The Evening Bulletin,* December 20, 1898. Also: 'The Keely Motor Company,' *New York Times,* December 21, 1898. This article stated: 'Charles S. Hill, attorney for the inventor's widow, stated that Keely's secret did not exist in manuscript, but that Keely had made a suggestion before his death that T.B. Kinraid of Boston was the one man who could succesfully carry out his idea.'

70. 'To Finish The Motor. Turbulent Meeting Of The Keely Company Stockholders,' *Public Ledger and Daily Transcript,* December 21, 1898.

71. 'Boston Man In Keely's Shoes. J.B. Kinraide Will Try To Make The Motor Mote,' *The Times,* December 21, 1898. There is some uncertainty concerning Smith's first name. Here he is called 'James J. Smith.'

72. 'To Finish The Motor. Turbulent Meeting Of The Keely Company Stockholders,' *Public Ledger and Daily Transcript,* December 21, 1898.

73. ibid. See also: 'Will Ship Keely's Devices,' *Public Ledger and Daily Transcript,* December 28, 1898.

74. 'Boston Man In Keely's Shoes. J.B. Kinraide Will Try To Make The Motor Mote,' *The Times,* December 21, 1898.

75. 'New Man On Keely's Motor. Boston Scientist to Take Up the Work on the Mysterious Engine,' *The Evening Bulletin,* December 21, 1898.

76. 'Keely Directors Sceptical,' *The Times,* December 22, 1898.

77. Short editorial in: *Public Ledger and Daily Transcript,* December 27, 1898.

78. 'To Test Keely's Theory. T.B. Kinrade Will Work One Year on the Motor,' *The Times,* December 27, 1898.

79. 'Will Ship Keely's Devices,' *Public Ledger and Daily Transcript,* December 28, 1898.

80. 'To Try Keely's Machines,' *The Times,* December 30, 1898.

81. 'Keely Inventory Filed,' *The Evening Bulletin*, January 4, 1899. Also: 'Keely Motor Inventory of Estate,' *New York Daily Tribune*, January 5, 1899.
82. 'Keely's Motor In Boston,' *New York Times*, January 4, 1899.
83. 'Official Statement To The Journal By A Director,' in: 'Keely Monumental Fraud Of The Century, *New York Journal*, January 29, 1899.
84. 'To Try Keely's Machines,' *The Times*, December 30, 1898.
85. 'Keely Sphere Was No Secret,' *Philadelphia Press*, January 8, 1899.
86. 'A Mysterious Globe,' *Public Ledger and Daily Transcript*, January 9, 1899.
87. 'Tesla Knew Of Keely's Secret,' *Evening Journal*, January 20, 1899.
88. 'Official Statement to the Journal by a Director,' in: 'Keely Monumental Fraud of the Century,' *New York Journal*, January 29, 1899.
89. 'Official Statement to the Journal by a Director,' in: 'Keely Monumental Fraud of the Century,' *New York Journal*, January 29, 1899. Hotel Stratford was the address where Kinraide resided during his stay in Philadelphia, according to documents of the West Laurel Hill Cemetery Company, dated December 20, 1898.
90. 'Tesla Knew Of Keely's Secret,' *Evening Journal*, January 20, 1899.
91. William Mill Butler, 'Keely and the Keely Motor,' *The Home Magazine*, 1898, page 114.
92. 'Will Ship Keely's Devices,' *Public Ledger and Daily Transcript*, December 28, 1898.
93. 'Keely Inventory Filed,' *The Evening Bulletin*, January 4, 1899.
94. 'Confession By Mrs. Keely's Attorney,' in: 'Keely The Monumental Fraud Of The Century,' *New York Journal*, January 29, 1899. See also: *Keely's Secrets, Public Ledger and Daily Transcript*, January 30, 1899.
95. 'Official Statement To The Journal By A Director,' in: 'Keely Monumental Fraud The Century, *New York Journal*, January 29, 1899.
96. 'The Keely Motor Secret Is Out,' *The Times*, January 30, 1899. See also: 'Light On Keely's Motor,' *The Record*, January 30, 1899.
97. *Keely's Secrets, Public Ledger and Daily Transcript*, January 30, 1899. Also: 'Keely Motor makes Kinraide Angry,' *New York Herald*, January 30, 1899, 'Did Keely Use Water?,' *Hartford Courant*, January 30, 1899.
98. 'Bridge Answers Kinraid,' *The Times*, January 31, 1899. Also: 'The Keely Motor Revelations,' *New York Times*, January 31, 1899.
99. 'Visiting Mr. Kinraide. Keely Motor People Start For Boston,' *Public Ledger and Daily Transcript*, February 9, 1899.
100. ibid. See also: 'Keely's Motor Again,' *The Evening Bulletin*, February 9, 1899.
101. 'Keely Motor Investigation In Boston,' *Public Ledger and Daily Transcript*, February 11, 1899.
102. 'Keely Directors Own Up,' *The Evening Bulletin*, March 1, 1899.
103. The newspapers do not refer to any relocation or disposal of Keely's equipment that was stored in Philadelphia.
104. 'End of the Keely Motor,' *Scientific American*, May 20, 1899, page 333.
105. 'Keely Motor Tricks Were All Reproduced,' *The Press*, July 16, 1899.
106. 'Kinraide On Keely,' *The Press*, July 16, 1899.
107. 'Keely Motor Tricks Were All Reproduced,' *The Press*, July 16, 1899.
108. 'Kinraide On Keely,' *The Press*, July 16, 1899.
109. 'Keely's Secret Known,' *The Times*, January 20, 1899.
110. 'Col. Astor Estate,' *New York Times*, June 22, 1913.
111. Seiffer speculates that the craft could have been equipped with Tesla's hydrofoil jet engine. In: Marc J. Seiffer, *Wizard, the Life and Times of Nikola Tesla*, Birch Lane Press, 1996, page 342.

112. The curator of the Smithsonian Institute wrote that 'It seems to be an early one - he operated from 1873 or 74 until his death in 1898...' In: 'Letter of the Smithsonian Institute to the American Precision Museum,' dated February 11, 1985. According to Pond it is a Hydro-vacuo Engine. In: 'Notes and Odds & ends,' August 9, 1985, not published.
113. The instrument was donated to the Franklin Institute in May 1932 by a Henry Howson of Philadelphia. The Franklin Institute sold the device in November 1973. I have the name of the current owner on file. Howson & Howson were at one time Keely's attorneys. Correspondence with Dale Pond, dated January 9, 1997.
114. According to Pond, the device is Keely's Disintegrator. The instrument was verified by the Smithsonian Institute. I have the name of the now deceased person who donated this device on file. In a telephone conversation with the person whose grandfather had found the device, he presumed that his grandfather had found the device in a junk-shop, since he was always on the lookout for things mechanical. Presumably this was in the thirties. Where and under what circumstances he had found the device, he did not know. In the letter of note 112, the curator wrote: 'Actually, I'm amazed to find that one of these things survives.' In a telephone conversation with the American Precision Museum, the Director expressed his wonder at the fine craftsmanship that went into the construction of the device. I also learned that some time ago a person from Florida had called the museum with the claim that he had purchased a device at an auction, of which the caller thought that it might be an original Keely device. Although I have been unable to check this story since the name and other details of the caller were lost, It is quite possible that more engines have survived.

Chapter 6. Anatomy of an Exposure

1. 'Keely's Mysterious Sphere. Strange Contrivance Found Hidden Under Floor of His Laboratory,' *The Evening Bulletin*, January 7, 1899. Also: 'Keely Mystery Unearthed. Queer Machine Found Under Floor of Inventor's Laboratory. Buried Iron Sphere Discovered,' *New Haven Evening Register*, January 7, 1899.
2. 'Keely's Secret,' unspecified clipping, December 16, 1881, in: 'John E. Worrell Keely, Chronology of Life and Times from 1874-1884,' Delta Spectrum Research, no date.
3. 'Keely Sphere Was No Secret,' *Philadelphia Press*, January 8, 1899.
4. ibid.
5. ibid.
6. ibid.
7. ibid.
8. 'Odd Find Under Keely Laboratory,' *New York Herald*, January 8, 1899.
9. Letter dated January 26, 1899 by E.H.C., 'Keely's Cylinder Not A New Discovery,' *Electrical Review*, February 1, 1899, vol.34, no.5, page 68.
10. 'Keely Sphere Was No Secret,' *Philadelphia Press*, January 8, 1899.
11. ibid.
12. ibid.
13. ibid.
14. ibid.
15. 'A Mysterious Globe,' *Public Ledger and Daily Transcript*, January 9, 1899.
16. 'The Keely Mystery,' *The Press*, January 9, 1899.
17. *Keely's Secrets*, Letter to the editor, *The Bulletin*, January 14, 1899.
18. 'Keely Investigation,' *Public Ledger and Daily Transcript*, January 19, 1899.
19. ibid.
20. 'The Extraordinary Story of John Worrell Keely,' *The Cosmopolitan*, April 6, 1899, vol. XXVI, no.6, page 634.
21. 'Keely Investigation,' *Public Ledger and Daily Transcript*, January 19, 1899.
22. 'Keely Enterprises,' *Public Ledger and Daily Transcript*, January 20, 1899.

23. As Moore himself stated: 'When Mrs. Moore was involved in litigation with me, the result of a written threat made by her to have me declared unfit to manage my father's estate...' in: 'Keely Enterprises,' *Public Ledger and Daily Transcript,* January 20, 1899.
24. 'Keely Investigation,' *Public Ledger and Daily Transcript,* January 19, 1899.
25. 'Keely's Secret Not Laid Bare,' *The Evening Bulletin,* January 19, 1899. Also: 'Keely, Knave or Fool,' *Washington Post,* January 20, 1899.
26. ibid.
27. 'Keely Motor A Humbug?,' *The Sun,* January 20, 1899.
28. 'Tesla Knew Of Keely's Secret,' *New York Evening Journal,* January 20, 1899.
29. 'Clarence B. Moore In Keely's Laboratory,' *The Times,* January 20, 1899.
30. See chapter 4 and chapter 4, note 80.
31. 'Traps For The Unwary. More Disclosures Of The Secrets Of Keely's Laboratory,' *Public Ledger and Daily Transcript,* January 21, 1899.
32. 'Clarence B. Moore In Keely's Laboratory,' *The Times,* January 20, 1899.
33. 'Tesla Knew Of Keely's Secret,' *New York Evening Journal,* January 20, 1899.
34. 'Investigation at The Keely Laboratory,' *Scientific American,* February 4, 1899, page 72.
35. 'Clarence B. Moore In Keely's Laboratory,' *The Times,* January 20, 1899.
36. 'Hidden Tubes Show How The Keely Motor Worked,' *New York Herald,* January 20, 1899. Also 'Keely's Secret Disclosed,' *New York Times,* January 20, 1899, 'Keely's Secret' and 'Keely Motor Deception,' both in: *New York Daily Tribune,* January 20, 1899, 'Hidden Compressed Air Ran the Keely Motor' and 'Scientists Expose Keely Motor Farce,' both in: *New York Herald,* January 20, 1899, 'Keely Secret Exposed,' *Hartford Daily Courant,* January 20, 1899. 'Probing Keely Motor Secrets In Inventor's House, Philadelphia,' *Chicago Tribune,* January 21, 1899.
37. 'Clarence B. Moore in Keely's Laboratory,' *The Times,* January 20, 1899.
38. ibid.
39. 'Traps For The Unwary. More Disclosures Of The Secrets Of Keely's Laboratory,' *Public Ledger and Daily Transcript,* January 21, 1899.
40. ibid.
41. ibid.
42. ibid.
43. ibid.
44. ibid.
45. ibid.
46. ibid. Around this time, newspaper reporters were given a tour of the workshop by Moore and his team. In: 'Keely Motor A Humbug?,' *The Sun,* January 20, 1899.
47. ibid. See also: 'Still Probing At Keely's Humbug,' *New York Herald,* January 21, 1899.
48. 'The Keely Investigation,' *The Record,* January 21, 1899.
49. ibid.
50. 'Keely's Spinning Motor,' *New York Times,* July 25, 1886.
51. 'The Keely Investigation,' *The Record,* January 21, 1899.
52. ibid.
53. 'Fortune was Squandered In Keely Motor,' *The Press,* January 22, 1899.
54. ibid.
55. ibid.
56. 'No Other Has Ever Been So Shrouded in the Mist of Publicity,' *New York Herald,* November 27, 1898.
57. 'Fortune was Squandered In Keely Motor,' *The Press,* January 22, 1899.
58. 'Hidden Tubes Show how the Keely Motor Worked,' *New York Evening Journal,* January 20, 1899.
59. 'Keely's Sphere Not His Secret,' *The Evening Bulletin,* January 25, 1899. 'The Keely Motor Defended,' *New York Daily Tribune,* January 26, 1899.

60. Megargee, 'Seen and Heard in Many Places,' *The Times,* November 26, 1898.
61. 'Keely's Power,' *Public Ledger and Daily Transcript,* January 26, 1899.
62. 'Tesla Knew Secret Of Keely Motor,' *New York Evening Journal,* January 20, 1899.
63. 'A Vote Of Confidence,' *The Times,* January 26, 1899. See also: 'Faith in Keely's Motor,' *The Evening Bulletin,* January 26, 1899.
64. 'Still Believe In The Motor,' *The Times,* January 28, 1899.
65. 'Keely The Monumental Fraud Of The Century,' *New York Journal,* January 29, 1899.
66. 'The Keely Motor Secret Is Out,' *The Times,* January 30, 1899.
67. ibid.
68. ibid.
69. 'Keely Monumental Fraud Of The Century,' *New York Journal,* January 29, 1899.
70. ibid.
71. See Philadelphia newspapers: 'The Keely Motor Secret Is Out,' *The Times,* January 30, 1899, 'Light On Keely's Motor,' *The Record,* January 30, 1899.
72. 'The Keely Myth,' *The Times,* January 31, 1899.
73. 'Grasping The Last Straw,' *The Times,* January 31, 1899.
74. 'The Keely Myth,' *The Times,* January 31, 1899.
75. 'Keely Directors Own Up,' *The Evening Bulletin,* March 1, 1899.
76. ibid.
77. 'Investigation Of The Keely Laboratory,' *Scientific American,* February 4, 1899, page 72.
78. 'The Extraordinary Life Story Of John Worrell Keely,' *The Cosmopolitan,* April 1899, vol. XXVI, no.6, page 634.
79. 'No Other Has Ever Been So Shrouded in the Mist of Publicity,' *New York Herald,* November 27, 1898.
80. Julius Moritzen, 'The Extraordinary Life Story of John Worrell Keely,' *The Cosmopolitan,* vol. XXVI, no.6, April 1899, page 636.
81. Megargee, 'Seen and Heard in Many Places,' *The Times,* March 24, 1898. Also in Megargee, 'Seen and Heard in Many Places,' *The Times,* November 22, 1898.
82. 'No Other Has Ever Been So Shrouded in the Mist of Publicity,' *New York Herald,* November 27, 1898.
83. Charles Fort, *Wild Talents,* Claude Kendall, 1932, pages 341-342.
84. In: 'Keely Monumental Fraud Of The Century,' *New York Journal,* January 29, 1899.
85. In: 'Investigation At The Keely Laboratory,' *Scientific American,* February 4, 1899, pg. 72.
86. Even Fort writes that 'the motor had been run by a compressed air engine, in the cellar,' in: *Wild Talents,* Claude Kendall, 1932, page 341. The 'New Encyclopedia Brittanica,' 'Macropaedia,' 15th edition, 1981, vol.14 lists Keely on page 105 in an article on perpetual motion and states about the exposure: 'They found concealed beneath the floor a giant metal tank evidently used as a reservoir of compressed air to the generator. In the walls was located a network of brass tubing that conducted the compressed air to the generator.' No mention is made about the water motor. In this, Fort was preceeded by the press. See 'Great Hoaxes of History,' *Wilkes Barre Times Leader,* December 28, 1909. The water motor was also noted by the *Hartford Courant:* 'Evidence that Rubber Tubes Were Connected With a Water Motor,' the lead of the article ran. In: 'Did Keely Use Water?,' *Hartford Courant,* January 30, 1899.
87. These were published in 'Keely The Monumental Fraud of the Century,' January 29, 1899.
88. 'Keely's Successor,' *Hartford Courant,* December 29, 1898. 'The Kinraide coil is described as being used for 'a portable power source for bedside examinations' and being 'highly succesful'. See: A Century of Radiology,' 'Section of Radiologic Computing and Imaging Science, Dept. of Radiology, Penn State University College of Medicine,' Internet. I have communicated with the Massachusetts Historical Society. On my request a search was conducted to uncover possible documentation on T. Burton Kinraide, but nothing was found.

89. Kinraide did suggest that persons who would produce evidence of 'being victimized,' could obtain an invitation for visiting his laboratory. In: 'Keely Motor Tricks Were All Reproduced,' *The Press*, July 16, 1899. Also chapter 5 and chapter 5, note 107. I have found no documentation of anybody actually having done so.
90. Charles Fort, *Wild Talents*, Claude Kendall, 1932, pages 341-342.
91. 'Tesla Knew Secret Of Keely Motor,' *New York Evening Journal*, January 20, 1899.
92. 'Use Found For Keely's Sphere,' *The Evening Bulletin*, March 19, 1900.

Chapter 7. To Understand the Art: Keely's Discoveries

1. William Mill Butler, 'Keely and the Keely Motor,' *The Home Magazine*, 1898, page 111.
2. Dale Pond, *Universal Laws Never Before Revealed: Keely's Secrets*, The Message Company, 1995, pages 84-85. The textbook for those wishing to study Keely's technology.
3. 'The Keely Motor Criticized,' page 10.
4. William Mill Butler, 'Keely and the Keely Motor,' *The Home Magazine*, 1898, page 106. About the accidental nature of his discoveries, see also: Clara Bloomfield-Moore, *Keely and His Discoveries, Aerial Navigation*, Kegan Paul, Trench, Trübner & Co., 1893, page 60.
5. The ' New Encyclopedia Brittanica,' 'Macropaedia,' 15th edition, 1981, vol.14, page 105.
6. 'The Keely Motor Criticized,' page 12.
7. 'The Keely Motor. Feats of Which it is Capable,' *New York Times*, July 3, 1875.
8. Clara Bloomfield-Moore, 'Aerial Navigation,' *The Arena*, 1894, page 391, reprinted by Delta Spectrum Research, no date.
9. 'The Keely Motor Criticized,' page 12.
10. Clara Bloomfield-Moore, *Keely and His Discoveries, Aerial Navigation*, Kegan Paul, Trench, Trübner & Co., 1893, page 11.
11. 'The Keely Motor,' *Public Ledger Almanac*, 1900, page 101.
12. 'Keely's Secret,' *The World*, May 11, 1890.
13. ibid.
14. Letter by H.O. Ward (Clara Bloomfield-Moore). In: 'The Keely Motor,' *The Evening Bulletin*, December 17, 1895.
15. 'Keely's Etheric Vapor,' *New York Times*, September 21, 1884.
16. 'Two Hours With Keely,' *Public Ledger and Daily Transcript*, November 11, 1895.
17. Clara Bloomfield-Moore, *Keely and His Discoveries, Aerial Navigation*, Kegan Paul, Trench Trübner & Co., 1893, page 75.
18. 'Keely's Secret,' *The World*, May 11, 1890.
19. Clara Bloomfield-Moore, *Keely and His Discoveries, Aerial Navigation*, Kegan Paul, Trench, Trübner & Co., 1893, page 369.
20. O.M. Babcock, 'The Keely Motor, Financial, Mechanical, Philosophical, Historical, Actual, Prospective,' privately printed, Philadelphia, June, 1881, page 25. Facsimile reprinted by Delta Spectrum research.
21. *H.P. Blavatsky Collected Writings*, T.P.S., 1982, vol. XIII, page 385.
22. O.M. Babcock, 'The Keely Motor, Financial, Mechanical, Philosophical, Historical, Actual, Prospective,' privately printed, Philadelphia, June, 1881, page 20. Facsimile reprinted by Delta Spectrum research.
23. ibid. page 22.
24. Clara Bloomfield-Moore, *Keely's Secrets*, T.P.S., 1888, page 17. Also in Clara Bloomfield-Moore, *Keely and His Discoveries, Aerial Navigation*, Kegan Paul, Trench, Trübner & Co., 1893, page 87.
25. ibid.
26. A. Wilford Hall, 'John Keely - A Personal Interview,' *Scientific Arena*, January 1887. Reprinted in: *Sympathetic Vibratory Physics*, December 1987, Vol. III, Issue 3, pages 3-7. See also: William Mill Butler, 'Keely and the Keely Motor,' *The Home Magazine*, 1898, page 110.

27. ibid.
28. See chapter 5 and chapter 5, notes 50 and 51.
29. Megargee, 'Seen and Heard in Many Places,' *The Times,* November 22, 1898.
30. 'The Keely Motor. What Is Claimed For It,' *New York Times,* June 10, 1875.
31. 'Fortune Was Squandered In Keely Motor,' *The Press,* January 22, 1899.
32. William Mill Butler, 'Keely and the Keely Motor,' *The Home Magazine,* 1898, page 111.
33. Company names mentioned in 'Keely's Power,' *Public Ledger and Daily Transcript,* January 26, 1899.
34. Companies names mentioned in *H.P. Blavatsky Collected Writings,* T.P.S., 1982, vol. XIII, page 385. The letter by H.R. Borle in note 28, chapter 1, came from the Delaware Iron Company's works at New Castle, Delaware. In: Megargee, 'Seen and heard in Many Places,' *The Times,* November 26, 1898.
35. 'Saw The Keely Motor,' *The Times,* January 31, 1893.
36. 'Keely Not Dead Yet,' *New York Times,* November 24, 1884.
37. 'Keely's Power,' *Public Ledger and Daily Transcript,* January 26, 1899.
38. ibid.
39. ibid.
40. ibid.
41. ibid.
42. ibid.
43. 'Keely Motor Company,' *Public Ledger and Daily Transcript,* December 12, 1895.
44. 'The Extraordinary Story Of John Worrell Keely,' *The Cosmopolitan,* April 1899, vol. XXVI, no.6, pages 638-639.
45. 'Is It A New Power?,' *Public Ledger and Daily Transcript,* April 16, 1896.
46. Clara Bloomfield-Moore, 'Aerial Navigation,' *The Arena,* 1894, page 391, reprinted by Delta Spectrum Research, no date.
47. William Mill Butler, 'Keely and the Keely Motor,' *The Home Magazine,* 1898, page 114.
48. Introduction of R. Harte in Clara Bloomfield-Moore, *Keely's Secrets,* T.P.S., 1888, page 4. See also: Clara Bloomfield-Moore, *Keely and His Discoveries, Aerial Navigation,* Kegan Paul, Trench, Trübner & Co., page 165.
49. Clara Bloomfield-Moore, *Keely and His Discoveries, Aerial Navigation,* Kegan Paul, Trench, Trübner & Co., 1893, page 147, 60.
50. John Milner, *Russian Revolutionary Art,* Bloomsbury Books, 1987, pages 26-27.
51. Gerald L'E. Turner, *Nineteenth-Century Scientific Instruments,* University of California Press, 1983, page 129.
52. ibid. page 135.
53. 'New Encyclopedia Brittanica,' 'Macropaedia,' 15th edition, 1981, vol.14, page 60.
54. Gerald L'E Turner, *Nineteenth-Century Scientific Instruments,* University of California Press, 1983, page 135.
55. 'Brockhaus Encyclopedie,' 19th edition, 1987, vol.4, page 522.
56. Gerald L'E Turner, *Nineteenth-Century Scientific Instruments,* University of California Press, 1983, pages 141-142.
57. John G. Burke, *Cosmic Debris: Meteorites in History,* University of California Press, 1986, page 167.
58. Gerald L'E Turner, *Nineteenth-Century Scientific Instruments,* University of California Press, 1983, page 143-144.
59. ibid. pages 144-146.
60. Clara Bloomfield-Moore, *Keely and His Inventions, Aerial Navigation,* Kegan Paul, Trench, Trübner & Co., 1893, page 259. About Koenig's presence, see also page 261.
61. ibid. pages 259-260.
62. ibid. page 261.
63. William Mill Butler, 'Keely and the Keely Motor,' *The Home Journal,* 1898, page 104.

64. 'Is It A New Power?,' *Public Ledger and Daily Transcript,* April 16, 1896. The cladna is also mentioned in a letter by L.N.S. Pasmore, dated November 6, London. In: Letters to the editor, *The Globe,* November 10, 1896.

65. A. Wilford Hall, 'A Visit To Mr. Keely,' *Scientific Arena,* July, 1886. Reprinted in *Sympathetic Vibratory Physics,* vol.III, no.4, January 1988, pages 4-12. See also chapter 3 for a description of this experiment.

66. 'Keely's Secret,' *The World,* May 11, 1890.

67. 'Possibilities of the Keely Motor,' *The Evening Bulletin,* July 8, 1875.

68. Vaporic substance mentioned in: 'The Keely Motor Criticized, a republication in pamphlet form of a series of editorials which appeared in the *Public Record* of Philadelphia, August 3rd, 4th, 5th and 6th, 1875,' no place, no date, but in all probability published in 1875 by the *Public Record.*

69. Clara Bloomfield-Moore, 'Aerial Navigation,' *The Arena,* 1894, page 387, reprinted by Delta Spectrum research, no date.

70. O.M. Babcock, 'The Keely Motor, Financial, Mechanical, Philosophical, Historical, Actual, Prospective,' privately printed, Philadelphia, June, 1881, page 23. Facsimile reprint by Delta Spectrum Research, no date.

71. Clara Bloomfield-Moore, *Keely's and his Discoveries, Aerial Navigation,* Kegan Paul, Trench, Trübner & Co., 1893, page 41.

72. 'Keely and his Motor,' *Evening Bulletin,* August 22, 1887. Compare this statement with a passage on page 132 of Blavatsky's *Isis Unveiled.*

73. 'Was Keely Ever Huss?,' *The Times,* January 19, 1889.

74. Editorial in: *The Times,* March 29, 1889.

75. 'Keely's Secret,' *The World,* May 11, 1890.

76. 'Two Hours With Keely,' *Public Ledger and Daily Transcript,* November 11, 1895.

77. 'New Life In Keely Patents,' *The Evening Telegraph,* April 28, 1896.

78. H.P. Blavatsky, *Isis Unveiled,* J.W. Bouton, 1877, pages XXV - XXVII, 58, 125.

79. ibid. page 134.

80. Sir Oliver Lodge, 'The Ether and its Functions,' lecture held at the London Institute, December 28, 1882, reprinted in *Modern Views of Electricity.* My source: 'William Kingsland, 'De natuurkunde in de Geheime Leer,' Theosophische Uitgeversmaatschappij, 1911, page 19. An interesting monograph on the concepts of the ether as seen by early theosophy.

81. H.P. Blavatsky, *Isis Unveiled,* J.W. Bouton, page 140.

82. 'Support Withdrawn,' *Public Ledger and Daily Transcript,* May 9, 1896.

83. Clara Bloomfield-Moore, *Keely and His Discoveries, Aerial Navigation,* Kegan Paul, Trench, Trübner & Co., 1893, page 124.

84. ibid. page 15.

85. ibid. page 124. For her treatment of MacVicar, see also ibid. pages 16, 17, 18, 19, 20, 27 , chapter 2, 63, 64, 67, 68, 116, 129. On Hughes, see pages 64, 89, 116.

86. Contemporary newspapers often referred to Charles Redheffer in connection with Keely; Redheffer built a self moving machine around 1810, but appeared to have been a charlatan. This connection of Keely with perpetual motion, more or less, has stayed the same; for instance in the 'New Encyclopedia Brittanica,' 15th edition, 1981, vol. 14, page 105, an entry on Keely is to be found in an article concerning perpetual motion.

87. Friedrich Klemm, *Perpetuum Mobile, ein unmöglicher Menschheithstraum,* Harenberg, 1983, pages 13-21. An excellent book on the subject.

88. ibid. pages 46-47.

89. ibid. pages 47-48.

90. ibid. pages 10-12.

91. Clara Bloomfield-Moore, *Keely's Secrets,* T.P.S., 1881, page 19.

92. 'Saw The Keely Motor,' *The Times,* January 31, 1893.

93. Clara Bloomfield-Moore, *Keely and His Discoveries, Aerial Navigation*, Kegan Paul, Trench Trübner & Co., 1893, page 324, page 107.
94. See chapter 1
95. 'Says He Knows Keely's Secret,' *The Times*, January 1, 1899.
96. Clara Bloomfield-Moore, *Keely and His Discoveries, Aerial Navigation*, Kegan Paul, Trench, Trübner & Co., 1893, page 323. Also cited in: Gaston Burridge, 'The Baffling Keely Free Energy Machines,' *Fate*, vol.10, no.7, 1957, pages 45-46.
97. ibid. page 111.
98. ibid. page 201.
99. A. Wilford Hall, 'John Keely - a Personal Interview,' *Scientific Arena*, January 1887, reprinted in: *Sympathetic Vibratory Physics*, vol.III, no.3, December 1987, page 3.
100. 'Keely's Secret,' *The World*, May 11, 1890.
101. 'Two Hours With Keely,' *Public Ledger and Daily Transcript*, November 11, 1895.
102. Clara Bloomfield-Moore, *Keely and His Inventions, Aerial Navigation*, Kegan Paul, Trench, Trübner & Co., 1893, page 42.
103. Several later writers on Keely have also complained about this. See: Gaston Burridge, 'The Baffling Keely Free Energy Machines,' *Fate*, vol.10, no.7, 1957, page 47, and Egerton Sykes, *The Keely Mystery*, Markham House, 2nd edition, 1972, page 2.
104. William Mill Butler, 'Keely and the Keely Motor,' *The Home Magazine*, 1898, page 114.
105. Correspondence with Dale Pond, dated January 9, 1997.
106. Gaston Burridge, 'The Baffling Keely Free Energy Machines,' *Fate*, vol.10, no.7, 1957, page 48.
107. Short editorial in *The Times*, October 27, 1893. On this, Bloomfield-Moore wrote: '...and that Professor D.G. Brinton, of the University of Pennsylvania has prepared a paper on the subject, and will publish it when Mr. Keely is ready to have his system made known.' In: Keely and his Discoveries, Aerial Navigation,' Kegan Paul, Trench, Trübner & Co., page 267. However, the 'translation' or 'paper' or parts of it appeared in the same book on pages 358-364.
108. 'Dashed Against the Rock' was advertised in the back pages of *Colville's John Worrall Keely, A Memorial Address*, Banner of Light Publishing Co., 1899, as: 'This wonderful story contains authentic reports of interviews with John W. Keely, and introduces in popular form amazing information concerning nature's mysteries.' Also correspondence with Dale Pond, dated August 29, 1995.
109. Dale Pond, *Universal Laws Never Before Revealed: Keely's Secrets*, The Message Company, 1995. A definitive must for those wanting to study Keely's discoveries in the light of modern technological achievements and science. Pond also wrote and published a massive 'SVP Compendium of Terms and Phrases,' including over 4,000 entries of all of Keely's terms and useages, including those of Tesla, Cayce and others.
110. Frank Edwards, *Strangest of All*, Citadel, 1956, Caroll Paperbacks, 1991, page 164.
111. 'Mr. Keely's Performances. Another Postponement of the Motor Test Announced,' *New York Times*, August 30, 1883.
112. In: Letter by Alfred Plum to *Philadelphia Enquirer*, January 14, 1894. Also chapter 3, text with chapter 3, footnote 92.
113. 'Keely Motor Company,' *Public Ledger and Daily Transcript*, December 12, 1895. Probably the 'sensitizing' process was the same as the 'vitalizing' process. Keely allegedly told a reporter that 'objects vitalized or synchronized so as to vibrate in this ether in a certain definite relation to each other will together exert a force which...will supersede all other forms of energy.' In: 'Two Hours With Keely,' *Public Ledger and Daily Transcript*, November 11, 1895.
114. 'There Was No Meeting,' *The Times*, November 27, 1898.
115. 'Keely's Etheric Force,' *The Evening Bulletin*, June 30, 1897.
116. 'Keely Motor Tested,' *New York Times*, June 19, 1897.

117. 'Two Hours With Keely. Power Without Cost - Engines Which Do Not Move,' *Public Ledger and Daily Transcript,* November 11, 1895. Also chapter 4 and chapter 4, note 35. Also references to disks and metallic powder in 'A Keely Motor Tested,' *New York Times,* June 20, 1897, 'Keely Motors tested,' *Wilkes Barre Record,* June 21, 1897.

118. 'What is the Force, Hidden and Unseen, of Keely's Motor?,' *New York Herald,* August 22, 1897. George W. Browne, an inventor residing in Brooklyn, made investigations on behalf of the *New York Herald* into Keely's inventions. Browne 'had been ill for a long time, and during his convalescence he thought a great deal of Keely's so-called discoveries as they were outlined in the daily newspapers.' Browne also went to visit Keely in Philadelphia. Anecdote found in: 'Scientists Expose Keely Motor Farce,' *New York Herald,* January 20, 1899. A number of large articles appeared in which Browne provided his mundane explanations for the working of Keely's engines, such as compressed air and other means of trickery. The articles appeared, with cross-section drawings of Keely's devices as Browne thought they were, in: 'Keely's Marvels Made Plain At Last,' *New York Herald,* October 18, 1891, 'Wizard Keely's Music Of The Mystic Sphere,' *New York Herald,* October 25, 1891, 'Artillery Branch of Keely's Service,' *New York Herald,* November 1, 1891, 'The Wizard Latest Highest Mystery Is Explored,' *New York Herald,* November 8, 1891.

119. Clara Bloomfield-Moore, *Keely and His Discoveries, Aerial Navigation,* Kegan Paul, Trench, Trübner & Co., 1893, pages 86-87.

120. Published in Dale Pond, *Universal Laws Never Before Revealed: Keely's Secrets,* The Message Company, 1995.

121. Clara Bloomfield-Moore, *Keely and His Discoveries, Aerial Navigation,* Kegan Paul, Trench, Trübner & Co., 1893, pages 37-38, about Keely's third line, page 85.

122. ibid. page 250.

123. ibid. pages 91-97.

124. Clara Bloomfield-Moore, *Keely and His Discoveries, Aerial Navigation,* Kegan Paul, Trench, Trübner & Co., 1893, page 10.

125. Letter by John Keely, dated July 15th, 1885, in: ibid. pages 35-36.

126. ibid. pages 48-49.

127. ibid. page 50.

128. ibid. pages 50-51.

129. ibid. pages 60, 147.

130. H.P. Blavatsky, *The Secret Doctrine,* T.P.S., 1888, book I, part II, page 559.

131. ibid. 558.

132. ibid.

133. ibid. page 559.

134. ibid. page 561.

135. H.P. Blavatsky, *The Secret Doctrine,* T.P.S., 1888, book I, part II, page 562.

136. Clara Bloomfield-Moore, *Keely and his Discoveries,* Kegan Paul, Trench, Trübner & Co., 1893, pages 81, 112. About his supposed paranormal abilities, see also ibid. pages 10, 49, 102, 161, 197.

137. ibid. page 250.

138. 'Keely's Secret,' *The World,* May 11, 1890.

139. 'Will He Become A Medium?,' unspecified newspaper clipping, January 19, 1896, Sympathetic Vibratory Physics Homepage, Internet.

140. A.E. Thierens, *Cosmologie, wetenschappelijke opstellen,* Luctor et Emergo, 1913, page 76.

141. Egerton Sykes, *The Keely Mystery,* Markham House, 2nd edition, 1972, page 7.

142. Charles Fort, *Wild Talents,* Claude Kendall, 1932, page 339. Later adopted by Gaston Burridge, in: 'The Baffling Keely 'Free Energy Machines,' '*Fate,* vol.10, no.7, 1957, page 45.

Chapter 8. Prisoners of the Neutral Point: Keely's Antigravity Experiments

1. R. Harte, introduction in Clara Bloomfield-Moore, *Keely's Secrets*, T.P.S., July 10, 1888, page 3.
2. 'Keely And His Motor,' *The Evening Bulletin*, August 22, 1887.
3. Blavatsky's short statements to be found in *The Secret Doctrine*, T.P.S., 1888, book I, part II, page 555, and in 'The Blessings of Publicity,' *Lucifer*, no.48, August, 1891.
4. Clara Bloomfield-Moore, *Keely and His Discoveries, Aerial Navigation*, Kegan Paul, Trench, Trübner & Co., 1893, page 130.
5. ibid. page 129.
6. R. Harte, introduction in Clara Bloomfield-Moore, *Keely's Secrets*, T.P.S., July 10, 1888, page 4.
7. Clara Bloomfield-Moore, Keely and His Discoveries, Aerial Navigation, Kegan Paul, Trench, Trübner & Co., 1893, pages 178-179.
8. ibid. pages 310-318.
9. Letter by Alfred H. Plum to *Philadelphia Enquirer*, January 14, 1894. See also chapter 3, and chapter 3, note 94.
10. 'Keely's Red Letter Day,' *New York Times*, June 7, 1885. At another time a newspaper referred to Keely's 'flying machine' and a device to lift heavy weights, suggesting that the latter was not an anti-gravity device. See also this chapter note 42. The original account of the device for a California party appeared in 'Keely Shows His Motor,' *New York Times*, October 19, 1881.
11. Clara Bloomfield-Moore, *Keely and His Discoveries, Aerial Navigation*, Kegan Paul, Trench, Trübner & Co., 1893, pages 31-32.
12. ibid. page 106.
13. ibid. page 155.
14. 'Keely's Aerial Navigation,' unspecified clipping, December 1, 1887, Sympathetic Vibratory Physics Homepage, Internet. Newspaper cites The Philadelphia newspaper *The Sun* as the source.
15. Helena P. Blavatsky, *The Secret Doctrine*, T.P.S., 1888, page 555.
16. R. Harte, introduction to Clara Bloomfield-Moore, *Keely's Secrets*, T.P.S. July 10, 1888, page 4.
17. ibid. page 4. Quoted in: Clara Bloomfield-Moore, *Keely and His Discoveries, Aerial Navigation*, Kegan Paul, Trench, Trübner & Co., 1893, page 106. On Ricarde-Seaver, see chapter 3, and chapter 3 notes 10, 11 and 81.
18. Clara Bloomfield-Moore, *Keely and His Discoveries, Aerial Navigation*, Kegan Paul, Trench, Trübner & Co., 1893, pages 122-123.
19. Passage cited in: C.W. Snell, 'The Snell Manuscript,' 1934, Delta Spectrum research, 1995, page 9, also in *A Modern Wizard, the Keely Motor and its Inventor*, Cheiro, *Mysteries and Romances of the World's Greatest Occultists*, Herbert Jenkins, 1935, page 249, where, according to Cheiro's garbled account, he and Ricarde-Seaver saw—and note how his first sentence mimics that of Bloomfield-Moore's account—'In demonstrating what appeared to be the overcoming of gravity for aerial navigation, Mr. Keely showed us a model of an airship weighing about eight pounds. When the differential wire was attached to it, it also rose, floated, or remained stationary, at whatever height he wished it to be. This remarkable demonstration of this model airship, it must be remembered, was shown us at the Keely laboratories in 1890, some thirteen years before the brothers, Orville and Wilbur Wright, flew the first aeroplane in France in 1903.' Egerton Sykes in turn cites Harte's account without referencing it, in: *The Keely Mystery*, Markham House, 2nd revised edition, 1972, page 11.
20. 'Keely's Mysterious Sphere,' *The Evening Bulletin*, January 7, 1899.

21. Clara Bloomfield-Moore, *Keely and His Discoveries, Aerial Navigation*, Kegan Paul, Trench, Trübner & Co., 1893, page 123. Also: 'Keely's Secret,' *The World*, May 11, 1890. These experiments did not convince everybody, who assumed them to be another example of trickery. See: 'Those Floating Weights,' *New York Herald*, November 4, 1891.

22. ibid. page 123.

23. ibid. page 282. Cheiro writes how he and Ricarde-Seaver also witnessed Keely's glass in jars experiments. In: *A Modern Wizard, the Keely Motor and its Inventor*, Cheiro, *Mysteries and Romances of the World's Greatest Occultists*, Herbert Jenkins, 1935, pages 248-249.

24. See, for instance, text of contract in 'Keely To Explain,' *Public Ledger and Daily Transcript*, December 14, 1895. Also this chapter, notes 50-51.

25. 'She Lays Down A Great Burden,' *The Evening Telegraph*, May 11, 1896. See chapter 3, and chapter 3, note 82, for the pamphlet that was published instead. It is quite possible that she printed this passage in her 1893 book on the pages 126-128, although this is not clear from her writings.

26. 'Keely's New Force,' *San Francisco Chronicle*, April 7, 1890.

27. 'Keely's Secret,' *The World*, May 11, 1890.

28. Clara Bloomfield-Moore, *Keely and His Discoveries, Aerial Navigation*, Kegan Paul, Trench, Trübner & Co., 1893, page 280. Also in C.W. Snell, 'The Snell Manuscript,' Delta Spectrum Research, 1995, page 38.

29. 'Saw the Keely Motor,' *The Times*, October 27, 1893.

30. 'Two Hours With Keely,' *Public Ledger and Daily Transcript*, November 11, 1895.

31. Charles Morris, 'Apergy: Power Without Cost,' *New Scientific Review*, no.10, 1895, pages 185-186.

32. 'Will He Become a Medium?,' unspecified clipping, January 19, 1896, Sympathetic Vibratory Physics Homepage, Internet.

33. Clara Bloomfield-Moore, *Keely and His Discoveries, Aerial Navigation*, Kegan Paul, Trench, Trübner & Co., 1893, pages 315-316.

34. ibid. page 317.

35. ibid. page 282. Also in C.W. Snell, 'The Snell Manuscript,' Delta Spectrum research, 1995, page 38.

36. W.J. Colville, *Dashed Against the Rock*, Banner Of Light Publishing Company, 1894. Not seen. Passage taken from Dale Pond, *Universal Laws Never before Explained: Keely's Secrets*, The Message Company, 1995, page 64. Colville's book was sold as far back as 1909 by Masonic publisher and bookseller Macoy of New York. In the advertisement in the back of his 'Memorial Address' Colville stated that *Dashed Against the Rock* contained 'authentic reports of interviews with John W. Keely.'

37. Pond wrote 'Supposedly these small spheres rose in the air (while Keely played some music) with the larger 'Sun' sphere and they all revolved around the larger sphere as do the planets! I have not seen anything to substantiate this story.' Letter from Dale Pond, dated August 29, 1995.

38. Letter by Clara Bloomfield-Moore, dated May 30, 1894. She copied Keely's portion out of a letter that she received two days before from her correspondent, marked 'for you alone,' and 'confidential.'

39. Charles Morris, 'Apergy: Power Without Cost,' *New Scientific Review*, no.10, 1895, pages 186-187.

40. Letter by Clara Bloomfield-Moore, dated May 16, 1897. Copy obtained through Dale Pond.

41. 'Traps For The Unwary,' *Public Ledger and Daily Transcript*, January 21, 1899.

42. 'Two Hours With Keely,' *Public Ledger and Daily Transcript*, November 11, 1895.

43. Charles Morris, 'Apergy: Power Without Cost,' *New Scientific Review*, no.10, 1895, page 186.

44. Clara Bloomfield-Moore, 'A Newton of the Mind, The Propeller Described,' *New Scientific Review*, vol.1, 1895, facsimile reprint by Delta Spectrum Research, no date.

45. Clara Bloomfield-Moore, *Keely and His Discoveries, Aerial Navigation*, Kegan Paul, Trench, Trübner & Co., pages 126-127.
46. William Mill Butler, 'Keely and the Keely Motor,' *The Home Magazine*, 1898, page 111.
47. Clara Bloomfield-Moore, *Keely and his Discoveries, Aerial Navigation*, Kegan Paul, Trench, Trübner & Co., 1893, page XXIII.
48. 'Keely To Explain,' *Public Ledger and Daily Transcript*, December 14, 1895.
49. Copy obtained through Dale Pond.
50. 'Keely To Explain,' *Public Ledger and Daily Transcript*, December 14, 1895. Text of contract also published in this article.
51. Zak Samuels was referred to as 'Astor's expert' in the source of note 47.
52. W. Scott-Elliott, *The Story Of Atlantis*, T.P.S., 1896, page 51. He left this passage unchanged in the second edition in 1909, only adding the year '1895' to it. In: *The Story of Atlantis*, T.P.S., 1909, page 63.
53. Julius Moritzen, 'The Extraordinary Life Story of John Worrell Keely,' *The Cosmopolitan*, April 1899, vol.XXVI, no.6, page 638.
54. Leonard G. Cramp, *Piece For a Jig-Saw, UFO's Astounding Scientific Evidence in the Flying Saucer Puzzle*, Somerton, 1966, page 65.
55. W. Raymond Drake, *Gods and Spacemen Throughout History*, Neville Spearman, 1975, page 236.
56. Vladimir Terziski, *Close Encounters of the Foo Fighter Kind*, TRZ Consultants, 1994, page 35. Also on page 34 wrong dating (1860s) of Keely's levitation experiments.
57. Dan Davidson, 'A Breakthrough to New Free Energy Sources,' KeelyNet BBS, Internet.
58. A.A.C. Belinfante, *De Duwkracht. Beschouwingen van stoffen en hun bewegingen als gevolgen van botsende- of duwkracht. Verwerping der aantrekkingskracht, zwaartekracht en moleculaire krachten, enz.* Westzaan, Van Dijk en Allan, 1905.
59. ibid. page 184.
60. ibid. page 46.
61. ibid. page 66.
62. ibid. pages 66-67.
63. Afra, 'John Worrell Keely,' *Theosophia*, 2e jaargang, mei 1893, no.13. the article also refers to Keely's anti-gravity research: '...that Keely researches the ether and one of these days hopes to make it useful for many means, for instance the lifting of heavy objects...'
64. A.A.C. Belinfante, *De Duwkracht. Beschouwingen van stoffen en hun bewegingen als gevolgen van botsende- of duwkracht. Verwerping der aantrekkingskracht, zwaartekracht en moleculaire krachten, enz.* Westzaan, Van Dijk en Allan, 1905, page 183.
65. The Dutch translation of Bulwer-Lytton's 1871 *The Coming Race* was titled *De Mensch der toekomst*, De Breuk & Smits, 1873.
66. Various, *Rudolf Steiner in Nederland*, Pentagon, 1994, page 244.
67. J.K. Rensburg, *Wereldbouw*, Van Loghum, Slaterus en Visser, 1923, page 36.
68. ibid. page 36. There is some confusion as to the exact title; while Rensburg lists its title as 'Es gibt keine Gravitation,' the Dutch translation lists the title as 'Die Gravitationslehre...ein Irrtum!' In: Th. Newest, *De leer der zwaartekracht een dwaling!* Hollandia-drukkerij, 1910. Both sources do not mention the year of the German publication.
69. Unfortunately without giving any further bibliographical data, the booklets are, according to the Dutch translation: Isenkrahe, 'Rätsel der Schwerkraft,' and Sahulka 'Erklärung der Gravitation.' In: Th. Newest, *De leer der zwaartekracht een dwaling!*, Hollandia-drukkerij, 1910, page 3.
70. Everett F. Bleiler, *Science Fiction the Early Years*, The Kent State University Press, 1990, page 300.
71. Introduction by Sam Moskowits in *Across The Zodiac*, Hyperion Press, 1974, pages 4-5.
72. ibid. pages 5-6.

73. Everett F. Bleiler, *Science Fiction the Early Years*, The Kent State University Press, 1990, page 301: 'Obviously a child of Bulwer-Lytton's *The Coming Race.*'
74. Percy Greg, *Across the Zodiac*, Trübner & Co., 1880, page 22.
75. ibid.
76. J.R. Zuberbauer, 'Philosophy of the Keely Drama,' *The Sun*, November 28, 1897.
77. William Mill Butler, 'Keely and the Keely Motor,' *The Home Magazine*, 1898, page 114, 'Keely, Motor Man, Dead, *Public Ledger and Daily Transcript*, November 19, 1898: 'The principle upon which the new discovery was based was that of intermolecular vibration of the ether, to which he gave the name Apergy,' same statement in: *American Machinist*, vol.21, no.47, November 24, 1898. Another reference appeared in: 'Personal,' *New York Times*, March 26, 1898: '...talk like that in which Keely indulges when discussing the force he calls apergy.'
78. Megargee, 'Seen and Heard in Many Places,' *The Times*, March 21, 1898. The article refers to 'Two Hours With Keely,' *Daily Ledger and Public Transcript*, November 11, 1895, but in this article there is no reference to apergy.
79. 'Will retain her Privileges,' *Public Ledger and Daily Transcript*, November 9, 1895. 'Mrs. Moore On The Keely Motor,' *New York Times*, December 9, 1895. Also in: 'The Keely Motor,' *Evening Bulletin*, December 27, 1895.
80. Other brief references appeared in 'An Expert On Keely's Motor,' *New York Herald*, January 19, 1896: '...the force which Mr. Keely produces - the reflex action of gravity, known to the Greeks as apergy and to the ancients as one of the triune currents of a polar flow of force, by which all the operations of the universe are regulated and controlled.' and '...when he proposed to have Tesla investigate the operation of apergy on a revolving wheel.' and in: 'Personal,' *New York Times*, March 26, 1898: '...talk like that in which Keely indulges when discussing the force he calls Apergy.' Gaston Burridge, 'The Baffling Keely Free Energy Machines,' *Fate*, vol.10, no.7, 1957, page 44.
81. Letter by Clara Bloomfield-Moore, dated September 14, 1895. Copy obtained through Dale Pond.
82. Charles Morris, 'Apergy: Power Without Cost,' *New Scientific Review*, no.10, 1895.
83. Letter by Clara Bloomfield-Moore, dated September 14, 1895. Copy obtained through Dale Pond.
84. Sam Moskowitz, *Across the Zodiac: A Major Turning Point in Science Fiction*, foreword in: Percy Greg, *Across the Zodiac*, Hyperion Press, 1974, page 11.
85. Percy Greg, *Across the Zodiac*, Trübner & Co. 1880, pages 30-31.
86. Sam Moskowitz, *Across the Zodiac: A Major Turning Point in Science Fiction*, foreword in: Percy Greg, *Across the Zodiac*, Hyperion Press, 1974, page 10.
87. For a literary analysis, see: Everett F. Bleiler, *Science Fiction: The Early Years*, The Kent State University Press, 1990, pages 300-301.
88. Percy Greg, *Across the Zodiac*, Trübner & Co., 1880, pages 23-25.
89. ibid. page 25.
90. ibid. page 29.
91. ibid. pages 29-30.
92. Everett F. Bleiler, *Science Fiction: The Early Years*, The Kent State University Press, 1990, page 355.
93. Chris Morgan, *The Shape of Future's Past*, Webb & Bower, 1980, page 116.
94. ibid. page 158.
95. Everett F. Bleiler, *Science Fiction: The Early Years*, The Kent State University Press, 1990, page 355.
96. For a literary analysis and description of Hay's novel, see: ibid. pages 355-356, and Chris Morgan, *The Shape of Future's Past*, Webb & Bower, 1980, pages 116-118.
97. Chris Morgan, 'The Shape Of Future's past, Webb & Bower, 1980, page 116.
98. ibid. page 117.

99. ibid. pages 158-159, page 117.
100. Martin Gardner, *Urantia: The Great Cult Mystery*, Prometheus Books, 1995, page 163. Story also told in: *Oahspe, A New Bible in the Word of Jehovah and the Angel Ambassadors*, Oahspe Publishing Association, reprint, Ray Palmer, 1972, pages 907-910.
101. ibid.
102. ibid.
103. ibid. page 164.
104. John Ballou Newbrough: *Oahspe, A New Bible in the Word of Jehovah and the Angel Ambassadors*, Oahspe Publishing Association, 1882, Ray Palmer, 1970, page 583.
105. Clara Bloomfield-Moore, *Keely and His Discoveries, Aerial Navigation*, Kegan Paul, Trench, Trübner & Co., 1893, page 365.
106. John Ballou Newbrough: *Oahspe, A New Bible in the Word of Jehovah and the Angel Ambassadors*, Oahspe Publishing Association, 1882, Ray Palmer, 1970, page 587.
107. ibid. page 586.
108. ibid. page 584.
109. Clara Bloomfield-Moore, *Keely and His Discoveries, Aerial Navigation*, Kegan Paul, Trench, Trübner & Co., 1893, page 155.
110. Martin Gardner, *Urantia: The Great Cult Mystery*, Prometheus Books, 1995, page 166.
111. ibid. page 163.
112. ibid. page 175.
113. Robert Cromie, *A Plunge into Space*, Frederick Warne & Co., 2nd edition, 1891, page 167.
114. ibid. pages 15-16.
115. Sam Moskowitz, *Across the Zodiac: A Major Turning Point in Science Fiction*, foreword in: Percy Greg, *Across the Zodiac*, Hyperion Press, 1974, page 11.
116. Robert Cromie, *A Plunge into Space*, Frederick Warne & Co., 2nd edition, 1891, pages 41, 43.
117. ibid. pages 12-13, 17.
118. ibid. pages 49-50.
119. ibid. pages 128-129.
120. ibid. page 227.
121. Derek Wilson, *The Astors*, Weidenfeld and Nicolson, 1993, page 200.
122. Sam Moskowitz, *Across the Zodiac: A Major Turning Point in Science Fiction*, foreword in: Percy Greg, *Across the Zodiac*, Hyperion Press, 1974, page 11.
123. Percy Greg, *Across the Zodiac*, Trübner & Co. 1880, page 153.
124. Robert Cromie, *A Plunge into Space*, Frederick Warne & Co., 2nd edition, 1891, page 156.
125. John Jacob Astor, *A Journey In Other Worlds* Appleton, 1894, page 20.
126. ibid. page 29.
127. Clara Bloomfield-Moore, 'The Veil Withdrawn,' no date, Sympathetic Vibratory Physics Homepage, Internet.
128. John Jacob Astor, *A Journal in Other Worlds*, Appleton, 1894, page 20.
129. ibid. page 30.
130. ibid.
131. For a literary analysis and summary, see: Everett F. Bleiler, *Science Fiction: The Early Years*, The Kent State University Press, 1990, page 24.
132. John Jacob Astor, *A Journal in Other Worlds*, Appleton, 1894, page 72.
133. Robert Cromie, *A Plunge Into Space*, Frederick Warne & Co., 2nd edition, 1891, page 173.
134. Marc J. Seiffer, *Wizard, the Life and Times of Nikola Tesla*, Birch Lane Press, 1996, pages 152-153.
135. His name appears in the list of May, 1890. In: *Proceedings of the Society for Psychical Research*, vol.V, 1888-89, Trübner & Co., page 591.
136. Marc J. Seiffer, *Wizard, the Life and Times of Nikola Tesla*, Birch Lane Press, 1996, page 153.

137. Letter by Clara Bloomfield-Moore, dated September 14, 1895.
138. Letter by J.R. Zuberbrauer, 'Philosophy of the Keely Drama,' *The Sun*, November 28, 1897.
139. See: Joscelyn Godwin, *Arktos the Polar Myth*, Phanes Press, 1993, John Michell, *At the Centre of the World*, Thames and Hudson, 1994.
140. *The Unknown World*, December 15, 1894, vol.I, No.5, page 237.
141. Daniel Cohen, *The Great Airship Mystery*, Dodd, Mead & Company, 1981, page 129.
142. Curtis C. Smith, editor, *Twentieth-Century Science Fiction Writers*, St. James Press, 1986, page 643.
143. Illustration of the cover of this particular issue of 'Frank Reade Library' appeared in *The Complete Book of Outer Space*, edited by Jeffrey Logan, Maco Magazine Corporation, 1953, page 113, and stems from Sam Moskowitz' collection.
144. Everett F. Bleiler, *Science Fiction: The Early Years*, The Kent State University Press, 1990, page 557.
145. Daniel Cohen, *The Great Airship Mystery*, Dodd, Mead & Company, 1981, page 129.
146. Everett F. Bleiler, *Science Fiction: The Early Years*, The Kent State University Press, 1990, page 665. Also: Curtis C. Smith, editor, *Twentieth-Century Science Fiction Writers*, St. James Press, 1986, page 645.
147. Peter Haining, *The Jules Verne Companion*, Souvenir Press, 1978, pages 15, 90.
148. Everett F. Bleiler, *Science Fiction: The Early Years*, The Kent State University Press, 1990, pages 630-631.
149. ibid. page 823.
150. William Kingsland, 'De Natuurkunde in de geheime leer,' *Theosophische uitgeversmaatschappij*, 1911, page 25.
151. The name was the house name of the American Stratemeyer Syndicate and was used mainly in the two *Tom Swift* series, a series of juvenile hardcover scientific invention novels. Howard Garis wrote the first 35 of the series. The second series was begun by Mrs. Harriet Adams, the daughter of Edward Stratemeyer. In: Peter Nicholls, *The Encyclopedia of Science Fiction*, Granada, 1979, pages 40, 244, 609.
152. Everett F. Bleiler, *Science Fiction: The Early Years*, The Kent State University Press, 1990, page 15. Bleiler remarks on the name of the inventor, Tesledy, 'undoubtedly invokes the inventor Nikola Tesla.'
153. H.P. Blavatsky, *The Secret Doctrine*, T.P.S., 1888, book I, part II, page 555.
154. *The Complete Index of Tom Swift Adventures*, SF-lovers Archives, Rutgers University, no date. Tale not seen.
155. Desmond Leslie and George Adamski, *Flying Saucers Have Landed*, Werner Laurie, 1953, pages 80, 102-108. Leslie also notes the similarity between Keely and Newbrough's *Oahspe*.
156. Jerome Clark, *The Emergence of a Phenomenon, the UFO Encyclopedia*, vol.2, Omnigraphics, 1992, page 3. Since Cromie was also Northern-Irish, it is possible that both had access to the same source. Perhaps this was in the form of a bound volume of newspapers in a local library. Both could have read Blavatsky's *The Secret Doctrine*.

Chapter 9. The Sorcerer's Apprentice: The Occult Connection

1.   George M. Eberhart, *A Geo-Bibliography of Anomalies*, Greenwood Press, 1980, pages 781-786. Also mentions Keely.
2.   Margery Silver, Introduction to *Gypsy Sorcery and Fortune Telling*, University Books, 1962, page xiii.
3.   Arthur Edward Waite, *The Brotherhood of the Rosicross*, William Rider, 1924, pages 601-610, Christopher McIntosh, *The Rosy Cross Unveiled*, Aquarian Press, 1980, page 129, Rosemary Ellen Guiley, *Harper's Encyclopedia of Mystical & Paranormal Experience*, Castle Books, 1991, pages 519-520, H. Spencer Lewis ed., *Rosicrucian Manual*, Rosicrucian Library vol.8, The Rosicrucian Press Ltd., 1918, 1955, pages 66, 128-129.

4. Francois Clavel, *De Geschiedenis der Vrijmetselarij en der geheimen genootschappen van vroegeren en lateren tijd,* Van Goor, 1843, page 73.
5. This was 'Memoirs of the Year Two Thousand Five Hundred' by Louis Sebastian Mercier. In: Neil Barron, *Anatomy of Wonder: Science Fiction,* R.R. Bowker, 1976, page 26.
6. *The Fringes of Reason,* Harmony Books, 1989, pages 11-12. Also James Webb, *The Occult Underground,* Open Court, 1974, page 117.
7. About this Golden Dawn temple, either in Chicago or Philadelphia, see: Israel Regardie, My Rosicrucian Adventure, *Aries Press, 1936, page 33, and R.A. Gilbert, 'The Golden Dawn Companion,'* The Aquarian Press, 1986, page 40.
8. Christopher McIntosh, *The Rosy Cross Unveiled,* Aquarian Press, 1980, page 132, Michael Howard, *The Occult Conspiracy,* Rider, 1989, page 91.
9. Joscelyn Godwin, Christian Chanel, John P. Deveney, *The Hermetic Brotherhood of Luxor,* Weiser, 1995, page 66.
10. Arthur Edward Waite, *The Brotherhood of the Rosicross,* William Rider, 1924, page 613.
11. Rosemary Ellen Guiley, *The Encyclopedia of Witches and Witchcraft,* Facts on File, 1989, page 200.
12. ibid.
13. Margery Silver, introduction to *Gypsy Sorcery and Fortune Telling,* University Books, 1962, pages xiii, xiv.
14. ibid. page xiv.
15. Charles Godfrey Leland, *The Poetry and Mystery of Dreams,* E.H. Butler & Co., 1856.
16. Margery Silver, introduction to *Gypsy Sorcery and Fortune Telling,* University Books, 1962, page xi.
17. Clara Bloomfield-Moore, *Keely and His Discoveries, Aerial Navigation,* Kegan Paul, Trench, Trübner & Co., 1893, page 49.
18. Short mention in Richard Cavendish, *A History Of Magic,* Weidenfeld & Nicholson, 1977, page 138.
19. Joscelyn Godwin, Christian Chanel, John P. Deveney, *The Hermetic Brotherhood of Luxor,* Weiser, 1995, page 40.
20. Arthur Edward Waite, *The Brotherhood of the Rosicross,* William Rider, 1924, page 614-615, Christopher McIntosh, *The Rosy Cross Unveiled,* Aquarian Press, 1980, page 132.
21. Jean-Pierre Bayard, *La Spiritualité des Rose-Croix: histoire, tradition et valeur initiatique,* St.-Jean-de-Braye, 1990, Dutch translation, 1994 , pages 183-184.
22. Michael Howard, *The Occult Conspiracy,* Rider, 1989, page 91. According to Howard, the order had a total of 773 members in 1980.
23. Sylvia Cranston, *The Extraordinary Life of Helena Blavatsky,* Putnam, 1993, page 131.
24. ibid. page 133.
25. H.P. Blavatsky, *A Modern Panarion,* vol.1, T.P.S., 1895, pages 15-34.
26. H.P. Blavatsky, *Isis Unveiled,* J.W. Bouton, 1877, pages 125-127.
27. H.P. Blavatsky, *The Secret Doctrine,* T.P.S., 1888, vol.1, pages 554-560.
28. In: 'Lucifer,' no.20, April, 1889.
29. Sylvia Cranston, *The Extraordinary Life of Helena Blavatsky,* Putnam, 1993, page 13.
30. Patrick Curry, *A Confusion of Prophets, Victorian and Edwardian Astrology,* Collins & Brown, 1992, page 39.
31. Nicholas Goodrick-Clarke, *The Occult Roots of Nazism,* The Aquarian Press, 1985, page 19.
32. Joscelyn Godwin, Christian Chanel, John P. Deveney, *The Hermetic Brotherhood of Luxor,* Weiser, 1995, page 47.
33. Adepts are instructed to read Bulwer-Lytton's *Zanoni* and *A Strange Story* with the remark: 'Valuable for its facts and suggestions about Mysticism' and 'Valuable for its facts and suggestions about Magick.' In: *The Equinox,* vol.III, no.1, page 23.

34. Jack Sullivan, *The Encyclopedia of Horror and Science Fiction*, Penguin Books, 1989, page 63, Lewis Spence, *An Encyclopedia of Occultism*, Routledge, 1920, page 256, Karl R. H. Frick, *Licht Und Finsternis, Gnostisch-theosophische und Freimaurerisch-okkulte Geheimgesellshaften bis an die Wende zum 20. Jahrhundert. Wege in die Gegenwart, Teil II: Geschichte ihrer Lehren. Rituale und Organisationen*, Akademische Druck- u. Verlagsanstalt Graz,' 1978, page 351.

35. Thomas H. Burgoyne, *The Light Of Egypt*, vol.II, 1900, 1963, page 66.

36. George Andrews, 'Blind Spots and Taboo Areas in UFO Research,' lecture held at the 32nd National UFO Conference, page 3. Also: Eugenia Macer-Story, 'Abduction and Fascination,' in: 'Magick Mirror, MUFON-NYC Newsletter,' Fall, 1995, pages 6-8.

37. Karl R. H. Frick, *Licht Und Finsternis, Gnostisch-theosophische und Freimaurerisch okkulte Geheimgesellshaften bis an die Wende zum 20. Jahrhundert. Wege in die Gegenwart, Teil II: Geschichte ihrer Lehren. Rituale und Organisationen*, Akademische Druck- u. Verlagsanstalt Graz,' 1978, page 351. Also: Ellic Howe, *The Magicians of the Golden Dawn*, Routledge & Kegan Paul, 1972, pages 31-32. Howe writes that 'no information about an alleged Rosicrucain lodge in Frankfurt is available,' however, the information on the identity of this Frankfurt lodge stems from Frick's monumental standardwork, written six years after Howes study.

38. Karl R. H. Frick, *Licht Und Finsternis, Gnostisch-theosophische und Freimaurerisch-okkulte Geheimgesellshaften bis an die Wende zum 20. Jahrhundert. Wege in die Gegenwart, Teil II: Geschichte ihrer Lehren. Rituale und Organisationen*, Akademische Druck- u. Verlagsanstalt Graz,' 1978, pages 25-26.

39. Peter Bahn, 'Im Zeichen des Rosenkreuzes,' *New Scientific Times*, no.2, 1996, page 31.

40. Karl R. H. Frick, *Licht Und Finsternis, Gnostisch-theosophische und Freimaurerisch okkulte Geheimgesellshaften bis an die Wende zum 20. Jahrhundert. Wege in die Gegenwart, Teil II Geschichte ihrer Lehren. Rituale und Organisationen*, Akademische Druck- u. Verlagsanstalt Graz, 1978, pages 351-352, about this order see: Karl H. Frick, *Die Erleuchteten, Gnostisch-theosophische und alchemisch-rosenkreuzerische Geheimgesellschaften bis zum Ende des 18.Jahrhunderts - ein Beitrag zur Geistescheschichte der Neuzeit*, Akademische Druck- u. Verlagsanstalt Graz, 1973, pages 562-566.

41. Ellic Howe, *The Magicians of the Golden Dawn*, Routledge & Kegan Paul, 1972, page 31.

42. Confirmed in *Comte De Gabalis by the Abbé N. de Montfaucon de Villars. Rendered out of French into English with a Commentary*, published by the Brothers, 1913, page X, and Christopher McIntosh, *The Rosy Cross Unveiled*, Aquarian Press, 1980, page 124. On the connection of Zanoni with the 17th century Rosicrucians, see also: Peter Bahn, *Der Vril-Mythos*, Omega verlag, 1997, pages 42-43.

43. Eugen Lennhoff and Oskar Posner write: '...he appears to have been a freemason.' *Freemasons Magazine and Masonic Mirror* presents him in 1861 as an authority in a Masonic question.' In: *Internationales Freimaurer Lexicon*, Amalthea Verlag, 1932, 1980, page 973.

44. Karl R. H. Frick, *Licht Und Finsternis, Gnostisch-theosophische und Freimaurerisch-okkulte Geheimgesellshaften bis an die Wende zum 20. Jahrhundert. Wege in die Gegenwart, Teil II: Geschichte ihrer Lehren. Rituale und Organisationen*, Akademische Druck- u. Verlagsanstalt Graz,' 1978, page 351.

45. Frick for instance writes that Bulwer-Lytton's connections with contemporary occult, Martinist, Masonic and Rosicrucian circles remain largely unknown and are speculative at best. In ibid, page 352.

46. Tautriadelta, 'Un Magicien Moderne,' *Le Voile d'Isis*, no.46, October 1924, page 673, Alec Maclellan, *The Lost World of Agharti, the Mystery of Vril Power*, Souvenir Press, 1996, page 89.

47. Patrick Curry, *A Confusion of Prophets, Victorian and Edwardian Astrology*, Collins & Brown, 1992, page 40.

48. For these novels, see: J.M. Roberts, *The Mythology of Secret Societies*, Secker & Warburg, 1972, pages 4-7.

49. ibid. page 7, also Michael Howard, *The Occult Conspiracy*, Rider, 1989, page 108.
50. Christopher McIntosh, *The Rosy Cross Unveiled*, Aquarian Press, 1980, page 124.
51. John Hamill editor, *The Rosicrucian Seer*, The Aquarian Press, 1986, page 22. It is argued elsewhere that from the brotherhood of the swastika the Hermetic Brotherhood of Luxor sprang forth. In: Joscelyn Godwin, *Arktos the Polar Myth*, Phanes Press, 993, pages 147-148.
52. ibid. pages 22-23.
53. ibid. Also Patrick Curry, *A Confusion of Prophets, Victorian and Edwardian Astrology*, Collins & Brown, 1992, page 49.
54. Ellic Howe, *The Magicians of the Golden Dawn*, Routledge & Kegan Paul, 1972, pgs 26, 32.
55. Karl R. H. Frick, *Licht Und Finsternis, Gnostisch-theosophische und Freimaurerisch okkulte Geheimgesellschaften bis an die Wende zum 20. Jahrhundert. Wege in die Gegenwart, Teil II: Geschichte ihrer Lehren. Rituale und Organisationen*, Akademische Druck- u. Verlagsanstalt Graz, 1978, page 350.
56. Eliphas Levi, *Transcendental Magic, Its Doctrine and Ritual*, translated and annotated by Arthur Edward Waite, William Rider & Son, 1923, page 151, also Eliphas Levi, *Dogme et Rituel de la Haute Magie*, tome premier, 1855, Félix Alcan, 1903, page 281.
57. Paul Chacornac, *Eliphas Levi, Rénovateur de l'Occultism en France*, Chacornac frères, 1926, pages 143-160, dates this visit in 1854. Karl R. H. Frick, *Licht Und Finsternis, Gnostisch-theosophische und Freimaurerisch okkulte Geheimgesellschaften bis an die Wende zum 20. Jahrhundert. Wege in die Gegenwart, Teil II: Geschichte ihrer Lehren. Rituale und Organisationen*, Akademische Druck- u. Verlagsanstalt Graz,' 1978, page 352, and Arthur Edward Waite, *The Mysteries of Magic*, Redway, 1886, page XV, date this visit in 1853.
58. Paul Chacornac, *Eliphas Levi, Rénovateur de l'Occultism en France*, Chacornac frères, 1926, page 194.
59. ibid.
60. Arthur Edward Waite, *The Mysteries of Magic*, Redway, 1886, page XV.
61. Karl R. H. Frick, *Licht Und Finsternis, Gnostisch-theosophische und Freimaurerisch okkulte Geheimgesellschaften bis an die Wende zum 20. Jahrhundert. Wege in die Gegenwart, Teil II: Geschichte ihrer Lehren. Rituale und Organisationen*, Akademische Druck- u. Verlagsanstalt Graz,' 1978, page 262.
62. ibid. page 296.
63. For an elaborate description of these rituals, see: Eliphas Levi, *Transcendental Magic, Its Doctrine and Ritual*, translated and annotated by Arthur Edward Waite, William Rider & Son, 1923, page 151, also Eliphas Levi, *Dogme et Rituel de la Haute Magie*, tome premier, 1855, Félix Alcan, 1903, page 281, Arthur Edward Waite, *The Mysteries of Magic*, Redway, 1886, pages 309-313, Paul Chacornac, *Eliphas Levi, Rénovateur de l'Occultism en France*, Chacornac frères, 1926, pages 143-160.
64. Paul Chacornac, *Eliphas Levi, Rénovateur de l'Occultism en France*, Chacornac Frères, 1926, page 194. Chacornac cites Arthur Edward Waite, *The Mysteries of Magic*, page 8 as the source; however, there is no such anecdote in Waite's book.
65. ibid. pages 194-195.
66. About Levi's concept of Astral Light, see: Arthur Edward Waite, *The Mysteries of Magic*, Redway, 1886, pages 74-79, Eliphas Levi, *Transcendental Magic, its Doctrine and Ritual*, translated and annotated by Arthur Edward Waite, William Rider & Son, 1923, Eliphas Levi, *Dogme et Rituel de la Haute Magie*, vols. 1 & 2, Germer Baillière, 1855, 1861.
67. Fred Gettings, *Encyclopedia of the Occult*, Rider, 1986, pages 28-30.
68. Karl R. H. Frick, *Licht Und Finsternis, Gnostisch-theosophische und Freimaurerisch okkulte Geheimgesellschaften bis an die Wende zum 20. Jahrhundert. Wege in die Gegenwart, Teil II: Geschichte ihrer Lehren. Rituale und Organisationen*, Akademische Druck- u. Verlagsanstalt Graz,' 1978, page 352.

69. Arthur Edward Waite, *The Mysteries of Magic*, Redway, 1886, page XV, Paul Chacornac, *Eliphas Levi, Renovateur de l'Occultism en France*, Chacornac Frères, 1926, page 198.

70. Richard H. Laars, *Eliphas Levi, der grose kabalist*, 1922, page 25.

71. Bulwer-Lytton, *The Coming Race* Tauchnitz, 1873, pages 52-54.

72. ibid. page 64.

73. ibid. page 89.

74. ibid. pages 131-132.

75. ibid. pages 136-137.

76. ibid. page 73.

77. ibid. page 79.

78. ibid. pages 194-195.

79. This novel was *Beneath Your Very Boots*, published in 1889. See: Curtis C. Smith editor, *Twentieth-Century Science-Fiction Writers*, St. James Press, 1986, page 370. Novel also treated in: Alec Maclellan, *The Lost World Of Agharti, The Mystery of Vril Power*, Souvenir Press, 1982, 1996, pages 18-20.

80. On the Tesla-Bulwer-Lytton-connection, see: Marc J. Seiffer, *Wizard, the Life and Times of Nikola Tesla*, page 65. Also in: Steven R. Elswick, editor, *Proceedings of the 1990 International Tesla Symposium*, The International Tesla Society Inc., 1991, pages 1-36-1-38.

81. H.P. Blavatsky, *Isis Unveiled*, J.W. Bouton, 1877, pages 125-126.

82. ibid. page 126.

83. Thomas H. Burgoyne, *The Light Of Egypt*, 1889, 1963, page 97.

84. Guenther Wachsmuth, *Die Ätherischen Bildekräfte in Kosmos*, Der Kommende Tag, 1924, page 4.

85. 'Keely And His Motor,' *Evening Bulletin*, August 22, 1887.

86. H.P. Blavatsky, *The Secret Doctrine*, T.P.S., 1888, vol.1, page 555.

87. W. Scott-Elliot, *The Story of Atlantis*, T.P.S., 1896, page 53.

88. Peter Bahn, *Der Vril-Mythos*, Omega Verlag, 1997, page 47.

89. Michel Lamy, *Jules Verne Initié et Initiateur*, 1984, 1994, Editions Payot et Rivages, page 191.

90. ibid. page 132.

91. Sykes writes: 'The description of air craft powered by Vril would indicate that Lytton was aware of the inventions of John Worrell Keely, the first public announcement was made in the same year as *The Coming Race* was published. I cannot as yet find any written evidence to confirm this but the similarity of ideas makes it highly probable.' In: Egerton Sykes, 'The Nature of Vril,' *Disc Digest*, vol.1, no.1, April, 1964, page 15. The article was a reprint from *Pendulum*, vol.13, no.12., February 1964. However, in 1972 Sykes wrote: 'At the time he (Keely) must have been influenced by the work of Reichenbach in 1862, and of Bulwer-Lytton in 1871.' In: Egerton Sykes, *The Keely Mystery*, Markham House, 2nd edition 1972, page 8. Gettings states Sykes' first theory as a fact without referencing it. In: Fred Gettings, *Encyclopedia of the Occult*, Rider, 1986, page 6.

92. See for instance: Jacques Vallee, *Passport to Magonia*, Regnery, 1969, pages 9-10. Cites as his source Montfaucon's, *Comte de Gabalis*, 1670 edition, page 297. My source was a 1742 edition, where the story appears on pages 167-169 of vol.1.

93. 'Mrs. Moore Dies Broken-Hearted' *The Times*, January 6, 1899. Also: 'Mrs. Bloomfield-Moore Dead,' *New York Times*, January 6,1899.

94. 'Her Anxiety Brought Death,' *Philadelphia Inquirer*, January 6, 1899. Also: 'Mrs. Bloomfield-Moore Dead,' *New York Times*, January 6, 1899.

95. 'Keely's Secret,' *The World*, May 11, 1890. Also in Clara Bloomfield-Moore, 'Robert Browning, written by request to be read before a meeting of the Massachusetts Browning Society which is to be held at Mosely Homestead, Westfield, Lippincott, 1890,' text on Sympathetic Vibratory Physics Homepage, Internet. Browning by the way met with theosophist Olcott in 1884 but it is unlikely that the two would have discussed Keely and his discoveries during

their meeting. In: Sylvia Cranston, *The Extraordinary Life & Influence of Helena P. Blavatsky*, Putnams, 1993, pages 257-258.

96. Clara Bloomfield-Moore 'Robert Browning, written by request to be read before a meeting of the Massachusetts Browning Society which is to be held at Mosely Homestead, Westfield, Lippincott, 1890,' text on Sympathetic Vibratory Physics Homepage, Internet. On this matter see also: Clara Bloomfield-Moore, *Keely and His Discoveries, Aerial Navigation*, Kegan Paul, Trench, Trübner & Co., 1893, pages 295-296.

97. 'Count Von Rosen Talks Of Keely,' *The Evening Bulletin*, March 15, 1899. According to a J.E. Wright, an electrician who had visited Keely's workshop, Bloomfield-Moore 'believes in the theory of the control of mind over matter.' In: 'An Expert On Keely's Motor,' *New York Herald*, January 19, 1896.

98. A photo of the plane with the swastika's is kept in the Tikkakoski Air Musueum, Finland.

99. Peter Nicholls, editor, *The Encyclopedia of Science Fiction*, Granada, 1981, pages 282-283, 229, 312, 494. Bloomfield-Moore quotes a passage by Hinton. See: *Keely and His Discoveries, Aerial Navigation*, Kegan Paul, Trench, Trübner & Co., 1893, page 329-331.

100. George M. Eberhart, *UFOs and the Extraterrestrial Contact Movement*, The Scarecrow Press, 1986, vol.2, page 1123.

101. See: *Equinox*, vol.III, no.1, page 22. Crowley comments: 'The textbook on this subject.'

102. 'Mrs. Bloomfield-Moore Dead,' unspecified London newspaper, January 5, 1899, Sympathetic Vibratory Physics Homepage, Internet.

103. Clara Bloomfield-Moore, *Keely and His Discoveries, Aerial Navigation*, Kegan Paul, Trench Trübner & Co., 1893, page 190.

104. Joscelyn Godwin, Christian Chanel, John P. Deveney, *The Hermetic Brotherhood of Luxor*, Weiser, 1995, pages 428-429.

105. Karl R. H. Frick, *Licht Und Finsternis, Gnostisch-theosophische und Freimaurerisch okkulte Geheimgesellshaften bis an die Wende zum 20. Jahrhundert. Wege in die Gegenwart,' Teil II: Geschichte ihrer Lehren. Rituale und Organisationen*, Akademische Druck- u. Verlagsanstalt Graz, 1978, page 263.

106. Sylvia Cranston, *The Extraordinary Life of Helena Blavatsky*, Putnam, 1993, page 145.

107. R.A. Gilbert noted this in his, 'Provenance unknown, a tentative solution to the riddle of the Cipher Manuscript of the Golden Dawn,' page 84. The word appears in Blavatsky's *Isis Unveiled*, vol.I, page 19.

108. Sylvia Cranston, *The Extraordinary Life of Helena Blavatsky*, Putnam, 1993, page 361.

109. Lewis Spence, *An Encyclopedia of Occultism*, George Routledge & Sons, page 388.

110. Cheiro, *True Ghost Stories*, The London Publishing Company, no date, page 163.

111. Paul Begg, Martin Fido and Keith Skinner. *Jack the Ripper A to Z*, Headline Book Publishing, 1994, page 445. Stephenson's pen-name was 'Tau Tria Delta,' under which he wrote about Bulwer-Lytton's occult doings. As he may be considered an early source about Bulwer-Lytton's more darker side, one must treat these strange anecdotes accordingly. Stephenson never divulged the origin of his highly curious nom-de-plume, although he hinted at some profound occult significance. For an example of his writing: see this chapter, note 46.

112. ibid. pages 447-449.

113. Lewis Spence, *An Encyclopedia of Occultism*, George Routledge & Sons, page 388.

114. Marc J. Seiffer, *Wizard, the Life and Times of Nikola Tesla*, Birch Lane Press, 1996, page 164.

115. Clara Bloomfield-Moore, *Keely and His Discoveries, Aerial Navigation*, Kegan Paul, Trench, Trübner & Co., 1893, page 186. She borrowed this quotation from Blavatsky's article 'Kosmic Mind,' *Lucifer*, April, 1890, page 89. More on Edison in ibid. pages 197-198.

116. In the letter with which Edison returned the signed forms, he wrote: 'Please say to Madame Blavatsky that I have received her very curious work and I thank her for the same. I SHALL READ BETWEEN THE LINES!' (Edison's capitals). The 'very curious work' was Blavatsky's *Isis Unveiled*. Letter in the Edison Archives, Madison, New Jersey. Copies obtained through

Dale Pond. See also: Sylvia Cranston, *The Extraordinary Life and Influence of Helena Blavatsky,* Putnam, 1993, page XX, pages 183-185.

117. Sylvia Cranston and Joseph Head, *Reincarnation, The Phoenix Fire Mystery,* Julian Press, 1986, page 419.

118. Gregory Little, *People of the Web,* White Buffalo Books, 1990, page 10.

119. Helena P. Blavatsky, 'The Blessings Of Publicity,' *Lucifer,* no.48, August, 1891.

120. 'Is Keely The New Mahatma?,' *New York Times,* April 29, 1896.

121. 'Is It The Vibratory Keely?,' *New York Times,* May 5, 1896.

122. Henry Steel Olcott, 'H.P.B. and the Keely Force,' *The Theosophist,* Vol. XX, no. 11, august 1899, pages 687-688.

123. Henry Steel Olcott, *Old Diary Leaves, The Only Authentic History of the Theosophical Society, Third Series, 1883 - 1887,* T.P.S., 1904, page 386.

124. Dr. A. Marques, *Scientific Corroborations of Theosophy,* The Theosophical Publishing Society, 1901, revised and greatly enlarged edition, 1908, pages 85-86.

125. Helmut Werner, *Lexicon der Esoterik,* Fourir, 1991, pages 188, 567.

126. James Webb, *The Occult Underground,* Open Court Publishing, 1974, pages 93-94. More biographical details in Nicholas Goodrick-Clarke, *The Occult Roots of Nazism,* the Aquarian Press, 1985, pages 24-27.

127. Leopold Engel, Mallona, *Der Untergang des Asteroiden-Planeten,* Turm Verlag, 1961, page 13.

128. Clara Bloomfield-Moore, *Keely's Secrets,* T.P.S., 1888, pages 25, 33-34.

129. Joël Rost and Jacques Ravatin, *Les Desintegrators,* editions L'Originel, 1994, page 272.

130. Clara Bloomfield-Moore, *Keely's Secrets,* T.P.S., 1888, page 23.

131. ibid. page 34, also pages 33-35.

132. Joscelyn Godwin, Christian Chanel, John P. Deveney, *The Hermetic Brotherhood of Luxor,* Weiser, 1995, page 429.

133. Thomas H. Burgoyne, *The Light Of Egypt,* vol.I, 1889, 1963, page 116.

134. Joscelyn Godwin, Christian Chanel, John P. Deveney, *The Hermetic Brotherhood of Luxor,* Weiser, 1995, page 419.

135. Emma Hardinge Britten, *Ghost Land or Researches into the Mysteries of Occultism,* Progressive Thinker Publishing House, 1897, page 10 of the catalogue at the back. The advertisement slogan on page 358. The identity of the writer of *Ghost Land* has never been established; rumor has it that Burgoyne was the writer. See also: Joscelyn Godwin, Christian Chanel, John P. Deveney, *The Hermetic Brotherhood of Luxor,* Weiser, 1995, page 57.

136. Sylvia Cranston, *The Extraordinary Life of Helena Blavatsky,* Putnam, 1993, page 124.

137. W.J. Colville, 'Autobiography of W.J. Colville,' in: *Universal Spiritualism, Spirit Communion In All Ages Among All Nations,* R.F. Fenno & Company, 1906, pages 15-38.

138. Leslie Shepard, *Encyclopedia of Occultism & Parapsychology,* Gale Research Inc., 1996, pages 307-308.

139. W.J. Colville, 'Autobiography of W.J. Colville,' in: *Universal Spiritualism, Spirit Communion in All Ages Among All Nations,* R.F. Fenno & Company, 1906, pages 15- 38.

140. William J. Colville, 'John Worrall (sic) Keely. A memorial address, delivered Sunday, Nov. 27, 1898, in Casino Hall, Thirteenth Street and Girard Avenue, Philadelphia, Penn.,' Banner of Light Publishing Co., 1899, page 5. On his lecture: short, unreferenced clipping in George Canby, 'Keely Motor Scraps,' no.2, not published.

141. ibid. page 15.

142. '...whose interest in the Occult had been cut short by his untimely death by drowning through the loss of the Titanic,' Cheiro notes. In: *True Ghost Stories,* the London Publishing Company, no date, page 34. On his account of his meeting with Blavatsky, see: Cheiro, *Mysteries and Romances of the World's Greatest Occultists,* Herbert Jenkins, 1935, pages 170-180.

143. Cheiro, *True Ghost Stories,* the London Publishing Company, no date, page 219.

144. Helmut Werner, *Lexicon der Esoterik,* Fourier, 1991, page 125.

145. Cheiro, *True Ghost Stories,* the London Publishing Company, no date, pages VII, 90. On his refusal to join the theosophical society, see: Cheiro, *Mysteries and Romances of the World's Greatest Occultists,* Herbert Jenkins, 1935, page 179.
146. Richard Cavendish, *Encyclopedia of the Unexplained, Magic, Occultism and Parapsychology,* Rainbird Reference Books, 1974, page 55.
147. Cheiro, *True Ghost Stories,* London Publishing Company, no date, pages 228-231.
148. The short story is part of the collection written by Cheiro, *True Ghost Stories,* the London Publishing Company, no date. Citation appears on page 235.
149. ibid. pages 239-240.
150. ibid. page 239.
151. ibid. page 236.
152. ibid. page 238.
153. ibid. pages 240-241.
154. ibid. page 245.
155. ibid. pages 245-246.
156. ibid. page 251.
157. 'The Keely Motor,' *Ledger And Transcript,* December 25, 1895.
158. ibid.
159. Clara Bloomfield-Moore, *Keely's Secrets,* T.P.S., 1888, page 25. See also Clara Bloomfield-Moore, *Keely and His Discoveries, Aerial Navigation,* Kegan Paul, Trench, Trübner & Co., 1893, page 250.
160. Dumas Malone editor, *Dictionary of American Biography,* vol. VII, Scribner's, 1934, page 199.
161. James Webb, *The Occult Underground,* Open Court Publishing, 1974, page 293.
162. Sylvia Cranston, *The Extraordinary Life & Influence of Helena P. Blavatsky,* Putnam, 1993, page 134.
163. Clara Bloomfield-Moore, *Keely's and His Discoveries, Aerial Navigation,* Kegan Paul, Trench, Trübner & Co., 1893, page 92. It is possible that Pancoast and Philadelphia cabbalist Meyer knew each other, both being deeply involved in cabbalistic researches. But while we have established that we know not much on Pancoast, we know even less, which amounts to nothing, of Meyer. In 1888 Meyer privately published the book *Qabbalah, The Philosophical Writings Of Solomon Ben Yeduda Ibn Gebirol or Avicebron and their Connection with the Hebrew Qabbalah and Sepher Ha Zohar...Also an Ancient Lodge of Initiates, translated from the Zohar & an Abstract of an Essay upon the Chinese Qabbalah, contained in the...Yih King.* The book was based upon early 19th century French translations of 12th century Hebrew versions of the main work of the Spanish-Jewish philosopher Ibn Gebirol, who introduced neoplatonic thought in the west. Possibly Meyer's cabbalist activities explain why it is alleged that Meyer held information on Keely while nothing was found, perhaps later writers have confused Pancoast with Meyer since both were cabbalists. It is inviting to speculate that for instance Pancoast's huge library held files on Keely, but Pancoast died before Keely and it is not known what became of his library and private documents. It also is possible that Meyer became interested in Keely's researches through Pancoast. I have studied Meyer's book - the only documentation on him available to us now. Although published at the time that Keely was well known in Philadelphia, the book is a treatise on the cabbala and there is nothing at the surface that is pertaining to, or hinting at Keely in its pages.
164. Letter from Clara Bloomfield-Moore, dated September 9, 1894.
165. *Beiträge zur Rudolph Steiner Gesamtausgabe,* Nr.107, Michaeli, 1991, pages 22-23.
166. R. Harte, introduction in Clara Bloomfield-Moore, *Keely's Secrets,* Theosophical Publishing House, July, 1888, pages 6-7.
167. H.P. Blavatsky, *The Secret Doctrine,* T.P.S., 1888, book I, part II, page 555.

168. Apart from having read about Keely in Blavatsky's *The Secret Doctrine*, Steiner once again read about Keely in C.G. Harrison's, *The Transcendental Universe*, published in 1893, in which a reference on Keely occurred, according to the source of note 169.

169. See: 'Vom Keely-Motor zur Strader Machine,' in: *Der Europäer*, Jg.1/Nr.6, April, 1997, pages 13-14.

170. Walter Johannes Stein, *Weltgeschichte im Lichte des Heiligen Gral, das Neunte Jahrhundert*, Orient-Occident Verlag, 1928.

171. 'Vom Keely Motor zur Strader Machine,' *Der Europäer*, Jg.1/Nr.6, April, 1997, page 13.

172. For title, see note 84. On Wachsmuth, Steiner and Keely, see: 'Vom Keely Motor zur Strader-Machine,' *Der Europäer*, Jg.1/Nr.6, April, 1997, page 12.

173. Although Steiner, during a lectur on May 11, 1924, told that he had modeled Strader on Gideon Spicker (1840 - 1912), a Professor of philosophy at the German town of Münster, this is only so in the first and second parts of his 'Mystery Plays.' In the third, Strader invents his device and in the fourth his device refuses to work in a factory. Strader obtains a number of characteristics which stem from Keely. On this, see also: 'Vom Keely Motor zur Strader Machine,' *Der Europäer*, Jg.1/Nr.6, April, 1997, pages 13-14.

174. Rudolf Steiner, *In Geänderter Zeitlage, Die soziale Grundforderung unserer Zeit* (G.A. 186), Cycle nr. 51, 6 lectures, Dornach, 1918. *Onder veranderde tijdsomstandigheden*, Zevenster, 1983, page 68.

175. ibid. page 64.

176. ibid. page 62.

177. ibid. page 65.

178. ibid. page 67.

179. ibid. page 66.

180. H.P. Blavatsky, *The Secret Doctrine*, T.P.S., 1888, book I, part II, page 560.

181. Rudolf Steiner, *In Geänderter Zeitlage, Die soziale Grundforderung unserer Zeit* (G.A. 186), Cycle nr. 51, 6 lectures, Dornach, 1918. *Onder veranderde tijdsomstandigheden*, Zevenster, 1983, pages 68-69.

182. Anthony C. Sutton, *America's Secret Establishment*, Liberty House Press, 1986.

183. See: Anonymous, 'A ritual and Illustration of Freemasonry, accompanied by numerous engravings, and a key to Phi Beta Kappa,' W. Reeves, no date (but 1880's), page 249.

184. Anthony C. Sutton, *America's Secret Establishment*, Liberty House Press, 1986, page 247.

185. ibid. page 24.

Chapter 10. The Sorcerer's Apprentice: The Occult Connection

1. Gerry Vassilatos' 11 part *The Vril Series*, published by Borderlands, is an admirable exploration and essential reading in this otherwise largely unmapped territory.

2. Helmut Werner, *Lexicon der Esoterik*, Fourir, 1991, page 395.

3. Eliphas Levi, *Histoire de la Magie*, Felix Alcan, 1862, 1892, pages 265-267, tale also told by Levi's brilliant pupil Stanislaus de Guaita in his *Essays de Sciences Maudites*, vol.1, au Seul du Mystère, George Carré, 1890, pages 53-54. I have also noted a less detailed account in Rene Noorbergen, *Secrets of the Lost Races*, The Bobbs-Merrill Company Inc., 1977, page 51. The account is found in: Eliphas Levi, *Histoire de la Magie*, Felix Alcan, 1862, 1892, pages 267-270, also in Stanislaus de Guaita, *Essays de Sciences Maudites*, vol.1, au Seul du Mystère, George Carré, 1890, page 54. Both cite the book *Le Grand and Petit Albert*, attributed to - but this is by no means certain - Albertus Magnus, as their source.

4. Helmut Swoboda, *Der Künstliche Mensch*, Heimeran, 1967, pages 45-48. See also: Helmut Werner, *Lexicon der Esoterik*, Fourir, 1991, page 21. For a popular treatment see: Paul Thompson, 'The Lore of the Homunculus.' In: *Fate*, vol.47, no.9, Issue 534, page 52.

5. ibid. page 46.

6. Eliphas Levi, *Histoire de Magie*, Felix Alcan, 1862, 1892, pages 267-268.

7. Helmut Swoboda, *Der Künstliche Mensch*, Heimeran, 1967, page 48.

8. Clara Bloomfield-Moore, *Keely's Secrets*, T.P.S., 1888, page 30.
9. Hanns Manfred Heuer, *Hax Pax Max*, Merlin Verlag, 1973, page 56. The chapter 'Das Geheimnis der ewig brennnenden Lampen' also holds a number of interesting accounts from other sources beside Jennings. Other interesting accounts are found in Rene Noorbergen, *Secrets of the Lost Races*, The Bobbs-Merrill Company Inc., 1977, pages 50-54.
10. Hargrave Jennings, *The Rosicrucians, their Rites and Mysteries*, Routledge, n.d., pages 13-16. About the tomb being illumined by an ever burning lamp, see: Anonymous (but written by Hargrave Jennings), *Mysteries of the Rosicross*, A. Reader, 1891, pages 25, 29-30. This booklet was part of the 'Nature Worship Series,' in 1905 translated by O.T.O. co-founder Theodor Reuss. In: Helmut Möller and Ellic Howe, *Merlin Peregrinus*, Königshausen & Neumann, 1986, page 150.
11. The tale was written by Edward Olin Weeks. Everett F. Bleiler, *Science Fiction: The Early Years*, The Kent State University Press, 1990, page 794.
12. Jacques Bergier, *Livres Maudites*, J'ai Lu, 1972, *Verdoemde Boeken*, Ankh Hermes, 1972, page 62. Trithemius decribed this in his 'Steganographia,' according to Bergier.
13. Hanns Manfred Heuer, *Hax Pax Max*, Merlin verlag, 1973, pages 85-86. Heuer gives a lengthy quote from the original French text, *Les admirables secrets de magie naturelle du Grand-Albert et de Petit Albert*, Paris, no date, also Levi's and de Guaita's source.
14. ibid. page 87.
15. Stephen Skinner, preface, in: Donald C. Laycock, The *Complete Enochian Dictionary*, Askin Publishers, 1978, pages 11-13.
16. My source: 'Wichtiger Brief von dem Philosopho Jesse von den Geheimnissen der güldenen Zeit, neu herausgegeben von Dr. H. Haase,' *Lotusblätter*, III Jahrgang, Nr.3/6, März/Juni 1923, pages 32-36. His source is Fr. Chr. Oetinger, *Die Philosophie der Alten wiederkommend in der güldenen Zeit*, 1762. Account also published in abbreviated form in Hans Manfred Heuer, *Hax Pax Max*, Merlin Verlag, 1973, pages 84-85. Cites as his source Siegmund Heinrich Güldenfalk, *Sammlung von mehr als hundert wahrhaften Transmutationsgeschichten*, Frankfurt, 1784.
17. Bulwer-Lytton, *The House and the Brain* is also often called *The Haunted and the Haunters*. It is found in most anthologies of ghost stories.
18. Manfred Heuer, *Hax Pax Max*, Merlin Verlag, 1973, pages 87-88.
19. Letter from H.S. Olcott to Thomas A. Edison, dated December 20, 1889. These strange tales possibly found their continuation in the rumor that Albert Pike and other early 19th century freemasons secretly made use of radio systems long before Marconi's patent in 1896. This is briefly mentioned in Neal Wilgus, *The Illuminoids*, Sun Books, 1978, page 116. Wilgus cites conspiracy researcher Carr as his source.
20. Rupert T. Gould, *Oddities, A Book of Unexplained Facts*, Bless, 1928, University Books, second ed. 1966, page 91. About Orffyreus, see also: Friedrich Klemm, *Perpetuum mobile, Ein 'unmöglicher' Menschheitstraum*, Harenberg, 1983, pages 59-71, and C.A. Crommelin, 'La roue d'Orffyreus,' *Janus*, no.49, 1960, pages 47-52.
21. ibid.
22. Eugen Lennhoff & Oskar Posner, *Internationales Freimaurer Lexikon*, Amalthea Verlag, 1932, 1980, page 1027.
23. Francois Clavel, *Geschiedenis der vrijmetselarij en der geheime genootschappen van vroegen en lateren tijd*, Van Goor, 1843, pages 178-179.
24. Arthur Edward Waite, *A New Encyclopedia of Freemasonry*, Rider, 1926, vol.II, page 147.
25. Eugen Lennhoff & Oskar Posner, *Internationales Freimaurer Lexikon*, Amalthea Verlag, 1932, 1980, page 1027.
26. Hans-Jürgen Glowka, *Deutsche Okkultgruppen 1875-1937*, Arbeitsgemeinschaft für Religions und Weltanschauungsfragen, 1981, page 92. The existence of this order is highly dubious; a source in Berlin who had handled much original and unique materials, including documents, diaries, seals, charters and ritual papers of the Fraternitas Saturni, who told me

that he had grave doubts as to its existence. He had, at one time, seen the rituals of this Free-Masonic Order of the Golden Centurium, but these originated from the Fraternitas Saturni. According to him the story of this 1840's order was concocted by Bardon who is quoted by Glowka and Flowers, who admits that 'most of the material having to do with the FOGC seems quite legendary and fantastic in tone.'

27. Stephen E. Flowers, *Fire and Ice*, Llewellyn, 1994, page 6.
28. ibid. Page 107.
29. U. Milankowitsch, 'Die vergötterte Idee Der organische oder positive Idealismus,' *Lotusblätter*, III Jahrgang, Nr. 1/2, January/February 1923, page 20.
30. Emma Hardinge, *Modern Spiritualism: A Twenty Year Record of the Communion Between Earth and the World of Spirits*, published by the author, fourth edition, 1870, pages 217-218.
31. ibid. page 219.
32. ibid. page 220.
33. ibid. page 221.
34. ibid. page 222.
35. ibid. pages 222-224.
36. Lewis Spence, *An Encyclopedia Of Occultism*, Routledge, 1920, page 293.
37. ibid. Also in Emma Hardinge, *Modern Spiritualism: A Twenty Year Record of the Communion Between Earth and the World of Spirits*, published by the author, fourth edition, 1870, page 226.
38. Emma Hardinge, *Modern Spiritualism: A Twenty Year Record of the Communion Between Earth and the World of Spirits*, published by the author, fourth edition, 1870, pages 228-229.
39. Except from the earlier mentioned sources, see: George Riland, *The New Steinerbooks Dictionary of the Paranormal*, Steinerbooks, 1980, page 202. George M. Eberhart, *A Geo-Bibliography Of Anomalies*, Greenwood Press, 1980, page 963. Of this inventor, Fort writes: 'The earliest fuelless motor 'crank' of whom I have record is John Murray Spear, back in the period of 1855...' In: *Wild Talents*, 'Claude Kendall,' 1932, page 335. Eberhart lists a 1970's study on Spear which I have not seen.
40. Emma Hardinge, *Modern Spiritualism: A Twenty Year Record of the Communion Between Earth and the World of Spirits*, published by the author, fourth edition, 1870, pages 309, 314, 317, 320.
41. ibid. page 313.
42. ibid. page 312.
43. ibid. page 317.
44. ibid. pages 314, 318.
45. ibid. page 319, 332.
46. Emma Hardinge Britten, *Ghostland, or Researches into the Mysteries of Occultism*, Boston, 1876, Progressive Thinker Publishing House, 1897, pages 260-269. For the description of the temple, page 267.
47. ibid. page 267. See also: Joscelyn Godwin, *Arktos the Polar Myth*, Phanes Press, 1993, pages 81-82.
48. Paul Chacornac, *Eliphas Levi, Rénovateur De L'Occultisme en France*, Chacornac Frères, 1926, page 137.
49. James Webb, *The Occult Underground*, Open Court, 1974, pages 273-274.
50. Paul Chacornac, *Eliphas Levi, Rénovateur De L'Occultisme en France*, Chacornac Frères, 1926, pages 136-137.
51. James Webb, *The Occult Underground*, Open Court, 1974, Pages 257-258.
52. ibid.
53. ibid. Page 259.
54. ibid. Page 260.
55. Stephen E. Flowers, *Fire and Ice*, Llewellyn, 1994, page 3.
56. James Webb, *The Occult Underground*, Open Court, 1974, Pages 262-264.

57. Paul Chacornac, *Eliphas Levi Rénovateur De L'Occultism En France,* Chacornac Frères, 1926, page 136.
58. ibid.
59. ibid. page 137-138.
60. ibid. page 137.
61. James Webb, *The Occult Underground,* Open Court, 1974, Page 273.
62. For an analysis of Saint-Yves d'Alveydre and Agarttha, see Joscelyn Godwin, *Arktos the Polar Myth,* Phanes Press, 1993, pages 83-87.
63. Albert L. Caillet, *Manuel Bibliographique des Sciences Psychiques ou Occultes,* Lucien Dorbon, 1913, tome III, page 464.
64. ibid. page 464.
65. James Webb, *The Occult Underground,* Open Court, 1974, page 273.
66. Albert L. Caillet, *Manuel Bibliographique des Sciences Psychiques ou Occultes,* Lucien Dorbon, 1913, tome III, page 464.
67. ibid. page 464.
68. James Webb, *The Occult Underground,* Open Court, 1974, page 273.
69. This is a mechanism, not unlike what happens in ufology, which I have often seen repeated while studying early free energy inventors; in Holland for instance, the same happened during the press coverage of Johannes Wardenier, inventor of a fuelless motor around 1933.
70. Clara Bloomfield-Moore, *Keely and His Discoveries, Aerial Navigation,* Kegan Paul, Trench, Trübner & Co., 1893, page 101. Also in: 'Keely's Sunday In Jail,' *New York Times,* November 19, 1888.
71. 'Keely's Marvels made Plain At Last,' unspecified clipping, 1891. There is uncertainty concerning Browne and he seemed to have been bluffing; he was an 'inventing' journalist, who had been 'dissapointed of gain.' In: Clara Bloomfield-Moore, *Keely and His Discoveries, Aerial Navigation,* Kegan Paul, Trench, Trübner & Co., 1893, page 263-264. See also chapter VII and chapter VII note 118.
72. Clara Bloomfield-Moore, *Keely and His Discoveries, Aerial Navigation,* Kegan Paul, Trench, Trübner & Co., 1893, page 89.
73. Very probably this was 'The Keely Motor Criticized,' although this pamphlet does not contain a quotation by A. Arnold. See chapter 1 for a discussion of this pamphlet and citations of its contents.
74. 'The Keely Motor. It Is Not A New Discovery - Another letter On The Subject,' *New York Times,* July 14, 1875. In a second letter, Arnold was quick to disengage his name with that of Keely, as he had been 'annoyed with a host of enquiries since the notice referred to in *The Times...*' Letter dated July 12, 1875: 'The New Motor,' *New York Times.* The copy I have is erroneously dated July 14, 1875.
75. 'A Rival to the Keely Motor,' *New York Times,* July 28, 1875.
76. 'The Keely Motor In China,' *The Practical American,* February, 1880. The article cites the Imperial decree, published in the official *Gazette of Peking,* of which it gives a translation. Perhaps meant as a hoax.
77. Egerton Sykes, *The Keely Mystery,* Markham House, 2nd revised edition, 1972, page 23. Original article not seen.
78. Richard Cavendish, *Encyclopedia of the Unexplained, Magic, Occultism and Parapsychology,* Rainbird Reference Books, 1974, page 250.
79. For a fuller account of this amazing entanglement and Steiner's role, see: Stephen Flowers, *Fire and Ice,* Llewellyn, 1994, pages 7-14, Francis King, *The Secret Rituals of the O.T.O.,* Samuel Weiser, 1973, pages 22-27, Ellic Howe, *The Magicians of the Golden Dawn,* Routledge, 1972, page 266, and Francis King, *Ritual Magic in England,* Neville Spearman, 1970, pages 99, 101-106, 205-207. For a history of the Hermetic Brotherhood of Luxor, see: Joscelyn Godwin, Christian Chanel, John P. Deveney, *The Hermetic Brotherhood of Luxor,* Samuel Weiser, 1995.

80. Nicholas Goodrick-Clarke, *The Occult Roots of Nazism*, The Aquarian Press, 1985, pages 52-53.
81. Wilfried Daim, *Der Mann, der Hitler die Ideen gab*, Isar Verlag, 1958, pages 162-163. James Webb, *The Occult Establishment*, Richard Drew, 1981, page 279.
82. Peter Bahn, *Der Vril-Mythos*, Omega Verlag, 1996, page 106.
83. Manfred Ach and Clemens Pentrop, *Hitler's Religion*, Arbeitsgemeinschaft für Religions- und Weltanschauungsfragen, 3rd edition 1982, page 12, Nicholas Goodrik Clarke, *The Occult Roots of Nazism*, The Aquarian Press, 1985, page 95.
84. Nicholas Goodrik-Clarke, *The Occult Roots of Nazism*, The Aquarian Press, 1985, page 95.
85. Wilfried Daim, *Der Mann, der Hitler die Ideen gab*, Isar Verlag, 1958, pages 110-112, Manfred Ach and Clemens Pentrop, *Hitler's Religion*, Arbeitsgemeinschaft für Religions- und Weltanschauungsfragen, 3rd edition 1982, pages 12-13. Of Liebenfels' patents, Webb writes that Liebenfels in fact filed 'numerous patents.' In: James Webb, *The Occult Establishment*, Richard Drew, 1981, page 512.
86. According to Liebenfels. In: Wilfried Daim, *Der Mann, der Hitler die Ideen gab*, Isar Verlag, 1958, pages 92-102, Manfred Ach & Clemens Pentrop *Hitler's Religion*, Arbeitsgemeinschaft für Religions- und Weltanschauungsfragen, 3d edition 1982, page 13.
87. Stephen Flowers, *Fire and Ice*, Llewellyn, 1994, pages 12-13.
88. Francis King, *Ritual Magic in England*, Neville Spearman, 1970, pages 206-207.
89. Helmut Möller and Ellic Howe, *Merlin Peregrinus*, Königshausen & Neumann, 1986, page 93.
90. Blavatsky's Masonic diploma is depicted as a fold-out frontispiece in: C. Jinarajadasa, *H.P.B. Speaks*, T.P.S., 1951. The date on the diploma is November 24, 1877. About her Masonic diploma, see: H.P. Blavatsky, *A Modern Panarion*, T.P.S., 1895, vol.I, pages 128-131.
91. Stephen E. Flowers, *Fire and Ice*, Llewellyn, 1994, page 12.
92. Francis King, *Sexuality, Magic and Perversion*, Neville Spearman, 1971, page 97.
93. Joscelyn Godwin, Christian Chanel, John P. Deveney, *The Hermetic Brotherhood of Luxor*, Samuel Weiser, 1995, page 37.
94. Lady Queenborough, *Occult Theocracy*, privately published, 1933, page 575. To be used with caution.
95. Clara Bloomfield-Moore, *Keely's Secrets*, T.P.S., 1888, pages 33-34.
96. Helmut Möller and Ellic Howe, *Merlin Peregrinus*, Königshausen & Neumann, 1986, pages 84-85.
97. Clara Bloomfield-Moore, *Keely's Secrets*, T.P.S., page 23. See also: Clara Bloomfield-Moore, *Keely and His Discoveries, Aerial Navigation*, Kegan Paul, Trench, Trübner & Co., 1893, page 89.
98. ibid. page 26.
99. 'Keely Outdone,' *Scientific American*, August 4, 1888, page 69.
100. Gerry Vassilatos, 'Earth Energy And Vocal Radio: Nathan Stubblefield,' *Borderlands*, vol.LI, Number 3, 1995, pages 6-10.
101. ibid.
102. ibid.
103. Albert Ross Parsons, *New Light From the Great Pyramid*, The Metaphysical Publishing Company, 1893, pages 287-388. Parsons mentions as his source the London newspaper *Spectator*.
104. Oskar Korschelt, *Die Nutzbahrmachung der lebendigen Kraft des Äthers*, Berlin, Lothar Volkmar, 1892.
105. Wilhelm Hübbe-Schleiden, 'Die Sonnenäther-Strahlapparate des Professors Oskar Korschelt,' *Sphinx, Monatschrift für Seelen- und Geistesleben*, 1892, VII Jahrgang, Dreizehnter Band, 73, März, page 88.

106. Letter of Oskar Korschelt, dated March 30, 1892, in: *Sphinx, Montaschrift für Seelen- und Geistesleben,* 1892, VII Jahrgang, Dreizehnter Band, 75, Mai, page 286. Letter also refers to an article about his devices in *Zur Guten Stunde,* 1891/1892, Heft 14. Not seen.

107. Wilhelm Hübbe-Schleiden, 'Korschelts Sonnenäther-Strahlapparate,' *Sphinx, Monat schrift für Seelen- und Geistesleben,* 1892, VII Jahrgang, Vierzehnter Band, 77, Juli, page 92.

108. Wilhelm Hübbe-Schleiden, 'Ein neuer Aether-strahlapparat,' *Sphinx, Monatschrift für Seelen- und Geistesleben,* 1893, VIII Jahrgang, Sechzehnter Band, 88, Juni, pages 333 334.

109. G.W. Surya, *Moderne Rosenkreuser, oder die Renaissance der Geheimwissenschaften,* 1907, 7th edition, Johannes Baum Verlag, 1930, page 224.

110. André Nataf, *The Wordsworth Dictionary of the Occult,* Wordsworth Reference, 1994, pages 195-196.

111. A.A. Belinfante, *De Duwkracht, beschouwingen van stoffen en hun bewegingen als gevolgen van botsende- of duwkracht. Verwerping der aantrekkingskracht, zwaartekracht en moleculaire krachten, enz.* Van Dijk & Allan, 1905, page 183.

112. Harold T. Wilkins, 'Magic Ray of Life,' *Fate* vol.4, no.4, 1950, page 39-44. My research has confirmed some elements of Wilkins' story.

113. Walter Kafton-Minkel, *Subterranean Worlds,* Loompanics, 1989, pages 99-100.

114. Clemente Figueras was a forester on the Canary Islands. The clipping came from the *New York Herald,* June 9, 1902. Not seen. In: Christopher Bird and Oliver Nichelson, 'Nikola Tesla,' *Bres,* no.67, November/December 1977, page 117.

115. Jacques Bergier, *Livres Maudits,* J'ai lu, 1972, *Verdoemde Boeken,* Ankh Hermes, 1973, pages 112-123.

116. Erik van Ree, 'Virtueel Het Eeuwige Leven,' *Volkskrant,* April 1, 1995.

117. ibid.

Chapter 11. Vril from Atlantis: Keely's Legacy

1. Richard Cavendish, *The Encyclopedia of the Unexplained, Magic, Occultism and Parapsychology,* Rainbird Reference Books, 1974, page 250. Hartmann had a bad reputation; Blavatsky described him as 'a bad lot...I cannot trust him,' while the English Theososphists nicknamed him 'dirty Franz' because of his 'greasy appearance.' In: ibid. page 121, and Francis King, *The Secret Rituals of the O.T.O.,* Samuel Weiser, 1973, pages 23-24.

2. The experimenter was Ehrenfried Pfeiffer who wrote in his unpublished memoirs that while meeting Steiner to discuss the ether-forces in 1920-1921, he also met Guenther Wachsmuth at the same time who then 'became a lifelong friend.' The three often discussed the forces and energies of the ether. As a consequence, Wachsmuth published his book about the etheric forces in 1924, while Pfeiffer was appointed to the task to begin with the experiments... I had to do certain experiments which I am not allowed to describe any closer.' In: 'Vom Keely Motor zur Strader Machine,' *Der Europäer,* Jg.1/N.6, April, 1997, page 12.

3. 'Der Strader Apparat,' *Beiträge zur Rudolph Steiner Gesamtausgabe,* Heft Nr.107, Michaeli, 1991, page 5.

4. ibid. page 9.

5. ibid. page 1.

6. ibid. page 9.

7. ibid. page 1.

8. ibid. page 5.

9. ibid. page 9.

10. ibid.

11. Munin Nederlander, *Sheleg Deel I Analyse van het Strader-hoofdapparaat,* Warmtegroep, 1992.

12. Friedrich Feerhow *Die Photographie des Gedankens,* Max Altmann Verlag, 1913.

13. Friedrich Feerhow, *Die Menschliche Aura, und ihre experimentelle Erforschung,* Max Altmann Verlag, 1913.

14. Friedrich Feerhow, *Der Einfluss der Erdmagnetischen Zonen auf den Menschen, mit eiener Theorie des Polarlichts*, Max Altmann Verlag, 1913.

15. Friedrich Feerhow, *N-Strahlen und Od, ein Beitrag zum Problem der Radioaktivität de Menschen*, Max Altmann Verlag, 1913.

16. Friedrich Feerhow, *Eine Neue Naturkraft, oder: eine Kette von Täuschungen*, Max Altmann Verlag, 1913.

17. Albert de Rochas, *Die Grenzen der Wissenschaft*, Max Altmann verlag, 1913.

18. Robert Sigerius, *Die Telepathie. Telästhesie, Telenergie, Mentalsuggestion, magische Gedankenübertragung etc.*, Max Altmann Verlag, 1913.

19. Fritz Giese, *Die Lehre von Gedankenwellen*, Max Altmann Verlag, 1913.

20. Adam Voll, *Die Wünschelrute und der Siderischen Pendel*, Max Altmann Verlag, 1913.

21. Robert Blum, *Die Vierte Dimension*, Band I. Der Dreiklang in der Natur, Band II. Die Irrtümer moderner Theosophie, Band III. Im reiche der Vibrationen, Max Altmann Verlag, 1913.

22. The book also appeared by Max Altmann Verlag, who had published quite a few materials on the new forces and energies.

23. Karl H. Frick, *Licht und Finsternis, Gnostisch-theosophische und freimaurerisch-okkulte Geheimgesellschaften bis an die Wende zum 20. Jahrhundert, Wege in die Gegenwart, Teil 2: Geschichte ihrer Lehren, Rituale und Organisationen*, Akademische Druck- u. Verlagsanstalt Graz, 1978, pages 309-310.

24. Kenneth Rayner Johnson, *The Fulcanelli Phenomenon*, Neville Spearman, 1980, page 174.

25. Geneviève Dubois, *Fulcanelli Devoilé*, Éditions Dervy, 1992, page 94-95.

26. Kenneth Rayner Johnson, *The Fulcanelli Phenomenon*, Neville Spearman, 1980, page 174.

27. Dr. Ing. Franz Philipp, *Deutscher Raumflug ab 1934, ein unbequemes Buch*, the author, 2nd edition, 1970, page 7.

28. See: *Franz Philipp und Nikolas Roerich, Zwei Eingeweite und ihr Vermächtnis*, I.E.G.-Schrift Nr.4, Innere Erde Gemeinschaft, 1996. On the New Church, see also Horst Knaut, *Das Testament des Bösen*, Seewald Verlag, page 291, photo of pope Clemens XV on page 295.

29. Scott Parker, 'Fortech.' In: *Strange Magazine*, vol.1, no.4, 1989, page 38. Citing *Denver Post*, August 8, 1921.

30. Stan Deyo, *Cosmic Conspiracy*, Adventures Unlimited, 1994, pages 208, 233.

31. *The Fringes Of Reason*, Harmony Books, 1989, page 124.

32. Gaston Burridge, 'The Hubbard Energy Transformer,' *Fate* vol. 9 no.7, July 1956, pages 36-42.

33. F. D. Flemming, 'The Hendershot Motor Mystery,' *Fate*, vol.3, no.1, January 1950, pages 8-12. Also: Gaston Burridge, 'The So-Called Hendershot Motor,' *Round Robin*, vol.XI, no.6., March/April, 1956, pages 1-5, Tom Brown, 'The Hendershott Motor Mystery,' *Borderlands*, 1988, Charles Fort, *Wild Talents*, Claude Kendall, 1932, pages 335-337.

34. Bert Grater, 'Heat From The Air,' letter in *Fate*, vol.9, no.11, November 1956, pages 126-128.

35. Thomas De Hartmann, *Our Life With Mr. Gurdjieff*, Cooper Square Publishers, Inc., 1964, pages 96-97.

36. Georgi Ivanovitch Gurdjieff, *Beelzebub's Tales to his Grandson*, 'Beelzebub's verhalen aan zijn kleinzoon, Miranda, 1977, pages 110-129.

37. Helmut Möller & Ellic Howe, *Merlin Peregrinus*, Königshausen & Neumann, 1986, page 269, Stephen E. Flowers, *Fire and Ice*, Llewellyn, 1994, page 13, Richard Cavendish, *Encyclopedia of the Unexplained, Magic and Parapsychology*, Rainbird Reference Books, 1974, pages 91-92, Aythos 'Die Fraternitas Saturni - eine saturn-magische Loge,' *Arbeitsgemeinschaft für Religions und Weltanschauungsfragen*, Hiram-Edition 7, 1979, page 3.

38. *Pansophia, Urquellen Inneren Lebens zum heile der Welt neu kundgegeben von einem Collegium Pansophicum*, Mystischer Feuerschein, Abteilung III, Bd.1,2,3, 'zweiten teil: die Pansophischen Schule,' 1925, page 67.

39. ibid. page 203.
40. ibid. pages 89-110.
41. ibid. page 103.
42. ibid. pages 112-113.
43. ibid. pages 114-115.
44. Stephen E. Flowers, *Fire and Ice*, Llewellyn, 1994, pages 106-107.
45. Karl Hans Strobl & Alf von Czibulka, 'Der Orchideengarten,' Heft 23, *Elektrodämonen*, Zweiter Jahrgang, 1920.
46. John Clute and Peter Nichols editors, *The Encyclopedia of Science Fiction*, Orbit, 1993, page 1054.
47. ibid. page 15. Flowers muses: 'Indeed, there seems to have been a good deal of occult involvement among the German filmmakers of the pre-1933 era...' ibid. page 16.
48. Accounts on these films and some stills can be found in every good overview of science fiction films; my source: Douglas Menville and R. Reginald, *Things to Come*, Times Books, 1977.
49. *Saturn Gnosis, Offizielles Publikations Organ der deutschen Gros-Loge Fraternitas Saturni, Orient Berlin*, Januar/März, Band 5, 1930.
50. Clara Bloomfield-Moore, *Keely and His Discoveries, Aerial Navigation*, Kegan Paul, Trench, Trübner & Co., 1893, page 67, 329.
51. James Webb, *The Occult Establishment*, Richard Drew Publishing, 1981, page 315. These were possibly these were Crowley's paintings; Crowley held an exposition in Berlin in the 1930s of his paintings that had 'but little resonance.' In: Helmut Möller & Ellic Howe, *Merlin Peregrinus*, Königshausen & Neumann, 1986, page 268.
52. Rudolf von Sebottendorff, *Bevor Hitler Kam*, Deukula Verlag, 1933, pages 31-32, Ellic Howe, *Rudolph Freiherr von Sebottendorff*, the author, 1968, not published, page 24. Von Sebottendorff was also technically inclined; at one time, probably around 1913-1914, he was involved in the financial side of the development of an ill-fated tank project, called the Göbel tank after its inventor, Friedrich Wilhelm Göbel. In: ibid. page 16. In this endeavor of Sebottendorff, Webb notes a similarity with Wronski. In: James Webb, *The Occult Establishment*, Richard Drew Publishing, 1981, page 512. According to Sebottendorff Göbel also invented a 'wheel and railless railroad.' The tank was demonstrated in the Berlin stadium in 1914, but was ultimately rejected by the German military authorities. In: Rudolf von Sebottendorff, *Bevor Hitler Kam*, Deukula Verlag, 1933, page 239.
53. Fra. Johannes, 'Psychisch-Magische Beeinflussung durch Hoch Frequenz- und Ätherströme,' *Saturn Gnosis, Offizielles Publikations Organ der deutschen Gros-Loge Fraternitas Saturni, Orient Berlin*, October, Band 2, 1928, pages 84-91.
54. Fra. Gregorius, 'Die Neue Astro Religion,' *Saturn Gnosis, Offizielles Publikations Organ der deutschen Gros-Loge Fraternitas Saturni, Orient Berlin*, April/Oktober, Band 4, 1929, page 182.
55. Fra Gregorius, 'Die Sakralen Kulte der Loge,' *Saturn Gnosis, Offizielles Publikations Organ der deutschen Gros-Loge Fraternitas Saturni, Orient Berlin*, Januar/März, Band 5, 1930, page 240.
56. Fra. Johannes, 'Psychisch-Magische Beeinflussung durch Hoch Frequenz- und Ätherströme,' *Saturn Gnosis, Offizielles Publikations Organ der deutschen Gros-Loge Fraternitas Saturni, Orient Berlin*, October, Band 2, 1928, page 87.
57. ibid. page 88.
58. ibid.
59. ibid. pages 66-67.
60. Fra. Pacitius, 'Der Sternenmensch, *Saturn Gnosis, Offizielles Publikations Organ der deutschen Gros-Loge Fraternitas Saturni, Orient Berlin*, October, Band 2, 1928, page 64.
61. See: Nicholas Goodrick-Clarke, *The Occult Roots of Nazism*, pages 218-221.

62. A fanciful account and the one that started the myth, is given in: Louis Pauwels and Jacques Bergier, *Le matin des magiciens,* Gallimard, 1960.
63. Peter Bahn, 'Das Geheimnis der Vril-Energie,' *Neue Horizonte in Technik und Bewusstsein,* Jupiter Verlag, 1996, page 141.
64. Johannes Taufer, *Vril, Die Kosmische Urkraft. Wiedergeburt von Atlantis,* Wilhelm Becker, 1930, page 6.
65. Eugen Georg, *Verschollene Kulturen. Das Menscheitserlebnis. Ablauf und Deutungsversuch,* R. Voigtländer Verlag, 1930, pages 254-255.
66. Karl H. Frick, *Licht und Finsternis, Gnostisch-theosophische und freimaurerisch okkulte Geheimgesellschaften bis an die Wende zum 20. Jahrhundert, Wege in die Gegenwart, Teil 2: Geschichte ihrer Lehren, Rituale und Organisationen,* Akademische Druck- u. Verlagsanstalt Graz, 1978, page 529.
67. Helmut Werner, *Lexicon der Esoterik,* Fourier, 1991, pages 649-650.
68. G.W. Surya, *Moderne Rosenkreuzer, oder Die Renaissance der Geheimwissenschaften,* Johannes Baum Verlag, 1930, page 37.
69. ibid. pages 153-157.
70. ibid. page 157.
71. ibid. pages 159-160.
72. ibid. page 160.
73. ibid. page 163. Possibly Professor Henri Hertz is meant, see chapter 3 and chapter 3 note 45, although from the text it is possible that Surya refers to Professor Zöllner.
74. ibid. page 181.
75. ibid. pages 181-182.
76. ibid. page 182.
77. ibid. page 183.
78. Percy Greg, *Across the Zodiac,* Trübner & Co., 1880, vol.II, page 2.
79. Peter Bahn, 'Das Geheimnis der Vril-Energie,' *Neue Horizonte in Technik und Bewusstsein,* Jupiter Verlag, 1996, page 141.
80. Peter Bahn, *Der Vril-Mythos,* Omega Verlag, 1997, page 92.
81. Patrick Curry, *A Confusion of Prophets, Victorian and Edwardian Astrology,* Colins & Brown, 1992, page 126.
82. ibid. page 159.
83. Bessie Leo, *The Romance of the Stars,* Modern Astrology Office, 1914, page 3.
84. In: *Lotusblätter,* III Jahrgang, Nr.1/2, Januar/Februar 1923, page 85.
85. Peter Bahn, 'Das Geheimnis der Vril-Energie,' *Neue Horizonte in technik und Bewusstsein,* Jupiter Verlag, 1996, page 142.
86. ibid. page 142.
87. A.P. Sinnett & W. Scott-Elliott, *Maan-Pitris,* Theosophische Bibliotheek no.6, Theosophische Uitgeversmaatschappij, maart 1905, pages 3-4. Also: Franz Hartmann, *Grundriss der geheimlehre von H.P. Blavatsky,* Bücher der Schatzkammer, no date, pages 36-47.
88. C. Aq. Libra, *Astrologie,* Luctor et Emergo, 1922, page 15.
89. 'Towards the end of 1922 appeared three new works of the in Germany so highly esteemed Dutch author C. Aq. Libra...' In: 'Lotus blätter,' III Jahrgang, Nr.1/2, January/February 1923, page 3. Two of his books appeared in the German language.
90. ibid.
91. C. Aq Libra, *Astrologie,* Luctor et Emergo, 1922, page 15.
92. ibid. page 16.
93. ibid.
94. Rudolf Steiner, *Die Apokalypse Des Johannes,* Verlag der Rudolf Steiner Nachlassverwaltung, 1962, page 83.
95. C. Aq. Libra, *Astrologie,* Luctor et Emergo, 1922, page 32-33.

96. James M. Pryse, *The Magical Message According To Iôannês, commonly called the Gospel according to St. John,* The Theosophical Publishing Company of New York,' 1909, page 77.
97. Rudolf Steiner, *Die Apokalypse Des Johannes,* Verlag der Rudolf Steiner Nachlassverwaltung, 1962, page 85.
98. Peter Bahn, 'Das Geheimnis der Vril-Energie,' *Neue Horizonte in Technik und Bewusstsein,* Jupiter Verlag, 1996, page 143.
99. ibid. page 144.
100. Rolf Schaffranke, 'Schappellers Kosmischer Energie-Extraktor,' *Raum & Zeit,* Special 7, 1994, page 89.
101. Franz Wetzel & L. Gföllner, *Raumkraft, Ihre Erschliesung und Auswertung durch Karl Schappeller,* Herold Verlag, 1928, page 8.
102. ibid. pages 8-9.
103. A.A.C. Belinfante, *De Duwkracht. Beschouwingen van Stoffen en hun bewegingen als gevolgen van botsende- of duwkracht. Verwerping der aatrekkingskracht, zwaartekracht en moleculaire krachten, enz.,* Westzaan, Van Dijk en Allan, 1905, page 54.
104. M. Gföllner, 'Die Erschliesung der Dynamischen Technik durch Karl Schappeller,' *Zeitschrift für Geistes- und Wissenschaftsreform,* 5. Jahrgang, 1930, Doppelheft 9/10, pages 201-206, Rolf Schaffranke, 'Schappellers Kosmischer Energie-Extraktor,' *Raum & Zeit,* Special 7, 1994, pages 90-91.
105. Herbert Reichstein 'Totgeschwiegene Forscher,' *Zeitschrift für Geistes- und Wissenschaftsreform,* 5. Jahrgang, 1930, Doppelheft 9/10, pages 201-206.
106. Frenzolf Schmid, *Die neue Strahlenlehre and Die neue Strahlenheilmethode,* both published in 1929.
107. Lanz von Liebenfels, Praktische Einführung in die Arisch-Christliche Mystik, II. Teil: Naturwissenschaftliche Begründung, *Ariomantische Bücherei,* Brief Nr.6, the author, 1934, page 3.
108. ibid. pages 15-16.
109. J.K. Rensburg, *Wereldbouw,* van Loghum Slaterus & Visser, 1923, page 38.
110. ibid. pages 38, 28.
111. Lanz von Liebenfels, Praktische Einführung in die Arisch-Christliche Mystik, V. Teil: Die Mystische Verzückung u. Hochzeit (Ecstasis u. Unio), *Ariomantische Bücherei,* Brief Nr.9, the author, 1934, page 1.
112. Lanz von Liebenfels, Praktische Einführung in die Arisch-Christliche Mystik, III. Teil: Die Mystische Vorbereitung (Preambulum), *Ariomantische Bücherei,* Brief Nr.7, the author, 1934, page 16.
113. Olof Alexandersson, *Living Water, Viktor Schauberger and the Secrets of Natural Energy,* Gateway Books, 1995.
114. ibid. page 141.
115. Peter Bahn, *Der Vril-Mythos,* Omega verlag, 1997, page 104.
116. ibid. page 106.
117. Conversation with Peter Bahn.
118. D.A. Kelly, *The Manual of Free Energy Devices and Systems,* Cadake Industries and Tri State Press, 1986, page 2.
119. Harold T. Wilkins, *Mysteries Solved and Unsolved,* Odhams Press, 1958, page 121. With John Andruss, Wilkins probably meant John Andrews, a Pennsylvania inventor who, in 1916, went to the New York Navy Yard in a Packard with a friend. They claimed to have driven from Pittsburgh and that their car was powered by a fuel developed by Andrews. The fuel consisted principally on water, and Andrews demonstrated his fuel to the Navy the next day. He poured a half gallon of water into the engine's tank, then added a few drops of green liquid from a vial he had taken from his vest pocket. The engine ran. A report was made to Secretary of the Navy Joseph Daniels who then ordered more tests, but Andrews could not be found and was unable to make the $2 million deal he wanted. In 1942, reporter James L.

Kilgallen discovered Andrews living on a Pennsylvania farm. Andrews, who was now trying to develop synthetic rubber, said he had forgotten the formula for his fuel. In 1953 Andrews died, his secret intact. Interestingly, in 1916 a Louis Enricht of Long Island called a press conference to show how he could turn water into fuel with a green pill. Enricht, who had a past of embezzlement, ended up in Sing Sing, and the secret of his green pill was lost forever. Curiously, as recently as 1973 a Guido Franch demonstrated to reporter Tom Valentine what he called 'mota' (atom spelled backwards) fuel, made from green granules and water. The fuel which would not mix with water, burned clean. Franch told Valentine that the fuel was made from coal with a process he had learned when he was 15 from a German scientist named Kraft, who also had taught Andrews the secret. In: Irving Wallace, David Wallechinsky, Amy Wallace, 'Water Into Gasoline?,' *Significa, Parade,* July 19, 1981, pae 21.

120. ibid. page 168.
121. James Webb, *The Occult Establishment,* Richard Drew Publishing, 1981, page 512.

Chapter 12. The Great 19th Century Airship Wave

1. Thomas A. Bullard, *The Airship File,* the author, Bloomington, 1982. A massive work and the definitive sourcebook on the subject, as it provides transcripts of the original newspaper accounts.
2. ibid. Also: Paris Flammonde, *UFO Exist!,* Putnam's, 1976, Julien Weverbergh, *UFOs in het verleden,* Ankh Hermes, 1980, Daniel Cohen, *The Great Airship Mystery, a UFO of the 1890s,* Dodd, Mead, 1981, Wallace O. Chariton, *The Great Texas Airship Mystery,* Wordware, 1991, W.A. Harbinson, *Projekt UFO,* Boxtree, 1995, Theo Paijmans, *Kosmisch Netwerk,* Ankh Hermes, 1996.
3. Thomas A. Bullard, *The Airship File,* the author, Bloomington, 1982, pages V-VI.
4. 'Voices In The Sky,' *Sacramento Evening Bee,* November 18, 1896.
5. 'Airship, Or What?,' *Sacramento Evening Bee,* November 19, 1896.
6. 'Strange Craft In The Sky,' *San Francisco Call,* November 19, 1896.
7. Editor's Call, *San Francisco Call,* November 23, 1896.
8. 'A Winged Ship In The Sky,' *San Francisco Call,* November 23, 1896.
9. John Keel, *UFOs: Operation Trojan Horse,* Souvenir Press, 1971, page 96.
10. 'A Lawyer's Word For That Airship,' *San Francisco Chronicle,* November 22, 1896.
11. 'Queer Things You See When,' *San Francisco Examiner,* November 23, 1896
12. 'A Lawyer's Word For That Airship,' *San Francisco Chronicle,* November 22, 1896.
13. 'Sticks To His Story,' *Sacramento Evening Bee,* November 23, 1896.
14. 'Collins Sticks To His Airship Story,' *San Francisco Chronicle,* November 23, 1896. Also in: 'Sticks To His Story,' *Sacramento Evening Bee,* November 23, 1896.
15. 'A Clue At Last,' *Oakland Tribune,* November 24, 1896.
16. 'A Winged Ship In The Sky,' *San Francisco Call,* November 23, 1896.
17. 'Queer Things You See When,' *San Francisco Examiner,* November 23, 1896.
18. 'The Apparition Of The Air,' *San Francisco Call,* November 24, 1896. Also in: 'Have You Seen It In The Sky?,' *San Francisco Examiner,* November 24, 1896, 'Coy Mr. Collins And His Airship,' *San Francisco Chronicle,* November 24, 1896, which already noted the contradictions in Collins' statements.
19. 'Have You Seen It In The Sky?,' *San Francisco Examiner,* November 24, 1896.
20. 'A Winged Ship In The Sky,' *San Francisco Call,* November 23, 1896.
21. 'Mission Of The Aerial Ship,' *San Francisco Call,* November 25, 1896.
22. 'Airships Now Fly In Flocks,' *San Francisco Examiner,* November 25, 1896.
23. 'As Large As A Big Whale,' *San Francisco Call,* November 27, 1896.
24. 'It Flitted Over San Jose,' *San Francisco Call,* November 28, 1896.
25. 'Hart's Inventor Has Three Aerial Fliers,' *San Francisco Call,* November 29, 1896.
26. 'Airships Now Fly In Flocks,' *San Francisco Examiner,* November 25, 1896.
27. John Jacob Astor, *A Journal in Other Worlds,* Appleton, 1894, pages 58-61.

28. 'Hart Stands By His Ship,' *San Francisco Call,* November 26, 1896.
29. 'It Flitted Over San Jose,' *San Francisco Call,* November 28, 1896.
30. 'As Large As A Big Whale,' *San Francisco Call,* November 27, 1896.
31. 'It Flitted Over San Jose,' *San Francisco Call,* November 28, 1896.
32. 'Hart's Inventor Has Three Aerial Fliers,' *San Francisco Call,* November 29, 1896.
33. Daniel Cohen, *The Great Airship Mystery,* Dodd, Mead, 1981, pages 38-39. Also in: Wallace Chariton, *The Great Texas Airship Mystery,* Wordware, 1991, page 28.
34. Jerome Clark, *The Emergence of a Phenomenon: UFOs from the Beginning through 1959,* Omnigraphics, 1992, page 21.
35. John Jacob Astor, *A Journal in Other Worlds,* Appleton, 1894, page 62.
36. Correspondence and conversations with P.G. Navarro. Also: James Ward and P.G. Navarro, 'Peter Mennis You Are Not Forgotten,' *Gray Barker's Newsletter,* no.17, December 1982, James Ward and P.G. Navarro, 'Dellschau and Other Aeronauts,' *Gray Barker's Newsletter,* no.19, 1983, Jerome Clark and Loren Coleman, 'Mystery Airships of the 1800's - part one,' *Fate,* vol.26, no.5, May, 1973.
37. Jerome Clark and Loren Coleman, 'Mystery Airships of the 1800's - part one, *Fate,* vol.26, no.5, May 1973, page 87-89. Possibly meant as an April Fools day joke, although the pamphlet was printed before that date (April 1, 1849), and must have been an expensive one for such an obscure firm.
38. 'Golden Haired Girl Is In It,' *St. Louis Post-Dispatch,* April 19, 1897.
39. Jerome Clark, *The Emergence of a Phenomenon: UFOs from the Beginning through 1959,* Omnigraphics, 1992, page 87.
40. Paris Flammonde, *UFO Exist!,* Punam's, 1976, page 93. In response to Flammonde's theory, Cohen speculates that in fact Hearst hated the *San Francisco Call* and 'would have done anything he reasonably could to discredit the rival paper and story it printed...' in: Daniel Cohen, *The Great Airship Mystery,* Dodd, Mead & Company, pages 19-22.

Chapter 13. Into the Realms of Speculation

1. Joscelyn Godwin, *Arktos the Polar Myth,* Phanes Press, 1993, page 177.
2. Peter Haining, *The Jules Verne Companion,* Souvenir Press, 1978, page 23.
3. ibid. pages 21-22.
4. Michel Lamy, *Jules Verne, Initié et Initiateur,* Payot, 1984, 1994, page 218.
5. ibid. page 15.
6. Peter Haining, *The Jules Verne Companion,* Souvenir Press, 1978, page 21.
7. Michel Lamy, *Jules Verne, Initié et Initiateur,* Payot, 1984, 1994, page 226.
8. ibid. page 221.
9. ibid. pages 116-117.
10. Nigel Pennick, *Hitler's Secret Sciences,* Victor Gollancz, 1981, page 169.
11. Clara Bloomfield-Moore, *Keely and his Discoveries, Aerial Navigation,* Kegan Paul, Trench, Trübner & Co., 1893, page 139.
12. Michel Lamy, *Jules Verne, Initié et Initiateur,* Payot, 1984, 1994, page 139.
13. Clara Bloomfield-Moore, *Keely and his Discoveries, Aerial Navigation,* Kegan Paul, Trench, Trübner & Co., 1893, page 81.
14. Francois Clavel, *Geschiedenis der vrijmetselarij en der geheime genootschappen van vroegen en lateren tijd,* Van Goor, 1843, page 73.
15. Eugen Lennhoff & Oskar Posner, *Internationales Freimaurer Lexikon,* Amalthea Verlag, 1932, 1980, page 1766.
16. ibid.
17. Francois Clavel, *Geschiedenis der vrijmetselarij en der geheime genootschappen van vroegen en lateren tijd,* Van Goor, 1843, page 73.
18. Eugen Lennhoff & Oskar Posner, *Internationales Freimaurer Lexikon,* Amalthea Verlag, 1932, 1980, page 1754.

19. Jacq. Ph. Levesque, *Apercu Generale Et Historique Des Sectes Macconiques*, Caillot, 1821, page 112. Arthur Edward Waite, *A New Encyclopedia of Freemasonry*, vol. II, Rider, 1925, page 275.

20. Gisela von Frankenberg, *Nommo, der wiederkehrende Sonnenmensch*, Aurum Verlag, 1981, page 191.

21. For instance Tones Brunés, *Energien der Urzeit*, Edition Sven Erik Bergh, 1977, pages 24-25, 159-164.

22. A.E. Thierens, *Wetenschappelijke Opstellen*, Luctor et Emergo, 1913, page 107.

23. Percy Greg, *Across the Zodiac*, Trübner & Co., 1880, page 141.

24. ibid. pages 279-285.

25. ibid. page 286.

26. Joscelyn Godwin, *Arktos the Polar Myth*, Phanes Press, 1993, page 176.

27. Eugen Lennhoff & Oskar Posner, *Internationales Freimaurer Lexikon*, Amalthea Verlag, 1932, 1980, pages 380-383.

28. Peter Nichols editor, *The Encyclopedia of Science Fiction*, Granada, 1981, page 266, 586.

29. Everett F. Bleiler, *Science Fiction: The Early Years*, The Kent State University Press, 1990, page 311.

30. ibid. page 1426-1427.

31. Arthur Edward Waite, *A New Encyclopedia of Freemasonry*, vol. II, Rider, 1925, pages 220-237.

32. ibid. pages 1329-1330. Also Arthur Edward Waite, *A New Encyclopedia of Freemasonry*, vol. II, Rider, 1925, pages 369-371.

33. Leslie Shepard, *Encyclopedia of Occultism & Parapsychology*, Gale Research Inc., 1996, page 308.

34. Everett F. Bleiler, *Science Fiction: The Early Years*, The Kent State University Press, 1990, page 583-584.

35. Sam Moskowitz, *Across the Zodiac: A Major Turning Point in Science Fiction*, foreword in: Percy Greg, *Across the Zodiac*, Hyperion Press, 1974, page 11.

36. Cheiro, *True Ghost Stories*, The London Publishing Company, no date, page 248.

37. ibid. pages 249-250.

38. Sam Moskowitz, *Across the Zodiac: A Major Turning Point in Science Fiction*, foreword in: Percy Greg, *Across the Zodiac*, Hyperion Press, 1974, page 11.

39. ibid. Also: Peter Haining, *The Jules Verne Companion*, Souvenir Press, 1978, page 48.

40. Peter Haining, *The Jules Verne Companion*, Souvenir Press, 1978, page 29.

41. Joscelyn Godwin, *Arktos the Polar Myth*, Phanes Press, 1993, page 223.

42. Michel Lamy, *Jules Verne, initié et initiateur*, Payot, 1984, 1994, also: David Wood and Ian Campbell, *Geneset, Target Earth*, Bellevue Books, 1994, pages 116-127. They refer to Lamy. About codes clearly visible in Verne's texts, and in which he was very interested, see: Peter Haining, *The Jules Verne Companion*, Souvenir Press, 1978, pages 31-34.

43. ibid. ibid. page 119.

44. Joscelyn Godwin, *Arktos the Polar Myth*, Phanes Press, 1993, pages 143-144.

45. Clara Bloomfield-Moore, *Keely and his Discoveries, Aerial Navigation*, Kegan Paul, Trench, Trübner & Co., 1893, page 369.

46. ibid. page 318.

47. ibid. page 369.

48. ibid. pages 354-355.

49. Joscelyn Godwin, *Arktos the Polar Myth*, Phanes Press, 1993, page 144.

50. ibid. pages 147, 149.

51. ibid. page 148.

52. ibid. page 144.

53. See for instance: Michael Baigent, Richard Leigh and Henry Lincoln, *Holy Blood, Holy Grail*, Jonathan Cape, 1982, David Wood, *Genisis, the First Book of Revelations*, The Baton Press, 1985.
54. Michel Lamy, *Jules Verne, initié et initiateur*, Payot, 1984, 1994, page 242.
55. Robert Cromie, *A Plunge Into Space*, Frederick Warne and Co., 1891, page 51.
56. Cheiro, *True Ghost Stories*, The London Publishing Company, no date, page 261.
57. Robert Cromie, *A Plunge Into Space*, Frederick Warne and Co., 1891, page 5.
58. Michel Lamy, *Jules Verne, initié et initiateur*, Payot, 1984, 1994, also: David Wood and Ian Cambell, *Geneset, Target Earth*, Bellevue Books, 1994, pages 116-127. They refer to Lamy.
59. Peter Haining, *The Jules Verne Companion*, Souvenir Press, 1978, page 48.
60. Robert Cromie, *A Plunge Into Space*, Frederic Warne & Co., 1891, page 46.
61. Albert Ross Parsons, *New Light on the Great Pyramid*, Metaphysical Publishing Co., 1893. With this theory he was not alone; see Joscelyn Godwin, *Arktos the Polar Myth*, Phanes Press, 1993, pages 181-222. Also, in Cromie's, *The track of Doom*, which was published in 1895, a strange force resembling atomic energy is threatening the world. The first test of the substance, made thousands of years earlier, had destroyed the fifth planet which created the asteroid belt.
62. ibid. page 387.
63. ibid. page 400.
64. Michel Lamy, *Jules Verne, initié et initiateur*, Payot, 1984, 1994.
65. Ellic Howe, *The Magicians of the Golden Dawn*, Routledge & Kegan Paul, 1972, page 35.
66. Mentioned in Daniel Cohen, *The Great Airship Mystery*, Dodd, Mead & Company, 1981, page 129. At least, according to Senarens who told in the 1920's that Verne had written him to praise his tales forty years earlier. Verne's letters to Senarens have not survived. In: Brian Taves and Stephen Michaluk, Jr., *The Jules Verne Encyclopedia*, The Scarecrow Press, 1996, pages 20-21.
67. Peter Nichols editor, *The Encyclopedia of Science Fiction*, Granada, 1981, page 472.
68. Conversation with Dale Pond. About Keely's devices hidden in a church in France; Pond obtained this unverified information through a psychic person.
69. Arthur Edward Waite, *A New Encyclopedia of Freemasonry*, Rider, 1925, vol. II, page 221. Waite only gives the slightest reference here.
70. Pierre Barbet, *Baphomet's Meteor*, DAW-books, 1972.
71. David Wood and Ian Cambell, *Geneset, Target Earth*, Bellevue Books, 1994, page 169.
72. ibid. page 119.
73. Jean Robin, *Operation Orth, ou l'incroyable secret de Rennes-le-Château*, Guy Trédaniel, 1989.
74. John Keel, *Disneyland of the Gods*, IllumiNet Press, page 91.
75. Peter Nichols editor, *The Encyclopedia of Science Fiction*, Granada, 1981, page 575.
76. Peter Haining, *The Dracula Centennary Book*, Souvenir Press, 1987, pages 25-27.
77. Michel Lamy, *Jules Verne, initié et initiateur*, Payot, 1984, 1994, pages 188-191.
78. Peter Haining, *The Jules Verne Companion*, Souvenir Press, 1978, page 66.
79. Paul Chacornac, *Eliphas Levi, rénovateur de l'occultism en France*, Chacornac, 1926, pages 242-245.
80. David Wood, *Genisis, the First Book of Revelations*, The Baton Press, 1985, Henry Lincoln, *The Holy Place*, Jonathan Cape, 1991.
81. For an account of Astor's death, see: Robert D. Ballard, *The Discovery of the Titanic*, Madison Press Books, 1995, pages 17, 37, 216. Also: Derek Wilson, *The Astors*, Weidenfeld & Nicholson, 1993, page 207.
82. Peter Nichols editor, *The Encyclopedia of Science Fiction*, Granada, 1981, page 518. More in: John Clute and Peter Nichols editors, *The Encyclopedia of Science Fiction*, Orbit, 1993, page 1054.

83. R. Père Martin, *Le Renversement, Ou La Boucane Contre L'Ordre Noir*, Guy Trédaniel, 1984, pages 243-244.

84. Rudolf Elmayer von Vestenbrugg, *Eingriffe aus dem Kosmos*, Herman Bauer Verlag, 1971, pages 422-423. The author also refers to Leslie.

85. John Clute and Peter Nichols editors, *The Encyclopedia of Science Fiction*, Orbit, 1993, page 1054.

86. Marc J. Seifer, 'Nikola Tesla: The History of Lasers and Particle Beam Weapons,' *Proceedings of the 1988 International Tesla Symposium*, International Tesla Society, 1988, page 24.

87. Peter Nichols editor, *The Encyclopedia of Science Fiction*, Granada, 1981, page 252.

88. ibid. page 521. Amongst others, Nikola Tesla, William Pickering, Astor's one-time mentor, published in Gernsback's magazines, as did such interesting persons as early rocket pioneer R.H. Goddard, early German rocket experimenter Max Valier, Herman Oberth of the V1 and V2 fame and inventor of the solar space mirror, and the late Donald H. Menzel. Writes Gernsback: 'For the record, it should be noted too, that the American rocket society was founded by the writers, employees and authors' and he further states that its first organ, the *Bulletin of the Interplanetary Society* was 'published and financed by the writer (Gernsback) in 1930.' In: Hugo Gernsback, 'Forecast 1958,' *Space Review*, 1958, the author, pages 2-3. Goddard worked on his rocket designs in the thirties near Roswell, New Mexico. Valier and Oberth are briefly mentioned in one of the pamphlets of the 'Reichsarbeitsgemeinschaft'. Oberth expressed his thoughts on the UFO phenomenon on several occasions. Menzel would later become notorious in UFO-research circles for his attempts to debunk the UFO phenomenon in several books.

89. ibid. page 252.

90. In: Chris Morgan, *The Shape of Future's Past*, Webb & Bower, 1980, page 109.

91. Sidney G.P. Coryn, *The Faith of Ancient Egypt*, T.P.S., 1913.

92. About the possible Lovecraft-Waite-Golden Dawn connection, see: Philip A. Shreffler, *The H.P. Lovecraft Companion*, Greenwood Press, 1977, pages 101, 164, 169, 170, 176, 179.

93. John Hamill editor, 'The Rosicrucian Seer,' *The Aquarian Press*, 1986, page 14.

94. Philip A. Shreffler, *The H.P. Lovecraft Companion*, Greenwood Press, 1977, page 67.

95. ibid. page 117.

96. See: David Wood & Ian Campbell, *Geneset, Target Earth*, Bellevue Books, 1994.

97. J.K. Rensburg, *Wereldbouw*, van Loghum Slaterus & Visser, 1923, page 36.

98. H.P. Lovecraft, *Dagon and other Macabre Tales*, Arkham House, 1965, pages 68-69.

List of Illustrations

Index